VOLUME ONE HUNDRED AND FIVE

CURRENT TOPICS IN DEVELOPMENTAL BIOLOGY

Developmental Timing

CURRENT TOPICS IN DEVELOPMENTAL BIOLOGY

"A meeting-ground for critical review and discussion of developmental processes"
A.A. Moscona and Alberto Monroy (Volume 1, 1966)

VOLUME ONE HUNDRED AND FIVE

Current Topics in DEVELOPMENTAL BIOLOGY

Developmental Timing

Edited by

ANN E. ROUGVIE and MICHAEL B. O'CONNOR
Department of Genetics, Cell Biology and Development
University of Minnesota
Minneapolis, MN, USA

ELSEVIER

AMSTERDAM • BOSTON • HEIDELBERG • LONDON
NEW YORK • OXFORD • PARIS • SAN DIEGO
SAN FRANCISCO • SINGAPORE • SYDNEY • TOKYO
Academic Press is an imprint of Elsevier

Academic Press is an imprint of Elsevier
525 B Street, Suite 1800, San Diego, CA 92101-4495, USA
225 Wyman Street, Waltham, MA 02451, USA
The Boulevard, Langford Lane, Kidlington, Oxford, OX5 1GB, UK
32 Jamestown Road, London NW1 7BY, UK
Radarweg 29, PO Box 211, 1000 AE Amsterdam, The Netherlands

First edition 2013

ISBN: 978-0-12-396968-2
ISSN: 0070-2153

CONTENTS

Contributors *xi*
Preface *xiii*

1. How Clocks and Hormones Act in Concert to Control the Timing of Insect Development **1**

Francesca Di Cara and Kirst King-Jones

1. General Considerations: The Problem of Time in Development 2
2. Evidence for Mechanisms Controlling the Timing of Developmental Events 3
3. Developmental Timing in Insects is Under Hormonal Control 4
4. Circadian Control and Developmental Transitions 7
5. Circadian Core Clock in Flies 9
6. Evidence of a Clock in the Insect Prothoracic Glands 12
7. Circadian Oscillations of EcR 13
8. Oscillations in the Prothoracic Gland: Input from the Brain 14
9. The Timing of CCAP Neuronal Differentiation is Ecdysone-Controlled 19
10. JHs and Developmental Timing 23
11. Outlook 26

References 28

2. Nutrient Signaling and Developmental Timing of Maturation **37**

E. Thomas Danielsen, Morten E. Moeller, and Kim F. Rewitz

1. Introduction 38
2. Interplay Between Nutrients, Systemic Growth, and Developmental Transitions 41
3. A Nutritional Checkpoint Based on Crosstalk Between Different Tissues 48
4. Fine-Tuning Systemic Insulin Signaling 54
5. A Developmental Checkpoint Couples Organ Growth and Tissue Regeneration to Timing 57
6. Conclusions and Perspectives 60

Acknowledgments 61
References 61

3. Temporal Patterning of Neural Progenitors in *Drosophila* **69**

Xin Li, Zhenqing Chen, and Claude Desplan

1. Introduction 70
2. Sequentially Expressed Transcription Factors in NBs Control Birth-Order-Dependent Neuron Fates 73

3. How are Temporal Transitions in NBs Controlled? 78
4. Relationship Between Temporal Sequence and NB Competence 81
5. Feedforward Loops Downstream of the Temporal Sequence Control Neuron Fates 84
6. Progression of the TF Sequence Required for the End of Neurogenesis 85
7. Integration of Temporal and Spatial Information Determines Lineages 87
8. Integration of Temporal Identity with Binary Fate Choice 88
9. Conclusions and Future Questions 90
References 91

4. MicroRNAs as Components of Systemic Signaling Pathways in *Drosophila melanogaster* 97

Arthur Luhur, Geetanjali Chawla, and Nicholas S. Sokol

1. Introduction 98
2. Overview of miRNA Biogenesis and Function 99
3. Insulin and Nutritional Pathways 99
4. Lipophilic Hormone Signaling Pathways 104
5. Light Modulated Circadian Rhythm Pathways 113
6. Stress Induced and Immune Response Pathways 116
7. Summary 118
References 118

5. Vegetative Phase Change and Shoot Maturation in Plants 125

R. Scott Poethig

1. Introduction 126
2. Terminology 126
3. Heteroblasty and Vegetative Phase Change 127
4. miR156 and miR157: Master Regulators of Vegetative Phase Change 129
5. miR156 Targets 130
6. Molecular Insights into the Phenomenology of Vegetative Phase Change 134
7. Conclusion 144
Acknowledgments 144
References 144

6. Developmental Transitions in *C. elegans* Larval Stages 153

Ann E. Rougvie and Eric G. Moss

1. Introduction 154
2. A Worm Well-Suited to Timing Studies 154
3. Heterochronic Mutants 156

4. microRNA Switches 159
5. LIN-28: A microRNA Regulator at the Pathway's Center 160
6. The Heterochronic Pathway 161
7. Developmental Transitions: The Molts 162
8. LIN-42: A Link Between Molting and Developmental Decisions 164
9. Reversible Interruption of Developmental Progression: DAF-12 167
10. Interrupted Development: Reprogramming and Resynchronization 169
11. Launching Larval Development 171
12. A Heterochronic Pathway for Larger Animals? 173
13. Conclusions and Future Directions 173
Acknowledgments 174
References 175

7. Steroid Regulation of *C. elegans* Diapause, Developmental Timing, and Longevity 181

Adam Antebi

1. Introduction 182
2. Endocrine Regulation of the Dauer Diapause 183
3. DAF-12 Regulates Developmental Timing 192
4. Hormonal Control of Longevity 201
5. Perspectives 203
Acknowledgments 205
References 205

8. Diapause: Delaying the Developmental Clock in Response to a Changing Environment 213

Luca Schiesari and Michael B. O'Connor

1. Introduction 214
2. Hormonal Control of Embryonic Diapause in Insects: The "Hatch-Ready" Example 215
3. Hormonal Control of Insect Diapause: The Mid-Embryonic Arrest Example 216
4. Developmental Control of Larval Diapause 219
5. Pupal Diapause 221
6. Diapause Can Modify Larval/Pupal Growth 226
7. Seasonal Polyphenisms Correlate with Diapause 227
8. The Killifish Diapausing Embryo: A Developmental Analogy in a Vertebrate 233
9. Summary and Perspective 236
Acknowledgments 237
References 237

9. Circannual Transitions in Gene Expression: Lessons from Seasonal Adaptations 247

Christine Schwartz and Matthew T. Andrews

1. Introduction 248
2. Seasonal Courtship and Territorial Behavior 249
3. Migration 251
4. Hibernation 253
5. Potential Pathways Involved in Circannual Timing 259
6. Recent Advances in Circannual Timing Research 266
7. Conclusions: Future Directions and Potential Applications 267

Acknowledgments 267

References 268

10. Unliganded Thyroid Hormone Receptor Regulates Metamorphic Timing via the Recruitment of Histone Deacetylase Complexes 275

Yun-Bo Shi

1. Introduction 276
2. TR Function During Frog Development 279
3. Unliganded TR Recruits Corepressor Complexes to Control Metamorphic Timing 282
4. Coactivator Recruitment for Metamorphic Progression 285
5. Temporal Regulation of Organ-Specific Metamorphosis 285
6. Conclusions 291

Acknowledgment 291

References 291

11. Keeping Puberty on Time: *Novel* Signals and Mechanisms Involved 299

Manuel Tena-Sempere

1. Puberty is a Key Developmental Transition 300
2. Neurobiology of Puberty: Essential Roles of GnRH and Kiss1 Neurons 301
3. Putative Roles of Kisspeptin Partners in the Control of Puberty: The Case of NKB 307
4. Early Development and the Timing of Puberty: Kiss1 Neurons in Brain Sex Differentiation 309
5. Metabolic Control of Puberty: Leptin, Kisspeptins, and Cellular Energy Sensors 312

6. Novel Mechanisms in the Control of Puberty: Roles of Epigenetics and microRNAs 317
7. Concluding Remarks 320
Acknowledgments 321
References 321

Index *331*

CONTRIBUTORS

Matthew T. Andrews
Department of Biology, University of Minnesota Duluth, Duluth, Minnesota, USA

Adam Antebi
Max Planck Institute for Biology of Ageing; Cologne Excellence Cluster on Cellular Stress Responses in Aging Associated Diseases (CECAD), University of Cologne, Cologne, Germany, and Department of Molecular and Cellular Biology, Huffington Center on Aging, Baylor College of Medicine, Houston, Texas, USA

Geetanjali Chawla
Department of Biology, Indiana University, Bloomington, Indiana, USA

Zhenqing Chen
Department of Biology, New York University, New York, New York, USA

E. Thomas Danielsen
Department of Biology, Cell and Neurobiology, University of Copenhagen, Copenhagen, Denmark

Claude Desplan
Department of Biology, New York University, New York, New York, USA

Francesca Di Cara
Department of Biological Sciences, University of Alberta, Edmonton, Alberta, Canada

Kirst King-Jones
Department of Biological Sciences, University of Alberta, Edmonton, Alberta, Canada

Xin Li
Department of Biology, New York University, New York, New York, USA

Arthur Luhur
Department of Biology, Indiana University, Bloomington, Indiana, USA

Morten E. Moeller
Department of Biology, Cell and Neurobiology, University of Copenhagen, Copenhagen, Denmark

Eric G. Moss
Department of Molecular Biology, Rowan University, Stratford, New Jersey, USA

Michael B. O'Connor
Department of Genetics, Cell Biology and Development, University of Minnesota, Minneapolis, Minnesota, USA

R. Scott Poethig
Department of Biology, University of Pennsylvania, Philadelphia, Pennsylvania, USA

Kim F. Rewitz
Department of Biology, Cell and Neurobiology, University of Copenhagen, Copenhagen, Denmark

Ann E. Rougvie
Department of Genetics, Cell Biology and Development, University of Minnesota, Minneapolis, Minnesota, USA

Luca Schiesari
Department of Biology, University of Padova, Padova, Italy

Christine Schwartz
Department of Biology, University of Minnesota Duluth, Duluth, Minnesota, USA

Yun-Bo Shi
Section on Molecular Morphogenesis, Program in Cellular Regulation and Metabolism, NICHD, NIH, Bethesda, Maryland, USA

Nicholas S. Sokol
Department of Biology, Indiana University, Bloomington, Indiana, USA

Manuel Tena-Sempere
Department of Cell Biology, Physiology and Immunology, University of Córdoba; CIBER Fisiopatología de la Obesidad y Nutrición, Instituto de Salud Carlos III, and Instituto Maimónides de Investigaciones Biomédicas (IMIBIC)/Hospital Universitario Reina Sofía, Córdoba, Spain

PREFACE

During development of multicellular organisms, the identity, behavior, and ultimate function of cells must be coordinated with those of their neighbors. The eventual outcome of developmental regulatory networks that control these processes is the formation of specialized tissues, each endowed with unique physiological and functional capabilities necessary for organismal survival and reproduction. Clearly, spatial control of gene expression is paramount to correct organismal development. Equally important, but less understood, is the issue of developmental timing. Without proper timing, tissue components might form in their proper spatial positions, but if this happens too early or too late, inappropriate proportions or faulty tissue architecture could result.

Developmental timing is regulated by a variety of temporal parameters including the rate at which an event occurs and the duration of the developmental window allotted to the event, the latter set by time of onset and completion of a particular process. Additional levels of temporal complexity arise when one considers control of developmental synchrony among tissues and organs within an organism, and the timing of repeated processes, such as molting cycles, that require organism-wide coordination of gene expression patterns and behaviors to execute developmental transitions at regular intervals. Ultimately, developmental timing mechanisms ensure correct tissue organization, patterning, and integration, as well as final body size since this is largely determined by mechanisms that regulate the rate and duration of tissue growth.

Intrinsic genetic programs have been described that coordinate developmental time through hard-wired regulatory networks. For example, the early synchronous cell cycles of *Drosophila* embryos are largely driven by maternally supplied products. Subsequently, waves of transcription factors that control early syncytial development achieve correct timing as a result of sequential inductions where one set of transcription factors activates the next and so forth, with feedback mechanisms refining the patterns of gene expression in both time and space. In this mechanism, timing is controlled by the rates of protein production, diffusion, and degradation.

Classical genetic studies in the nematode revealed the heterochronic gene regulatory network that times lineage-specific cell fate decisions in the epidermis and other tissues. This pathway is now famous for the landmark discovery of microRNAs, which have emerged as important regulators in timing mechanisms of plants and animals.

More complex timing programs involve systemic signaling systems that coordinate developmental processes among tissues throughout the organism. In humans, for example, passage from adolescence to adulthood is accompanied by rapid changes in growth and acquisition of sexual maturity. Likewise, in frogs, a wholesale transformation remodels the immature swimming tadpole into the sexually active air-breathing adult. Perhaps the most dramatic examples, however, can be found in insects and other arthropods where developmental transitions occur at regularly defined intervals. These transitions include molting, a process whereby the rigid exoskeleton is shed and resynthesized to accommodate increased body size due to cell growth, and metamorphosis, a stunning remodeling in which the immature larva changes into the reproductively mature adult.

In some cases, the duration of these developmental intervals can be regulated in response to nutritional or environmental cues. Diapause in insects and *Caenorhabditis elegans*, vernalization in plants, and hibernation in mammals are all examples of induced developmental and physiological states that aid an organism in overcoming adverse environmental conditions such as starvation and seasonal changes in temperature and diurnal cycle. Dietary deficiencies or excesses are also known to delay or accelerate puberty, respectively, in humans. In some cases, developmental delay is also programmed directly into the normal lifecycle of the organism. One well-known example is that of the periodic cicadas where nymphal stages are prolonged to either 13- or 17-year intervals, perhaps as a means to avoid predation. Determining how these developmental timing programs are regulated at the molecular level is therefore of fundamental importance in deciphering how an organism interacts with and adapts to its environment.

In this volume, we have assembled selected reviews that exemplify several of the processes described above. This is by no means a complete set of topics and examples, but rather the intent is to highlight some of the regulatory mechanisms that have been identified for several well-characterized developmental timing processes and to provide the reader inroads into the corresponding literature. Chapters 1 and 2 focus on hormonal regulatory processes that time metamorphosis in *Drosophila*. These chapters lay out the core molecular pathway and emphasize the importance of environmental variables, such as the circadian clock and nutrient availability, in regulating the timing of systemic hormonal pulses. In Chapter 3, the focus turns to intrinsic temporal waves of transcription factor activation and shutdown that drive patterning of neuronal progenitors in the *Drosophila* brain. The discussion then switches to consider the roles of microRNAs in timing developmental transitions in flies (Chapter 4), plants (Chapter 5), and worms (Chapter 6). The *C. elegans* discussion is extended by examination of how the complex miRNA regulatory network impacts steroid hormone production to time transitions and regulate diapause and longevity (Chapter 7). Consideration of how external cues can induce developmental and physiological transitions continues with broader discussion of diapause (Chapter 8) and mammalian hibernation (Chapter 9). Here, the focus shifts to how seasonal environmental changes, such as temperature, diurnal cycles, and moisture, induce changes that help organisms survive through periodic adverse conditions. Finally, the last two chapters echo many of the previous themes in considering vertebrate examples of developmental timing. Chapter 10 reviews the roles of ligand-bound versus -unbound thyroid receptors in controlling the timing of metamorphosis onset and its rate of progression in *Xenopus*, while Chapter 11 concludes with a synopsis of the molecular mechanisms that help time puberty onset in mammals. The focus here is on neuroendocrine and metabolic controls and intriguing ties to *C. elegans* timing mechanisms via miRNAs. We hope that these selected examples provide a useful overview of the mechanisms that have evolved to properly time developmental progression in a variety of organisms.

Ann E. Rougvie
Michael B. O'Connor

CHAPTER ONE

How Clocks and Hormones Act in Concert to Control the Timing of Insect Development

Francesca Di Cara, Kirst King-Jones[1]

Department of Biological Sciences, University of Alberta, Edmonton, Alberta, Canada

[1]Corresponding author: e-mail address: kirst.king-jones@ualberta.ca

Contents

1. General Considerations: The Problem of Time in Development 2
2. Evidence for Mechanisms Controlling the Timing of Developmental Events 3
3. Developmental Timing in Insects is Under Hormonal Control 4
4. Circadian Control and Developmental Transitions 7
5. Circadian Core Clock in Flies 9
6. Evidence of a Clock in the Insect Prothoracic Glands 12
7. Circadian Oscillations of EcR 13
8. Oscillations in the Prothoracic Gland: Input from the Brain 14
9. The Timing of CCAP Neuronal Differentiation is Ecdysone-Controlled 19
10. JHs and Developmental Timing 23
11. Outlook 26

References 28

Abstract

During the last century, insect model systems have provided fascinating insights into the endocrinology and developmental biology of all animals. During the insect life cycle, molts and metamorphosis delineate transitions from one developmental stage to the next. In most insects, pulses of the steroid hormone ecdysone drive these developmental transitions by activating signaling cascades in target tissues. In holometabolous insects, ecdysone triggers metamorphosis, the remarkable remodeling of an immature larva into a sexually mature adult. The input from another developmental hormone, juvenile hormone (JH), is required to repress metamorphosis by promoting juvenile fates until the larva has acquired sufficient nutrients to survive metamorphosis.

Ecdysone and JH act together as key endocrine timers to precisely control the onset of developmental transitions such as the molts, pupation, or eclosion. In this review, we will focus on the role of the endocrine system and the circadian clock, both individually and together, in temporally regulating insect development. Since this is not a coherent field, we will review recent developments that serve as examples to illuminate this complex topic. First, we will consider studies conducted in *Rhodnius* that revealed how

Current Topics in Developmental Biology, Volume 105
ISSN 0070-2153
http://dx.doi.org/10.1016/B978-0-12-396968-2.00001-4

circadian pathways exert temporal control over the production and release of ecdysone. We will then take a look at molecular and genetic data that revealed the presence of two circadian clocks, located in the brain and the prothoracic gland, that regulate eclosion rhythms in *Drosophila*. In this context, we will also review recent developments that examined how the ecdysone hierarchy delays the differentiation of the crustacean cardioactive peptide (CCAP) neurons, an event that is critical for the timing of ecdysis and eclosion. Finally, we will discuss some recent findings that transformed our understanding of JH function.

1. GENERAL CONSIDERATIONS: THE PROBLEM OF TIME IN DEVELOPMENT

It appears that everything must exist within the framework of time, and thus it comes as no surprise that life itself is inexorably intertwined with the passing of time. While this may apply to both for our daily lives as well as the grand scale of evolution, we will deal in this review with a more narrow question: How do organisms keep track of time so that they can ensure the proper chronology of distinct developmental events?

When we consider how time intersects with biology, two major categories come to mind. First, the ability of organisms to adapt to natural rhythms, such as earth's rotation that results in day and night (circadian rhythms), or the changes of the seasons, forcing plants and animals alike to adapt to annual cycles of the climate and different lengths of daylight (photoperiodism). Very different from these cyclical processes is the second category, which only applies to developing organisms: How are developmental programs controlled in such a fashion that they are turned on and off only when they are supposed to? Given the complexity of coordinating multiple parallel programs during the development of higher organisms, one would expect that any given sequence of events must follow a specific order, and that the precocious or delayed execution of developmental events could have potentially fatal consequences.

At first glance one might object and argue that developmental programs are "hardwired" and that the sequence of these events may be genetically programmed, much like the domino effect. If true, this would eliminate the need for a time-keeping mechanism during development. While many developmental events will follow indeed such a genetically determined sequence, a developing organism must have the ability to integrate environmental variables, such as nutrient availability, temperature, and the daily changes in daylight, which in turn may affect the timing of certain developmental events. In the section below, which we shall briefly discuss examples of how various factors may influence the timing of development.

2. EVIDENCE FOR MECHANISMS CONTROLLING THE TIMING OF DEVELOPMENTAL EVENTS

Indirect evidence that developmental events are temporally controlled comes from a plethora of evolutionary studies showing that heterochrony constitutes an important evolutionary mechanism. Heterochrony ascribes morphological differences between (usually closely) related clades to relative changes in developmental timing, causing—for instance—acceleration or retardation of organ growth, or the retention of juvenile features into adulthood (Gould, 1977). While evolutionary changes caused by heterochronic constraints are obviously fixed genetically, they still illustrate the plasticity of certain developmental traits, giving rise to the idea that gene programs underlying these traits are easily adaptable not because these programs are altered per se, but rather because their temporal appearance or duration are "fine-tuned."

More direct evidence for developmental timing comes from a large body of genetic studies. If temporal regulators of development do exist, one would hypothesize the existence of "temporal" mutations that affect primarily the timing of certain development events, but leave the nature of the event itself largely unaffected. The early 1980s saw the first examples of such "temporal" mutations in the form of the "lineage" (lin) genes that were linked to corresponding heterochronic mutations in *Caenorhabditis elegans*. Specifically, loss-of-function mutations of the *lin-14* and *lin-28* genes caused animals to skip certain cell fates, resulting in the precocious expression of developmental events. In contrast, loss-of-function mutations in the *lin-29* gene caused animals to re-iterate specific cell fates at later developmental stages, instead of just during a single larval stage, and larval cell lineages continued to be active during the adult stage (Ambros, 2003; Ambros & Horvitz, 1984; Lee, Feinbaum, & Ambros, 1993).

In other organisms, however, mutants with such well-defined temporal defects are either remarkably scarce or simply not recognized as such. Given that many "temporal" mutations would presumably wreak havoc in a developing organism, it stands to reason that most mutations that fall into this category go unrecognized as such. However, some exceptions exist. In humans, for instance, it has been argued that many morphological traits associated with Down syndrome are heterochronic in nature. For instance, patients often exhibit clinodactyly, the bending of the small finger towards the other fingers, which represents the survival of more primitive fetal structures to adulthood. Down syndrome also causes delays in liver and bone tissue development and is associated with a range of malformations affecting facial features; heart formation and brain development, all of which are consistent with the idea that heterochronic aberrations are the underlying theme of these symptoms (Wilson, 1988).

In this review we will focus on developmental timing in insects, and in particular, we will explore the relationship between endocrine control and circadian rhythms. Heterochronic phenotypes have been described for some mutations in insect genes, but they are not linked to a group of genes such as the "lineage" genes in *C. elegans*. Instead, the timing of developmental transitions in insects is associated with hormonal and circadian control. Studying developmental timing in insect model organisms provides certain advantages over other model systems. In particular, the presence of clearly delineated developmental transitions, in combination with well-described stage-specific events throughout the insect's life cycle, provides a unique platform for isolating and characterizing mutations that affect developmental timing. While *C. elegans* development is also characterized by molts, they lack the well-characterized endocrine and circadian systems found in insects. Here, we will review a string of recent advances from disparate research fields to highlight how endocrine control and circadian rhythms affect developmental timing in *Drosophila* and other model insects.

3. DEVELOPMENTAL TIMING IN INSECTS IS UNDER HORMONAL CONTROL

In insects, three classes of hormones act as key determinants of developmental timing: (1) Ecdysteroids, exemplified by the insect molting hormone ecdysone, which are closely related steroid hormones that promote developmental transitions such as the molts. (2) The prothoracicotropic hormones (PTTH), which are neurosecretory peptide hormones that stimulate the synthesis of ecdysteroids. (3) The juvenile hormones (JHs), which when present at high levels ensure that molts result in another immature stage, such as larvae or nymphs. For simplicity, we will refer to these hormones generically as ecdysone, PTTH, and JH.

JH, as the name suggests, keeps the developing insect in a juvenile state by repressing metamorphosis. A decline of JH titers during the last instar allows a subsequent pulse of ecdysone to trigger metamorphosis, defining the onset of insect maturation that leads to the completion of all adult structures (Fig. 1.1A) (Mizoguchi, 2001; Riddiford, Truman, Mirth, & Shen, 2010; Rountree & Bollenbacher, 1986). As one would expect, genetic manipulations that cause (a) JH titers to prematurely regress, (b) interfere with the expression of the JH receptor, and (c) mutations in enzymes necessary for JH synthesis all trigger premature metamorphosis in a range of insects (Daimon et al., 2012; Gu & Tsai, 2012; Konopova & Jindra, 2007; Minakuchi, Namiki, Yoshiyama, & Shinoda, 2008; Tan, Tanaka, Tamura, & Shiotsuki, 2005; Wang, Ali, Moriyama, Iwanaga, & Kawasaki, 2012). We will discuss the role of JH and developmental timing in a separate section in further detail.

During insect development, pulses of ecdysone initiate all developmental transitions, including the molts, wandering behavior, and metamorphosis. Ecdysone is produced as a prohormone in the prothoracic glands, and further processed in target tissues to achieve its biologically active form. Our understanding of how the onset, amplitude, and duration of ecdysone pulses are controlled is still incomplete. However, we do know that the rhythmic release of PTTH, a brain-derived neuropeptide, elicits the onset of ecdysone pulses (Rybczynaki, 2005), similar to what was described for adrenocorticotropic hormone (ACTH) and glucocorticoids in vertebrates (Lightman & Conway-Campbell, 2010). PTTH transcripts display ultradian rhythmicity in *Drosophila* (McBrayer et al., 2007), which in turn is under control of brain neurons that are critical for establishing circadian behaviors, and we will explore this relationship in more detail in the next section.

To initiate a transition between two developmental stages, the brain releases PTTH (Fellner, Rybczynski, & Gilbert, 2005; Kawakami et al., 1990; McBrayer et al., 2007; Westbrook & Bollenbacher, 1990), which induces a signaling cascade that activates ecdysone biosynthesis in PG cells (Gilbert, 2004; Rewitz, O'Connor, & Gilbert, 2007) and its secretion into the hemolymph (Huang, Warren, & Gilbert, 2008). Once ecdysone reaches its target tissues, it is converted into its active form, 20-hydroxyecdysone (20E) and binds the heterodimer of two nuclear receptors, ecdysone receptor (EcR) and ultraspiracle (USP) (King-Jones & Thummel, 2005; Koelle et al., 1991; Yao et al., 1993) resulting in the activation of a transcriptional response that drives behavioral and physiological changes as well as the molts (Bollenbacher, Agui, Granger, & Gilbert, 1979; Bollenbacher, Smith, Goodman, & Gilbert, 1981; Jiang, Lamblin, Steller, & Thummel, 2000; King-Jones & Thummel, 2005; Riddiford, 1993; Sullivan & Thummel, 2003; Thummel, 1996, 2002). The nature of a molt

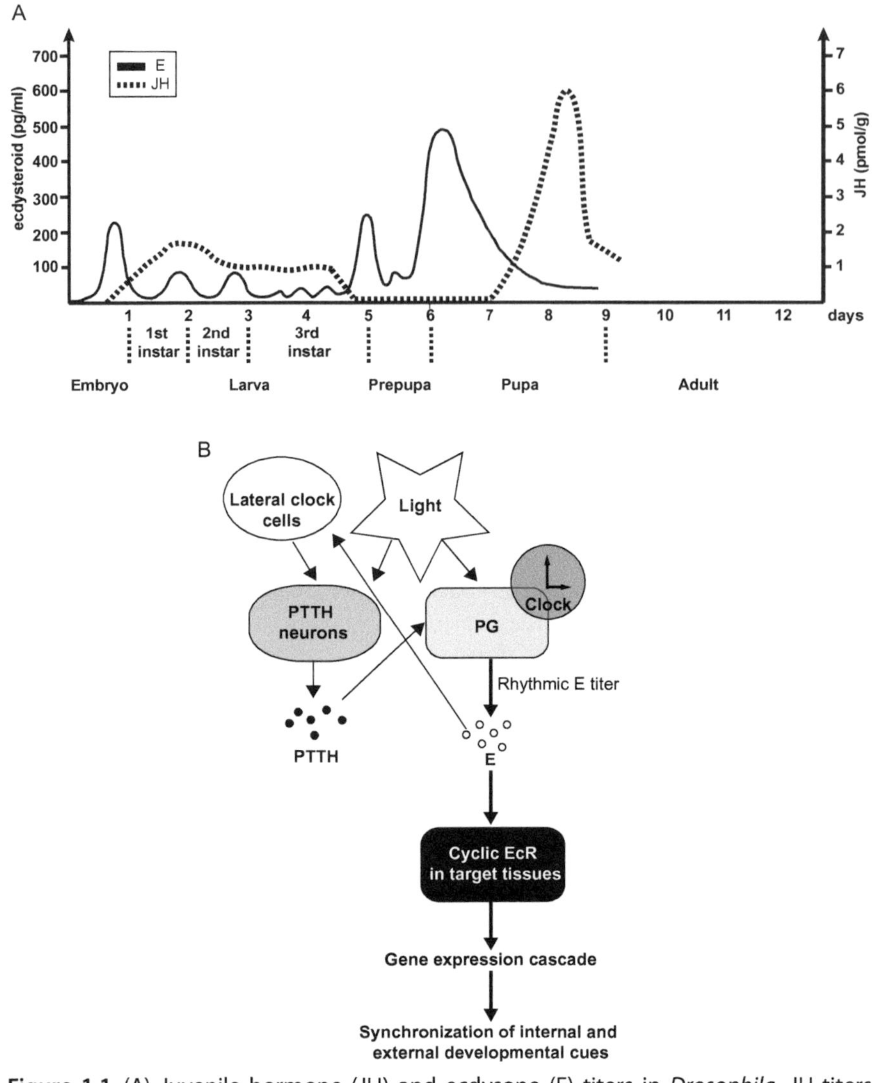

Figure 1.1 (A) Juvenile hormone (JH) and ecdysone (E) titers in *Drosophila*. JH titers decline in third instar larvae, prior to the large ecdysone peak that triggers puparium formation. (B) Daily ecdysone (E) titer rhythms in *Rhodnius*. Summary of the circadian clocks that regulate developmental hormones in *Rhodnius*. Lateral ventral clock neurons (LNv) signal to PTTH-producing neurons. Light and PTTH signals entrain the PG cells for ecdysone secretion, resulting in rhythmic release of the hormone. The ecdysone receptor, EcR, anticipates these daily rhythms by undergoing cyclic nucleocytoplasmic oscillations in target cells. PG: prothoracic gland, PTTH: prothoracicotropic hormone.

depends on the presence or absence of JH, where larval to larval molts require high levels of JH, but larval-pupal transitions necessitate low JH titers (Doane, 1973; Mizoguchi, 2001; Riddiford, 1970a, 1970b; Truman, Riddiford, & Safranek, 1974).

An apparently complex network of signaling events regulates the production of ecdysone (Caceres et al., 2011; Fellner et al., 2005; Gibbens, Warren, Gilbert, & O'Connor, 2011; Ou, Magico, & King-Jones, 2011; Rewitz, Yamanaka, Gilbert, & O'Connor, 2009; Song & Gilbert, 1995, 1997) and JH (Huang et al., 2011; Kinjoh et al., 2007; Wigglesworth, 1936). The downstream effects of these hormones result in a highly regulated succession of developmental gene programs, and we refer the reader to the following references for more detail (Ou & King-Jones, 2013; Rewitz, Yamanaka, & O'Connor, 2013; Song & Gilbert, 1998; Thummel, 1995, 2001; Yamanaka, Rewitz, & O'Connor, 2013). Despite the progress in dissecting the regulatory pathways that drive insect development, the mechanisms controlling the timing of developmental transitions are still largely unknown.

4. CIRCADIAN CONTROL AND DEVELOPMENTAL TRANSITIONS

Although the timing of developmental transitions in insects is not directly determined by circadian clocks, interesting links do exist between molting, eclosion, and circadian circuitries. For instance, under standard laboratory conditions, the major ecdysone pulses in *Drosophila* show a clear daily rhythm, where the embryonic and first two larval pulses occur ~24 h apart, followed by a third instar pulse 48 h later (Fig. 1.1A). Similar rhythms have been described in other insects as well (Cymborowski, Smietanko, & Delbecque, 1989; Mizoguchi, Dedos, Fugo, & Kataoka, 2002; Mizoguchi, Ohashi, Hosoda, Ishibashi, & Kataoka, 2001; Richter, 2001; Schwartz & Truman, 1983). At lower temperatures, however, insect development slows down, whereas circadian rhythms are temperature-compensated (not affected by temperature fluctuations), suggesting that the link between circadian clocks and molting is not based on a simple linear relationship. Rather, it appears that circadian clocks provide a frame of reference—or gate—during which a developmental transition might occur, thus linking the timing of molting and eclosion to daily cycles.

Four decades ago, James Truman demonstrated that in *Manduca sexta*, the secretion of PTTH is regulated by a photosensitive circadian clock located in the brain, thus introducing the concept of circadian "gating" during development (Truman & Riddiford, 1974). The photoperiodic gating mechanism ensures that PTTH can be released every day only during a specific 8-h window (Truman, 1972; Truman & Riddiford, 1974). During this photoperiodic gate, larvae determine whether they have achieved critical weight, a checkpoint that ensures that larvae have sufficient resources to survive metamorphosis. Larvae that have not achieved critical weight during this time window continue feeding until the next photoperiodic gate is reached the following day.

More direct evidence linking circadian clocks to molting comes from the observation that the PG of the blood-sucking insect *Rhodnius* harbors an endogenous circadian clock. In particular, the circadian core proteins TIMELESS and PERIOD (see Section 5) have been shown to co-localize and oscillate in PG cells of *Rhodnius* and *Drosophila*, suggesting that circadian control over ecdysone biosynthesis may also directly occur in the PG (Steel & Vafopoulou, 2006).

In *Drosophila*, the circadian clock neurons (aka LN or lateral neurons) innervate the dendrites of the PTTH producing neurons (McBrayer et al., 2007; Siegmund & Korge, 2001). Interestingly, PTTH transcript levels oscillate, but they do so with an ultradian rather than circadian rhythm. Specifically, PTTH mRNA displays an 8-h rhythmicity in *Drosophila*, at least when determined via semi-quantitative PCR based on total RNA (McBrayer et al., 2007). Also, it is unknown whether this oscillatory transcript behavior corresponds to similar cycles of secreted PTTH peptides. However, there is evidence that ecdysone titers underlie daily rhythms in *Bombyx* and *Rhodnius*, raising the possibility that *Drosophila* ecdysone levels fluctuate in accordance with the transcriptional PTTH oscillations (Ampleford & Steel, 1985; Satake, Kaya, & Sakurai, 1998). *Drosophila* larvae with ablated pigment dispersing factor (PDF) neurons display both an increase in PTTH periodicity and higher PTTH transcript levels.

Additional supporting evidence for circadian clocks controlling developmental transitions comes from studies on eclosion in *Drosophila*. Eclosion within a *Drosophila* population typically follows a rhythm, where most animals eclose during dawn. Interestingly, *Drosophila* pupae harbor a functional clock in the PG at the time of eclosion, and disrupting this clock by overexpressing *timeless* in this tissue results in a disruption of normal eclosion timing (Myers, Yu, & Sehgal, 2003). For the PG clock to function properly, circadian LNs must be present in order to maintain accurate eclosion rhythms in the population, indicating that the PG acts as a peripheral clock.

Next, we will take a more detailed look at circadian clocks in *Drosophila*, and follow with an overview of the evidence that insect prothoracic glands harbor their own peripheral clocks.

5. CIRCADIAN CORE CLOCK IN FLIES

Most organisms have evolved a circadian clock in order to anticipate daily environmental changes in light, temperature, and mate availability. Thus, they play pivotal roles in synchronizing and integrating physiological and biochemical processes (Allada & Chung, 2010). Although circadian rhythms are endogenous ("built-in," self-sustained), they are adjusted (entrained) by local environmental cues called zeitgebers (German for "synchronizer"), such as daylight or temperature.

In order to be considered a circadian clock, a biological system needs to have four fundamental characteristics: (1) an oscillatory event that occurs approximately every 24 h; (2) these cycles occur even in constant conditions (e.g. constant darkness), indicating the presence of a self-sustained clock; (3) the rhythm must be entrainable by environmental cues, most commonly daylight; and (4) the oscillatory rhythm is temperature-compensated, meaning that shifts in the external temperature do not alter the periodicity of the clock. Much of the work that contributes to our current understanding of circadian clocks is derived from studies in *Drosophila* melanogaster (Collins & Blau, 2007; Dubruille & Emery, 2008; Stanewsky, 2003; Zheng & Sehgal, 2008). The main components of the clock are transcription factors that act in an auto-regulatory feedback loop (Nitabach & Taghert, 2008).

The heterodimer CLOCK/CYCLE binds to specific consensus sequences (E-BOX) in the regulatory regions of their target genes. Among these target genes are two circadian genes: *timeless* and *period*. Both *timeless* and *period* are actively transcribed during the day, but by early night, TIMELESS and PERIOD protein levels have risen in the cytoplasm, which allows them to dimerize and translocate to the nucleus. There, TIMELESS/PERIOD dimers bind to CLOCK/CYCLE, thereby blocking their transcriptional activity. In the canonical core clock, CYCLE/CLOCK heterodimers act as positive regulators of *period* and *timeless* transcription (Meyer, Saez, & Young, 2006; Yu, Zheng, Houl, Dauwalder, & Hardin, 2006), which activates their own repressors, thus generating a transcriptional feedback loop that builds the foundation of circadian periodicity (Kadener, Menet, Schoer, & Rosbash, 2008) (Fig. 1.2). The molecular core components CLOCK, CYCLE, TIMELESS, and PERIOD are modified at the post-transcriptional level allowing the system to fine-tune cycle

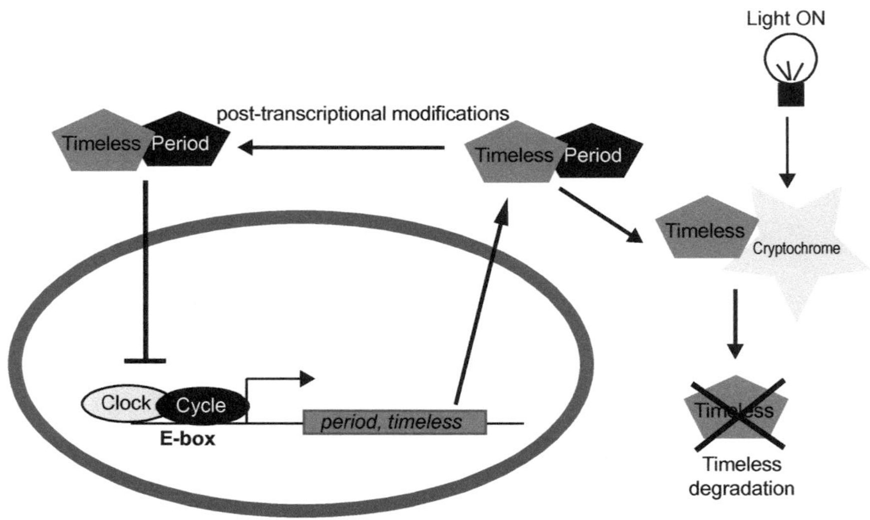

Figure 1.2 The core clock in *Drosophila*. (A) The transcription factors CYCLE and CLOCK transcriptionally activate *period* and *timeless*. The *period* and *timeless* mRNAs are transported to the cytoplasm in the early evening where they are translated into their respective proteins. Once PERIOD and TIMELESS proteins are modified by phosphatases and kinases, they are translocated as a dimer to the nucleus in the late evening. Once in the nucleus, PERIOD–TIMELESS dimers inhibit the transcriptional activity of CLOCK–CYCLE, resulting in a decline in *period* and *timeless* mRNA levels. Nuclear TIMELESS and PERIOD are then degraded, allowing the derepression of *period* and *timeless* promoters, which starts the cycle all over again. (B) In the presence of light, the protein CRYPTOCHROME binds TIMELESS, preventing its dimerization with PERIOD and leading to its degradation. This cycle continues in constant darkness but is amplified in light–dark cycles, when the blue light receptor CRYPTOCHROME is activated. When the PERIOD–TIMELESS heterodimer is accumulating in the early evening, light retards the accumulation of the PERIOD–TIMELESS dimer and sets the clock back. When PERIOD–TIMELESS are degrading late in the night, light accelerates this breakdown and sets the clock ahead.

times to roughly 24 h (Chiu, Vanselow, Kramer, & Edery, 2008; Fang, Sathyanarayanan, & Sehgal, 2007; Kim & Edery, 2006; Kim, Ko, Yu, Hardin, & Edery, 2007; Kivimae, Saez, & Young, 2008; Meissner, Kilman, Lin, & Allada, 2008). CRYPTOCHROME (CRY) is a cell-autonomous blue light photoreceptor that mediates the entrainment of the core clock by light, through a mechanism that involves the degradation of TIMELESS upon stimulation by light (Dubruille & Emery, 2008).

Most cells harbor their own circadian machinery, however, these independent clocks are usually aligned with the activity of a central clock that acts on an organismal level. Central clocks are also referred to as pacemaker neurons, such as the suprachiasmatic nuclei in humans and the LNs in *Drosophila*. In particular, the adult *Drosophila* brain harbors ~150 neurons that express the molecular clock components and comprise the central pacemaker neural networks. These neurons consist of three clusters of dorsal neurons (DN1, DN2, and DN3), one cluster of the dorsal lateral neurons (LN_ds), one of large ventral lateral neurons (l-LN_vs), one cluster of small ventral lateral neurons (s-LN_vs), and the lateral posterior neurons (LPN). A subset of the LNs express a neuropeptide called PDF and are considered to represent the actual pacemaker cells in the *Drosophila* brain (Collins & Blau, 2007; Dubruille & Emery, 2008; Nitabach & Taghert, 2008). The s-LN_vs appear to indirectly project to the prothoracic gland by innervating neurosecretory neurons of the lateral protocerebrum, PG–LP, which in turn directly project to the prothoracic gland (Siegmund & Korge, 2001).

In *Drosophila*, circadian clocks are also found in peripheral tissues (Giebultowicz, 2001; Saunders, 1982). Although there is evidence that the central pacemaker clock can synchronize peripheral clocks through secretion of humoral factors, emerging evidence supports the idea that peripheral clocks can entrain their own rhythm independently (Liu, Lorenz, Yu, Hall, & Rosbash, 1988; Plautz, Kaneko, Hall, & Kay, 1997). While CRY is expressed in some of these peripheral clocks, it appears to function in a photoreceptor-independent manner, suggesting fundamental differences to oscillator mechanisms found in core clocks (Krishnan et al., 2001). The presence of peripheral clocks independent from the central pacemaker neurons has been reported in multiple *Drosophila* tissues, including the fat body (So et al., 2000; Xu, Zheng, & Sehgal, 2008) and antennae (Krishnan et al., 2001) where local clocks control metabolism and feeding behavior in the adult fly. Other peripheral clocks in *Drosophila* seem to operate in the *Corpus allatum* (CA) and the PG, both structures of the central larval endocrine system, the ring gland (RG) (Allada & Chung, 2010; Emery,

Noveral, Jamison, & Siwicki, 1997). The peripheral clock in PG cells has been described in detail in *Rhodnius prolixus*, which we will be the focus of the next section.

6. EVIDENCE OF A CLOCK IN THE INSECT PROTHORACIC GLANDS

The endocrine system is believed to be the principal route through which the pacemaker cells transmit temporal information to the target tissues (Handler & Konopka, 1979). Therefore, the secretion of hormones seems to be driven by pacemaker neural cells. A daily rhythm of steroid hormones has been reported in several insects (Richter, 2001; Satake et al., 1998), and circadian control has been proposed for *Samia cynthia* (Mizoguchi & Ishizaki, 1982), *Galleria mellonella* (Cymborowski, Porcheron, & Cassier, 1991) and demonstrated in *R. prolixus* (Steel & Vafopoulou, 2006).

Vafopoulou and Steel were the first to demonstrate the circadian regulation of an invertebrate hormone, based on studies of explanted prothoracic glands isolated from *Rhodnius* larvae. They showed that ecdysone synthesis was 3–5-fold higher at night, consistent with the observation that ecdysone hemolymph titers in living larvae display a similar pattern (Vafopoulou & Steel, 1991). The results also exhibited a 24-h periodicity in constant darkness (DD), which suggested that these daily changes in ecdysone biosynthesis are controlled by a free-running clock. Immunostaining of *Rhodnius* PGs not only revealed the presence of the clock proteins PERIOD and TIMELESS in tissue, but also demonstrated that both proteins undergo synchronous nucleocytoplasmic shuttling with a 24-h periodicity. Specifically, at night, when intense ecdysone synthesis occurs, TIMELESS and PERIOD shuttle to the nucleus, while both proteins are absent from both nucleus and cytoplasm during daytime. These data demonstrated that the prothoracic glands in *Rhodnius* harbor a circadian clock, which presumably controls daily rhythms of steroid hormone synthesis in this species.

However, PTTH stimulates ecdysone synthesis in all insects, which raised the question whether PTTH is under circadian control as well. In *Rhodnius*, the release of PTTH was found to peak at night in both *in vivo* and *in vitro* experiments. The exposure for several hours to constant light (LL) conditions tends to block the secretion of PTTH, while a shift from LL to DD conditions restored rhythmic PTTH release (Vafopoulou & Steel, 1996). These data suggested the presence of a circadian clock that drives rhythmic PTTH release in the *Rhodnius* brain. Double-immunostaining larval *Rhodnius* brains with anti-PTTH and anti-PERIOD (or anti-TIMELESS) antibodies revealed that dendrites originating from PERIOD and TIMELESS-positive ventral-lateral and dorsal neurons extended very close to the axons of the PTTH-positive neurons (Vafopoulou, Steel, & Terry, 2005). These observations suggested that a subset of neurons in the brain of *Rhodnius* larvae, thought to be the equivalent to the *Drosophila* lateral clock neurons, harbor photosensitive clocks and establish synaptic interactions with the PTTH-producing neurons, thereby controlling the periodicity of PTTH release.

Since PG cells possess their own local photosensitive clock and rhythmically secrete ecdysone as well, it may seem that the role of PTTH in stimulating ecdysone synthesis is redundant. However, abolishing PTTH function, either by genetic ablation of PTTH neurons in *Drosophila* (McBrayer et al., 2007), or, in other insects, by decapitation and neurotoxic treatments with tetrodotoxin (TTX) all negatively affected ecdysteroid synthesis. Rather, it appears that PTTH is required for entraining the PG clock (Vafopoulou & Steel, 2006). The experiments described above provided the first molecular and physiological evidence for the existence of a peripheral clock in the PG. While the PG maintains its own circadian rhythm, the clock has to be calibrated with external cues that act as zeitgebers, such as environmental signals (light) and neurosecretory input (PTTH) (Fig. 1.1B). Next, we will discuss how these daily ecdysone rhythms are transduced in the developing organism.

7. CIRCADIAN OSCILLATIONS OF EcR

In insects, the heterodimer of the nuclear receptors EcR and Usp binds to 20E, the principal step in transducing the ecdysteroid signal in target tissues (Riddiford, 1993; Russel, 1996). In the section above, we have discussed daily oscillations of ecdysone production, which leads us to the question as to whether the ecdysteroid receptor (EcR/Usp) would also exhibit some oscillatory behavior. Indeed, in *Rhodnius*, EcR exhibits circadian shuttling in specific tissues and thus appears to respond to or anticipate the daily peaks of ecdysone. Specifically, immunostains with EcR antibodies in epidermis cells showed that during daytime, EcR protein levels increase and tend to accumulate in the cytoplasm until early night, while around midnight, when ecdysone levels are high, EcR protein localizes to the nucleus (Vafopoulou & Steel, 2006, 2012a). Similar nucleocytoplasmic shuttling of EcR protein was observed in the fat body, the rectal epithelium, oenocytes, the PG, and the salivary glands. EcR shuttling was also observed in the lateral clock neurons, where EcR was shown to co-localize in the nuclei with the circadian protein PERIOD during the night, albeit most of the PER signal appeared to be cytoplasmic. Other tissues, however, showed no circadian shuttling of EcR, including the Malpighian tubules and spermatocytes.

Taken together, these findings suggest that hormones act as circadian outputs to coordinate periodic changes in the physiology of the animal (Fig. 1.1B). Ecdysone is periodically released, while EcR is similarly cycling in anticipation of maximum hormone levels. Recently, it was shown that the circadian system itself is reorganized during metamorphosis in response to ecdysone, suggesting that the endocrine system not only responds to circadian cues, but also signals back to it (Vafopoulou & Steel, 2012b).

EcR protein exhibits circadian cycling in a range of tissues (Vafopoulou & Steel, 2006, 2012a), and many nuclear receptors appear to underlie circadian regulation at the transcript level. In particular, large-scale approaches examining all 49 mouse nuclear receptor genes revealed that roughly half of them are rhythmically expressed and linked to basal metabolic functions (Yang et al., 2006). In addition, nuclear receptor-signaling pathways have been shown to play a direct role in central and peripheral circadian clocks in mammals. For instance, RORα plays a direct role in the suprachiasmatic nuclei as an activator of *Bmal1* transcription, the vertebrate ortholog of the *Drosophila cycle* (Sato et al., 2004). In addition, the REV-ERB/retinoid-related orphan receptor (ortholog of the insect *E75*, an early ecdysone response gene) regulates the expression of core clock genes and contributes to the robustness of the clock mechanism (Preitner et al., 2002; Solt et al., 2012). For a detailed discussion of the link between nuclear receptors and circadian rhythms we refer the reader to a review by Teboul, Guillaumond, Grechez-Cassiau, and Delaunay (2008).

8. OSCILLATIONS IN THE PROTHORACIC GLAND: INPUT FROM THE BRAIN

In a recent study, Morioka, Matsumoto, and Ikeda (2012) demonstrated that transcriptional rhythms of *period* in *Drosophila* PG cells are autonomous, but influenced by and dependent on the central nervous system (CNS). To study circadian rhythms on a cellular level in prothoracic glands, the authors used video imaging to follow the expression of a *period*-driven *luciferase* reporter gene. To test whether the CNS influences these oscillations, *ex vivo* organ cultures of isolated RGs were compared to RGs that were still attached to the CNS (CNS-RG). In either case, the oscillation of *period* transcript levels was stable for many days. Interestingly, in CNS-RGs, *period* oscillations started in the proximal part of the RG, corresponding to where projections from the brain innervate the RG. Subsequently, these oscillations then propagate in a wave-like manner to the distal part of the RG. However, when RGs were not attached to the CNS, this spatial phasing was lost, and instead oscillations occurred uniformly and synchronized across PG cells. These findings indicate that *period* transcript oscillations in PGs are self-sustained and that signals from the brain determine how these transcriptional rhythms are spatially synchronized across PG cells.

In the same report, the CNS was also shown to be required for relaying a light-dark transition response to the prothoracic glands. When CNS-RGs were moved from LD to DD conditions, the first 12-h cycle resulted in a high amplitude of *period* transcript levels, which then rapidly declined to a lower but stable level under constant darkness (DD). This light-dark transition response did not occur when brain-derived signals where blocked either by administering the Na^+ channel inhibitor TTX or the L-type Ca^{2+} channel inhibitor nimodipine, or when RGs were cultured in isolation from the CNS. Clearly, exposing CNS-RGs to light relayed photic information to the RG, and indeed, a mutation (*cry*^{b}) in the blue light photoreceptor gene *cryptochrome* also abolished the light-dark transition response.

To further characterize the oscillatory response of *period* to light, the authors exposed CNS-RGs to 30 min light pulses either shortly before a peak of *period* expression, or shortly after such a peak. This treatment affected the phase of the following period peaks, resulting in delayed or advanced peaks, respectively. Similar to the above series of experiments, the effect of the light pulse was abolished using pharmacological inhibitors of synaptic transmission, or treating RGs in isolation. Likewise, removing *cryptochrome* function resulted in strongly diminished phase shifts, demonstrating that the CNS plays a key role in receiving and processing light information that is relayed to the PG. However, this does not appear to generally true for all peripheral clocks, since the authors go on to show that the Malpighian tubules are in fact photoresponsive and thus independent from the CNS.

How do these findings relate to developmental timing? The authors argue that the phase shifts observed after administering light pulses correspond well to earlier findings that showed light-induced shifts in eclosion rhythms of *Drosophila* populations (Pittendrigh, 1964; Winfree, 1970). As discussed in an earlier section, PG cells are thought to gate eclosion timing, and this gating in turn is linked to circadian-controlled outputs from the PG, which is strongly supported by this study. It follows that exposing pupae to light causes phase-shifting of *period* pulses, which in turn distorts the gating of eclosion rhythms through a mechanism that is still unclear.

The above experiments used a *period-luciferase* transgene as a transcriptional readout, but how do these transcript peaks relate to the nuclear oscillations of PERIOD–TIMELESS heterodimers? After all, the circadian circuit in insects is based on the nuclear translocation of PERIOD and TIMELESS, which then act as negative regulators of *clock* and *cycle* (Fig. 1.2). When explanted, CNS-RGs were incubated under LD conditions, PERIOD showed rhythmic nuclear localization, similar to what has been observed previously (Emery et al., 1997), however, this was not observed in constant darkness. In contrast, rhythmic accumulation of nuclear PERIOD was evident in cultured LNs under DD conditions, suggesting rhythmic accumulation of PERIOD in PG nuclei does not rely on LN input alone.

Interestingly, previous work had shown that TIMELESS protein, the heterodimeric partner of PERIOD, does exhibit nuclear rhythms in the PG in constant darkness (Myers et al., 2003), raising the possibility that PERIOD behaves differently from TIMELESS. To test this directly, the authors examined the nuclear rhythm of TIMELESS in CNS-RGs or RGs alone. Indeed, nuclear TIMELESS rhythms were maintained in PG cells of CNS-RGs and even in RGs not attached to a brain, regardless of whether the tissues were cultured in LD or DD environments. This data suggest that TIMELESS behaves differently from its dimerization partner PERIOD, since rhythmic accumulation of TIMELESS in PG nuclei is not driven by signals originating from the CNS, and therefore appears to be self-sustained. The difference between TIMELESS and PERIOD regulation is intriguing, given their close functional relation, however the physiological significance of this novel finding remains unclear. It is possible that TIMELESS is also part of non-canonical circadian transcriptional complex in PG cells where PERIOD is not required, in which case the dependency on the CNS would reflect the presence of different TIMELESS protein complexes.

Intracellular Ca^{2+} signaling has long been known to play an important role in ecdysteroidogenic signal transduction (Henrich, 1995; Venkatesh & Hasan, 1997). Given the dependency of oscillatory PG components on the CNS, Morioka and colleagues examined how Ca^{2+} levels fluctuated in PG cells of cultured CNS-RGs. They found that intracellular Ca^{2+} concentrations spiked spontaneously, and that the frequency of these spikes could be reduced by exposing the CNS to 20 min of blue light, while red light had little effect. In addition, the blue light treatment was ineffective in the presence of TTX, indicating that synaptic transmission from the CNS to the RG was necessary for this inhibitory signal. Finally, *cryptochrome* mutants displayed overall higher Ca^{2+} spike frequencies, however a partial reduction was still observed under blue light exposure, suggesting that some blue light sensor activity remained in these mutants.

As expected Ca^{2+} spike frequencies showed daily rhythms, with higher frequencies occurring during the night, and were sensitive to TTX and nimodipine. In addition, Ca^{2+} spiking was significantly less frequent at all times in cultured RGs that were severed from the CNS. Taken together these data demonstrated that CNS signaling to the PG is critical for the generation of Ca^{2+} spiking rhythms, and that blue light blocks this signal in a cryptochrome-dependent manner (Fig. 1.3A).

Drosophila eclosion rhythms are known to peak at dawn, and the authors therefore suggest that the first appearance of light might directly lower intracellular calcium concentration in PG cells, thus triggering the decline of ecdysone production that is necessary for activating the endocrine cascade necessary for eclosion to occur (Mesce & Fahrbach, 2002).

However, the situation is likely more complex *in vivo*. Clearly, while the proposed model provides an elegant mechanism for the gating of eclosion rhythms, it does not explain the problem of competence, namely why the same signal 24 h prior to these events does not trigger eclosion (when the pupa is not yet ready to eclose). In addition, and outlined in an earlier section, ablation of the circadian LNs results in arrhythmic eclosion behavior, and while LNs do innervate PTTH-producing neurons, they also signal to dorsal neurons that release eclosion hormone (EH) (Blanchardon et al., 2001), and a different set of pacemaker neurons projects to the cells that release crustacean cardioactive peptide (CCAP) (Park, Schroeder, Helfrich-Forster, Jackson, & Ewer, 2003). Both hormones, EH and CCAP, regulate eclosion behavior downstream of the initial event, a drop in ecdysone levels (Fig. 1.3B). A decline in ecdysone levels not only triggers the release of EH, but also provides competence of the CNS to respond EH (Morton & Truman, 1988; Truman, 1981; Truman, Rountree, Reiss, & Schwartz, 1983). Similarly, the decrease of ecdysone concentration under a critical threshold allows Inka cells to release Ecdysis triggering hormone (ETH), another hormone required

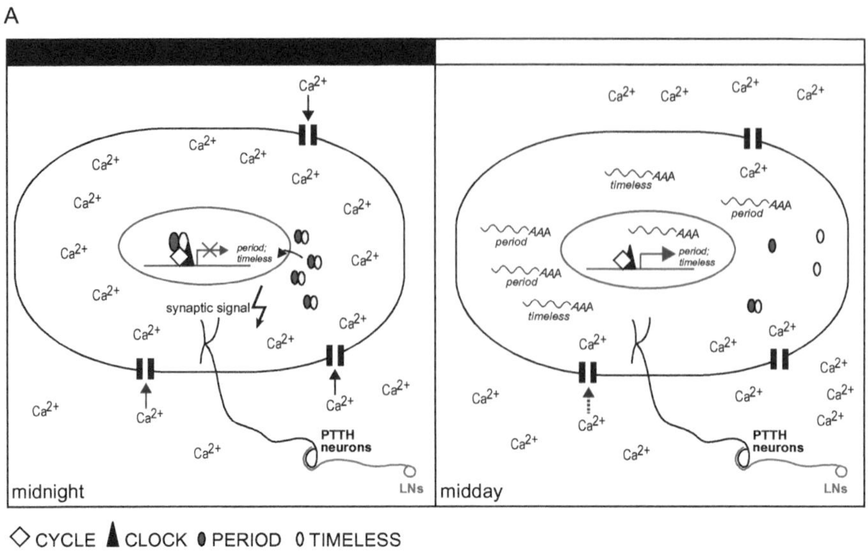

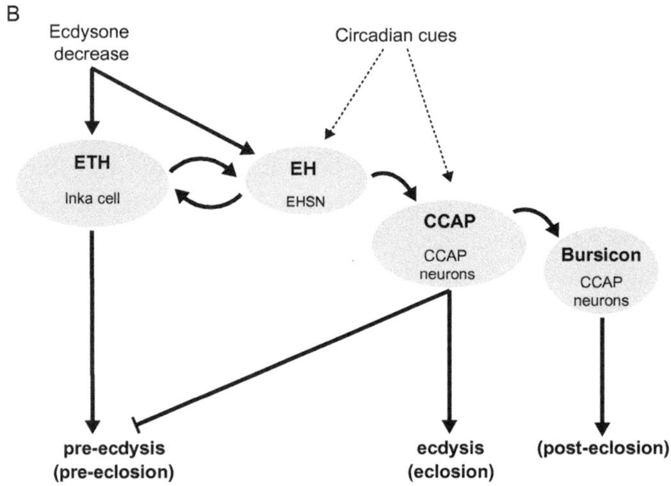

Figure 1.3 (A) Diagram of the signaling events occurring in PG cells shortly before eclosion. The peripheral clocks in PG cells are coordinated by synaptic input coming from the circadian lateral neurons (LNs) that project onto the dendritic field of PTTH-producing neurons, both of which are located in the CNS. The central clock present in LNs is entrained by light. During the night, synaptic input from the LNs cause intracellular [Ca^{2+}] spikes in PG cells that trigger the nuclear translocation of PERIOD–TIMELESS heterodimers. This event may trigger the endocrine cascade underlying eclosion. During daylight, intracellular [Ca^{2+}] concentrations are lower than during the night, and PERIOD–TIMELESS protein levels are very low, however, their transcripts accumulate in the cytoplasm. (B) Signaling cascade controlling eclosion. Two days prior to eclosion, ecdysone levels start to decline. This decrease stimulates the secretion of eclosion hormone (EH) and ecdysis triggering hormone (ETH), which stimulate the production of each other in a positive feedback loop. ETH stimulates pre-ecdysis (here pre-eclosion, since this is the pupal-adult transition) behaviors, while EH increases the secretion of two other peptide hormones, Crustacean cardioactive peptide (CCAP) blocks pre-ecdysis and simultaneously triggers ecdysis, while bursicon regulates post-eclosion behaviors such as cuticle tanning.

for eclosion (Kingan & Adams, 2000). EH and ETH stimulate each other's production in a positive feedback loop. ETH appears to be primarily involved in stimulating pre-ecdysis behaviors, while EH stimulates secretion of CCAP, which triggers ecdysis behaviors. These hormonal cascades were originally characterized in *Bombyx*, however, similar events have been observed during *Drosophila* eclosion (Mesce & Fahrbach, 2002). Clearly, given the complexity of the neuro-endocrine network that controls eclosion, more work is required to identify the exact events that determine the timing by which the onset of eclosion is regulated. In the next section, we will discuss the CCAP neurons in more detail and in particular, we will consider the temporally tuned differentiation of a sub-group of the CCAP neurons as a model for temporally controlled neuronal differentiation during development.

9. THE TIMING OF CCAP NEURONAL DIFFERENTIATION IS ECDYSONE-CONTROLLED

CCAP neurons are critical for proper larval and pupal ecdysis, as well as maintaining eclosion rhythms in *Drosophila* and *Manduca* (Ewer & Truman, 1996; Gammie & Truman, 1997; Jackson et al., 2001; Kim, Zitnan, Cho, & Adams, 2006; McNabb et al., 1997; Mesce & Fahrbach, 2002; Park et al., 2003). The ablation of CCAP neurons prolongs larval ecdysis and results in a failure to complete pupal ecdysis, thus causing lethality. CCAP neurons express the neuropeptides CCAP and the subunits of the bursicon peptide hormone, bursicon-α (Burs-α) and its partner bursicon-β (Burs-β) (Kim et al., 2006; Veverytsa & Allan, 2011).

CCAP neurons are located in the ventral nerve cord, and are comprised of a single CCAP interneuron (CCAP-IN) per hemi-segment T1-A7, as well as a single CCAP efferent (CCAP-EN) per hemisegment T3-A4 (Fig. 1.4A) (Martinez-Padron & Ferrus, 1997; Prokop, 2006; Zhao et al., 2008). Earlier work had demonstrated that CCAP neurons undergo morphological remodeling after pupal ecdysis (Zhao et al., 2008). In a recent study, Veverytsa and Allan closely examined CCAP neuron organization during *Drosophila* prepupal development and found that the CCAP network is also remodeled just before pupal ecdysis. In particular, they identified 12 late emerging CCAP neurons that are spatially restricted to the segments A5–A9, and they go on to show that this subset of CCAP neurons is sufficient to ensure normal pupal ecdysis (Veverytsa & Allan, 2012).

The late-emerging CCAP neurons appear during early pupariation before pupal ecdysis, and can be subdivided into two groups each

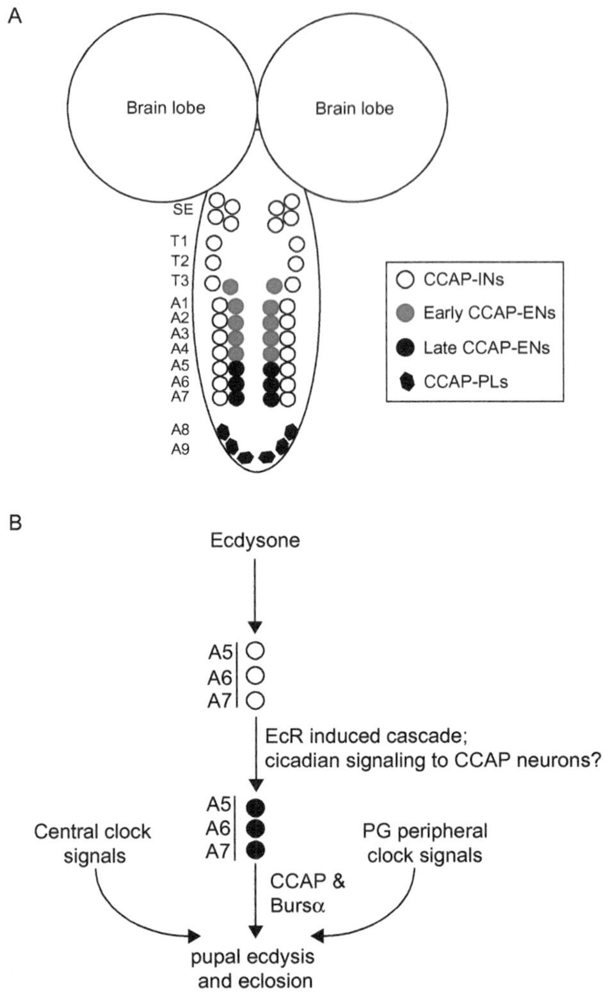

Figure 1.4 (A) Schematic summary of CCAP neurons (modified from Veverytsa & Allan, 2012). Distribution of the CCAP interneurons and efferents (CCAP-IN and CCAP-EN) subesophageal (SE), thoracic (T), and abdominal (A) segments of *Drosophila* larval ventral nerve cord (VNC). CCAP-PLs: posterior lateral CCAP neurons. (B) Diagram summarizing the late differentiation of CCAP-ENs, their role in pupal ecdysis and eclosion, as well as circadian inputs. Both, signaling from the PDF neurons and the peripheral clock in the prothoracic gland are likely necessary to regulate eclosion rhythms (Myers et al., 2003).

comprising six neurons, the late CCAP-ENs (A5–A7) and the CCAP-PLs ("posterior lateral CCAPs," A8–A9), determined by the presence or absence of the *Dachshund* marker, respectively (Fig. 1.4A). The late CCAP neurons do not differentiate and do not extend out of the ventral nerve cord until early pupariation. After differentiation they become competent to secrete

CCAP, Burs-α and Burs-β at around 10–12 h after puparium formation, and continue to express the peptides up to the pharate adult stage. Surprisingly, of all CCAP neurons, the 12 late CCAP neurons are sufficient for completing eclosion in all animals. When the authors ablated all CCAP neurons up until to the mid-L3 stage, and then allowed for the late CCAP neurons to differentiate at pupariation, all animals eclosed successfully, suggesting that the late CCAP neurons exert an essential role during pupal development. In contrast, ablation of all CCAP neurons including the late CCAP neurons until after pupal ecdysis resulted in 100% failure of eclosion. Likewise, when the late CCAP neurons were ablated after pupal ecdysis, that is after they are required, all the animals eclosed with normal timing. The only defect that was not rescued by the presence of late CCAP neurons was wing inflation, which was consistent with the earlier finding that wing inflation requires the presence of a subset of the CCAP-IN neurons, which were also ablated in this experiment (Davis, O'Keefe, Primrose, & Hodgetts, 2007). Interestingly, when the authors used a stochastic ablation approach, they found that the two subclasses of late CCAP neurons, the CCAP-ENs and the CCAP-PLs, correlate with distinct phenotypes, and are therefore functionally separable. Specifically, they showed that the six late CCAP-ENs correlated with a failure to properly execute leg extension, while the CCAP-PL neurons corresponded to head eversion defects, both commonly observed phenotypes in mutants with defects in prepupal development.

The authors were curious as to how the delayed appearance of the late CCAP neurons was controlled. When they tracked postembryonic neuroblast lineages using BrdU, they never observed BrdU incorporation into late CCAP neurons, suggesting these neurons derive from different cells. Next, they followed early CCAP neurons by genetically marking these cells, but they did not find late CCAP neurons expressing this marker, indicating that the late CCAP neurons did not derive from their early cousins either. These negative data prompted them to test the idea that the late CCAP neurons are embryonic in origin, but stay undifferentiated until pupariation. Indeed, through the elegant use of a series of genetic markers, the authors demonstrated that the CCAP neurons are born in the embryonic segments T3-A9, and that an unidentified mechanism delayed the differentiation of the CCAPs in segments A5–A9, i.e. the late CCAP neurons.

Puparium formation is triggered by a large pulse of ecdysone at the end of the third instar, which is followed by a smaller pulse ~10–12 h later that causes head eversion and pupal ecdysis. This raised the possibility that the large peak of ecdysone at the end of larval development triggers the onset of late

CCAP neuronal differentiation, and that the production of CCAP and bursicon peptides might be linked to the second pulse. Indeed, blocking the ecdysone hierarchy specifically in CCAP neurons by expressing dominant negative forms of EcR resulted in a loss of CCAP and bursicon expression in the late CCAP neuron population. Similarly, expressing RNAi against EcR in early and late CCAP-EN neurons caused strongly reduced expression of CCAP and Burs-α. On a whole animal level, these disruptions of EcR function resulted in leg extension defects, but not head eversion phenotypes. This however, was consistent with the observation that manipulating EcR function in CCAPs did not significantly affect CCAP expression in the six CCAP-PL neurons, which correspond to the head eversion phenotype, and the authors reason that the GAL4 driver used in this experiment has much weaker expression in CCAP-PL neurons, which would result in reduced RNA interference.

While these results nicely demonstrated the requirement of EcR for the onset of late CCAP neuronal differentiation, it was still unclear why this event did not occur in response to an earlier ecdysone pulse. A body of work has examined the issue of stage-specific ecdysone responses. In short, the expression of βFTZ-F1, a nuclear receptor isoform acting directly downstream of EcR, provides temporal specificity to the prepupal stage, ensuring that the responses to the late-larval ecdysone pulse are different from the responses to the later pulse that triggers head eversion (Broadus, McCabe, Endrizzi, Thummel, & Woodard, 1999; Woodard, Baehrecke, & Thummel, 1994). This prompted the authors to ask whether ectopic expression of *βFTZ-F1* would be sufficient to trigger the differentiation of late CCAP neurons. Indeed, when *βFTZ-F1* was ubiquitously expressed in first instar larvae, they observed precocious production of CCAP and bursicon peptides in late CCAP-EN neurons only 4 h after induction of the transgene. Thus, it appears that CCAP neurons are competent throughout larval development to receive an inductive ecdysone signal to undergo terminal differentiation into peptide producing CCAP neurons, however, this inductive signal is dependent on the presence of βFTZ-F1, allowing the cells to put the ecdysone pulse into temporal context (Fig. 1.4B). While the data is compelling, its interpretation is complicated by the fact that βFTZ-F1 is expressed in most tissues during earlier stages (Yamada et al., 2000), giving rise to the possibility that additional distinguishing factors must be involved in triggering the differentiation of CCAP neurons. Alternatively, it is possible that these neurons simply do not express βFTZ-F1 at earlier stages, which can only be resolved by a careful expression analysis.

The delayed terminal differentiation of late CCAP neurons may reflect the need to discriminate between larval and pupal ecdysis, both of which require functional CCAP-EN neurons (Veverytsa & Allan, 2011). By differentiating late CCAP neurons just before they are needed for pupal ecdysis, this developmental delay may act as a sophisticated switch to ensure that CCAP neurons can distinguish between larval and pupal ecdysis behavior. While the late CCAP neurons are sufficient for pupal ecdysis, animals still fail to inflate their wings, suggesting that the other CCAP neurons have some functionality during pupal development. For this reason, it seems likely that the switch from larval to pupal ecdysis is accomplished by integrating the late CCAPs into the existing CCAP network, as opposed to a true switch where existing CCAP neurons would surrender all functionality to the late-emerging CCAP neurons.

10. JHs AND DEVELOPMENTAL TIMING

In this last section, we will discuss recent findings that explore the relationship between JHs and developmental timing as well as circadian rhythms. JHs function in a wide range of biological processes, including morphological processes, behavioral changes, caste determination in social insects, and diapause and vitellogenesis in adult insects (Ishikawa et al., 2012; Saunders, Richard, Applebaum, Ma, & Gilbert, 1990; Sroka & Gilbert, 1974). JH blocks the differentiation of imaginal discs, but promotes their proliferation, consistent with its role in inhibiting metamorphosis and promoting juvenile stages. Multiple lines of evidence from the last two decades show that JH antagonizes ecdysone production by repressing PTTH secretion in the final instar of holometabolous insects. In the *Manduca sexta*, during the last larval instar, JH appears to control the timing of pupal commitment by regulating the gated release of PTTH (Berger & Dubrovsky, 2005; Mizoguchi, 2001; Rountree & Bollenbacher, 1986; Sakurai, Okuda, & Ohtaki, 1989). However, the role of JH in regulating PTTH secretion is not well defined in *Drosophila*.

The discovery of the JHs started with a series of elegant experiments carried out by Vincent B. Wigglesworth in the 1930s, who identified a brain-derived hormone activity that could suppress molting to the adult form in last (i.e. 5th) stage *Rhodnius* nymphs (Wigglesworth, 1934, 1936). Essentially, Wigglesworth joined immature nymphs from an earlier stage to 5th stage nymphs using a tube that allowed the hemolymph of the two animals to mix, a technique referred to as parabiosis. Using this approach, he tracked the source of this metamorphosis-blocking activity, now known as JH, back to a pair of small glands, the *corpora allata*, which are connected to the brain by a nerve. However, the prospects for identifying the chemical nature of the JHs were limited by the fact that the *corpora allata* remained the sole known JH source for many years, and that these glands only produced minute amounts of the hormone. This situation changed when Carrol M. Williams identified large amounts of JH in the abdomens of *Cecropia* and *Cynthia* moths (Williams, 1963), which spurred a new era of JH research that led to the identification of the chemical structure of JH (now known as JH-I) by Röller, Dahm, Sweeley, and Trost (1967) a few years later. The hormone turned out to be an unusual lipophilic sesquiterpenoid derivative of farnesoic acid, not dissimilar from retinoic acid, a metabolite of vitamin A that acts as a morphogen in vertebrate development. The similarity to retinoic acid spawned the idea that JH binds to USP, the insect ortholog of the vertebrate nuclear receptor for 9-*cis*-retinoic acid, RXRα. The compelling concept that USP could double-function as JH receptor and as heterodimeric partner for EcR to form the functional EcR has lost some of its attractiveness due to the difficulty of identifying natural JH ligands for USP. Instead, methyl farnesoate, a compound secreted from *Drosophila corpora allata* and required for normal development, was shown to bind to USP with nanomolar affinities (Jones et al., 2010; Jones, Jones, Teal, Sapa, & Wozniak, 2006).

Due to the lack of a receptor, the mode of JH action remained unclear for many years. The first insight into the nature of the JH receptor was the identification of the *Drosophila Met* mutant (*Methoprene-tolerant*) in a screen aimed at identifying *Drosophila* mutants resistant to the JH mimic Methoprene, which normally causes toxicity and morphological changes when applied topically to third instar larvae (Wilson & Fabian, 1986). However, *Met* mutant flies displayed mainly normal development and were homozygous viable, which was inconsistent with the idea that Met represented a bona fide JH receptor, unless there was some functional redundancy with another protein. Indeed, the later identification of a bHLH-PAS paralog of Met, Germ cell-expressed (Gce), led to the discovery that Gce and Met can form heterodimers (Godlewski, Wang, & Wilson, 2006) and are partially redundant in transducing JH signaling (Abdou et al., 2011; Baumann, Barry, Wang, Fujiwara, & Wilson, 2010). Similar to *Met*, *gce* mutants have increased resistance to Methoprene, but are homozygous viable, while double mutants homozygous for *gce* and *Met* die as prepupae. Both proteins were shown to bind to JH with high affinity (Charles et al., 2011; Miura, Oda, Makita, & Chinzei, 2005), corroborating the idea that both Gce and Met are functionally redundant JH receptors in *Drosophila*. *Gce* and *Met* are represented by a single gene outside the *drosophilid* branch, due to a recent gene duplication in this lineage.

While there is little doubt that Gce/Met function as the principal JH receptors in *Drosophila*, the phenotypes that are usually associated with JH deficiency in other insects, such as precocious development or a decreased number of larval instars, are largely absent in *Drosophila gce/Met* double mutants, suggesting that the role of *Drosophila* JH deviates from most other insects (Baumann, Fujiwara, & Wilson, 2010). Despite this, *gce/Met* double mutants displayed precocious induction of the *broad* gene, a key player in the ecdysone-induced gene hierarchy controlling the larval-prepupal transition (Abdou et al., 2011). This precocious induction is caused by the loss of *kruppel-homolog-1* (*kr-h1*) expression, which encodes a JH-inducible transcription factor (Minakuchi, Namiki, & Shinoda, 2009; Minakuchi, Zhou, & Riddiford, 2008; Zhu, Busche, & Zhang, 2010) that is normally expressed during larval stages, but then sharply turned off in mid-prepupae to allow for the upregulation of *broad* (Pecasse, Beck, Ruiz, & Richards, 2000). A recent study identified a functional JH-response-element ~2 kb upstream of the *Bombyx kr-h1* gene, to which Met2, one of the two Met orthologs found in *Bombyx*, was able to bind to when co-expressed with another bHLH-PAS protein, SRC (Kayukawa et al., 2012). A similar JH response element was also identified in the *kr-h1* gene from *Aedes aegypti* (Shin, Zou, Saha, & Raikhel, 2012). These data strongly suggest that Kr-h1 is a key mediator of juvenile stages, and that precocious absence of this factor initiates developmental programs that normally occur much later. Finally, *gce/Met* double mutants exhibited precocious induction of programmed cell death of larval fat body cells, a phenotype that is also consistent with reduced JH signaling. Taken together, when JH receptor function is abolished in *Drosophila*, phenotypes are not nearly as dramatic as expected, but still consistent with JH mediating juvenile stages in this species.

Both, the redundancy of the *Drosophila* Gce and Met receptors and the subtle defects on developmental timing limit the usefulness of *Drosophila* as a model for studying JH and Met function. Recently, investigators focused their attention to the red flour beetle *Tribolium castaneum* to study JH pathways. Disrupting *Met* function in *Tribolium* via RNAi not only caused resistance to topical application of JH, but—in stark contrast to *Drosophila*—also caused early larval stages to enter precocious metamorphosis, which provided strong genetic evidence that Met might be the long-sought-after JH receptor (Konopova & Jindra, 2007). In addition, the authors observed a range of heterochronic phenotypes, including the precocious presence compound eyes and a pupal-like cuticle. The *Tribolium* Met protein binds JH with high affinity, and forms heterodimers in a ligand-dependent manner with another bHLH-PAS protein, Taiman, which is orthologous to the above-mentioned SRC (aka FISC) (Charles et al., 2011; Li, Mead, & Zhu, 2011).

Taiman is not the only known partner for Met. The Raikhel laboratory recently identified the circadian clock protein CYCLE as a binding partner for Met in *Aedes aegyti*, using yeast-two-hybrid screening (Shin et al., 2012). Cycle encodes, like Met, a member of the bHLH-PAS protein family, raising the interesting possibility that circadian circuits intersect with JH response pathways by direct protein–protein interactions. The interaction of these two proteins required the presence of JH, similar to what has been observed for Taiman. Interestingly, Taiman was also shown to be important for the JH-dependent action of the Cycle-Met heterodimer. In particular, the expression of two JH-responsive genes, *Kr-h1* and *Hairy*, was impaired when RNAi against Met, Cycle, or Taiman was used. Interestingly, the transcripts of both *kr-h1* and *hairy* showed clear periodicity when animals were kept on a LD cycle, but this oscillatory behavior was lost in DD conditions, indicating that *kr-h1* and *hairy* are regulated by both light and JH. Consistent with this, the expression levels of both transcripts were highly elevated, but still rhythmic when JH was applied topical (Fig. 1.5). This observation raises the interesting question whether JH levels change rhythmically during a particular developmental window, and whether JH could act as an internal zeitgeber for peripheral clocks.

The authors then tested whether CLOCK (Yu & Hardin, 2006), the heterodimeric partner of Cycle in the canonical circadian pathway, modulates JH induction of *Kr-h1* and *hairy*. RNAi-mediated knockdown of *clock* did not affect the rhythmic expression of *kr-h1* and *hairy*, suggesting that the canonical function of CYC/CLK might not be directly involved in the JH-dependent regulation of *kr-h1* and *hairy*. Deciphering the precise crosstalk between JH-mediated regulation and the circadian clock represents an exiting goal for future experiments, which could pave the way for a better understanding how the endocrine system interacts with the circadian clock.

11. OUTLOOK

Since the first evidence of circadian gating of developmental transitions in *Manduca*, significant progress has been made that allow for a better

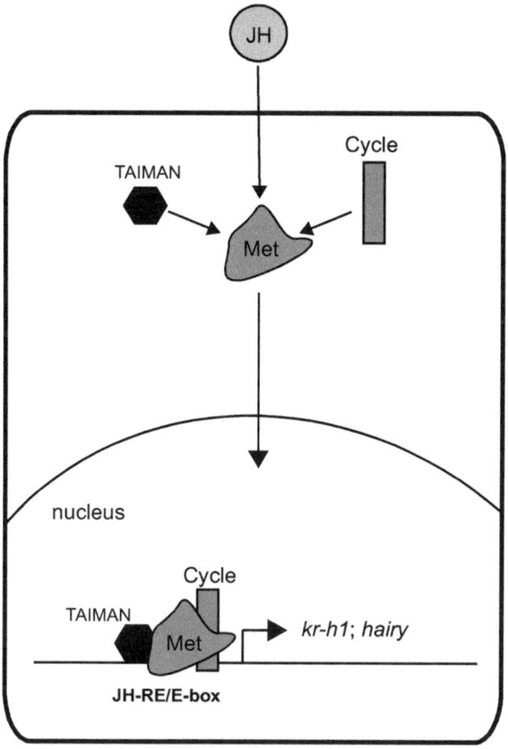

Figure 1.5 The JH receptor, Methoprene-tolerant (Met) interacts with CYCLE and Taiman (aka FISC or SRC), which are, like Met, members of the bHLH-PAS protein family (Kewley, Whitelaw, & Chapman-Smith, 2004; Li et al., 2011). These proteins form a functional complex in fat body cells following JH signaling during mosquito oocyte development.

understanding of strictly timed processes that govern insect development. It is becoming increasingly clear that the circadian machinery plays a pivotal role in regulating many of the key developmental events that drive the remarkable transformation of an egg into a complex adult. This notion is not just restricted to insect development but similar concepts have been unearthed in *C. elegans* (Banerjee, Kwok, Lin, & Slack, 2005), zebrafish (Dekens et al., 2003), and in mammals (Kennaway, 2005; Pilorz & Steinlechner, 2008).

However, the idea that peripheral clocks dictate the timing, at least in part, of developmentally active hormones and their corresponding responses has gained some considerable traction, but more research is required to understand how endocrine systems interact with peripheral and central clocks. Specifically, a systematic disruption of circadian rhythm genes in endocrine tissues such as the prothoracic glands would be a fruitful approach to genetically dissect the requirement of circadian control for hormonal output.

In the future, the ever-sinking costs of next-generation sequencing could help to employ RNA-seq in a wide-spanning strategy to obtain highly reliable tissue-specific expression profiles. This could be done for a range of animal models, factoring in complex parameters such as different developmental stages, LD vs. DD cycles, gender, zeitgeber times, and different genetic backgrounds. This and similar high-throughput approaches will ultimately improve our repertoire of circadian-regulated gene networks, and pave the way for linking clock functions to the regulation of developmental pathways.

REFERENCES

Abdou, M. A., He, Q., Wen, D., Zyaan, O., Wang, J., Xu, J., et al. (2011). *Drosophila* Met and Gce are partially redundant in transducing juvenile hormone action. *Insect Biochemistry and Molecular Biology, 41*, 938–945.

Allada, R., & Chung, B. Y. (2010). Circadian organization of behavior and physiology in *Drosophila. Annual Review of Physiology, 72*, 605–624.

Ambros, V. (2003). MicroRNA pathways in flies and worms: growth, death, fat, stress, and timing. *Cell, 113*, 673–676.

Ambros, V., & Horvitz, H. R. (1984). Heterochronic mutants of the nematode *Caenorhabditis elegans. Science, 226*, 409–416.

Ampleford, E. J., & Steel, C. G. (1985). Circadian control of a daily rhythm in hemolymph ecdysteroid titer in the insect *Rhodnius prolixus* (Hemiptera). *General and Comparative Endocrinology, 59*, 453–459.

Banerjee, D., Kwok, A., Lin, S. Y., & Slack, F. J. (2005). Developmental timing in *C. elegans* is regulated by kin-20 and tim-1, homologs of core circadian clock genes. *Developmental Cell, 8*, 287–295.

Baumann, A., Barry, J., Wang, S., Fujiwara, Y., & Wilson, T. G. (2010a). Paralogous genes involved in juvenile hormone action in *Drosophila* melanogaster. *Genetics, 185*, 1327–1336.

Baumann, A., Fujiwara, Y., & Wilson, T. G. (2010b). Evolutionary divergence of the paralogs Methoprene tolerant (Met) and germ cell expressed (gce) within the genus Drosophila. *Journal of Insect Physiology, 56*, 1445–1455.

Berger, E. M., & Dubrovsky, E. B. (2005). Juvenile hormone molecular actions and interactions during development of Drosophila melanogaster. *Vitamins and Hormones, 73*, 175–215.

Blanchardon, E., Grima, B., Klarsfeld, A., Chelot, E., Hardin, P. E., Preat, T., et al. (2001). Defining the role of Drosophila lateral neurons in the control of circadian rhythms in motor activity and eclosion by targeted genetic ablation and PERIOD protein overexpression. *European Journal of Neuroscience, 13*, 871–888.

Bollenbacher, W. E., Agui, N., Granger, N. A., & Gilbert, L. I. (1979). In vitro activation of insect prothoracic glands by the prothoracicotropic hormone. *Proceedings of the National Academy of Sciences of the United States of America, 76*, 5148–5152.

Bollenbacher, W. E., Smith, S. L., Goodman, W., & Gilbert, L. I. (1981). Ecdysteroid titer during larval–pupal–adult development of the tobacco hornworm, *Manduca sexta*. *General and Comparative Endocrinology*, *44*, 302–306.

Broadus, J., McCabe, J. R., Endrizzi, B., Thummel, C. S., & Woodard, C. T. (1999). The Drosophila beta FTZ-F1 orphan nuclear receptor provides competence for stage-specific responses to the steroid hormone ecdysone. *Molecular Cell*, *3*, 143–149.

Caceres, L., Necakov, A. S., Schwartz, C., Kimber, S., Roberts, I. J., & Krause, H. M. (2011). Nitric oxide coordinates metabolism, growth, and development via the nuclear receptor E75. *Genes & Development*, *25*, 1476–1485.

Charles, J. P., Iwema, T., Epa, V. C., Takaki, K., Rynes, J., & Jindra, M. (2011). Ligand-binding properties of a juvenile hormone receptor, Methoprene-tolerant. *Proceedings of the National Academy of Sciences of the United States of America*, *108*, 21128–21133.

Chiu, J. C., Vanselow, J. T., Kramer, A., & Edery, I. (2008). The phospho-occupancy of an atypical SLIMB-binding site on PERIOD that is phosphorylated by DOUBLETIME controls the pace of the clock. *Genes & Development*, *22*, 1758–1772.

Collins, B., & Blau, J. (2007). Even a stopped clock tells the right time twice a day: Circadian timekeeping in Drosophila. *Pflugers Archiv: European Journal of Physiology*, *454*, 857–867.

Cymborowski, B., Smietanko, A., & Delbecque, J. P. (1989). Circadian modulation of ecdysteroid titer in Galleria mellonella larvae. *Comparative Biochemistry and Physiology. Part A, Physiology*, *94*, 431–438.

Cymborowski, B. M.-P. M. , Porcheron, P., & Cassier, P. (1991). Hemolymph ecdysteroid titers controlled by a circadian clock mechanism in larvae of the wax moth, *Galleria mellonella*. *Journal of Insect Physiology*, *37*, 35–40.

Daimon, T., Kozaki, T., Niwa, R., Kobayashi, I., Furuta, K., Namiki, T., et al. (2012). Precocious metamorphosis in the juvenile hormone-deficient mutant of the silkworm, *Bombyx mori*. *PLoS Genetics*, *8*, e1002486.

Davis, M. M., O'Keefe, S. L., Primrose, D. A., & Hodgetts, R. B. (2007). A neuropeptide hormone cascade controls the precise onset of post-eclosion cuticular tanning in Drosophila melanogaster. *Development (Cambridge, England)*, *134*, 4395–4404.

Dekens, M. P., Santoriello, C., Vallone, D., Grassi, G., Whitmore, D., & Foulkes, N. S. (2003). Light regulates the cell cycle in zebrafish. *Current Biology*, *13*, 2051–2057.

Doane, W. W. (1973). Role of hormones in insect development. In S. J. C. a. C. H. Waddington (Ed.), *Developmental systems: Insects*. London: Academic Press.

Dubruille, R., & Emery, P. (2008). A plastic clock: How circadian rhythms respond to environmental cues in Drosophila. *Molecular Neurobiology*, *38*, 129–145.

Emery, I. F., Noveral, J. M., Jamison, C. F., & Siwicki, K. K. (1997). Rhythms of Drosophila period gene expression in culture. *Proceedings of the National Academy of Sciences of the United States of America*, *94*, 4092–4096.

Ewer, J., & Truman, J. W. (1996). Increases in cyclic 3',5'-guanosine monophosphate (cGMP) occur at ecdysis in an evolutionarily conserved crustacean cardioactive peptide-immunoreactive insect neuronal network. *The Journal of Comparative Neurology*, *370*, 330–341.

Fang, Y., Sathyanarayanan, S., & Sehgal, A. (2007). Post-translational regulation of the Drosophila circadian clock requires protein phosphatase 1 (PP1). *Genes & Development*, *21*, 1506–1518.

Fellner, S. K., Rybczynski, R., & Gilbert, L. I. (2005). Ca^{2+} signaling in prothoracicotropic hormone-stimulated prothoracic gland cells of *Manduca sexta*: Evidence for mobilization and entry mechanisms. *Insect Biochemistry and Molecular Biology*, *35*, 263–275.

Gammie, S. C., & Truman, J. W. (1997). Neuropeptide hierarchies and the activation of sequential motor behaviors in the hawkmoth, *Manduca sexta*. *Journal of Neuroscience*, *17*, 4389–4397.

Gibbens, Y. Y., Warren, J. T., Gilbert, L. I., & O'Connor, M. B. (2011). Neuroendocrine regulation of Drosophila metamorphosis requires TGFbeta/Activin signaling. *Development (Cambridge, England)*, *138*, 2693–2703.

Giebultowicz, J. M. (2001). Peripheral clocks and their role in circadian timing: Insights from insects. *Philosophical Transactions of the Royal Society of London. Series B, Biological Sciences*, *356*, 1791–1799.

Gilbert, L. I. (2004). Halloween genes encode P450 enzymes that mediate steroid hormone biosynthesis in Drosophila melanogaster. *Molecular and Cellular Endocrinology*, *215*, 1–10.

Godlewski, J., Wang, S., & Wilson, T. G. (2006). Interaction of bHLH-PAS proteins involved in juvenile hormone reception in Drosophila. *Biochemical and Biophysical Research Communications*, *342*, 1305–1311.

Gould, S. J. (1977). *Ontogeny and phylogeny*. Cambridge: Belknap Press of Harvard University Press.

Gu, S. H., & Tsai, W. H. (2012). Decreased JH biosynthesis is related to precocious metamorphosis in recessive trimolter (rt) mutants of the silkworm, *Bombyx mori*. *Archives of Insect Biochemistry and Physiology*, *79*, 235–246.

Handler, A. M., & Konopka, R. J. (1979). Transplantation of a circadian pacemaker in Drosophila. *Nature*, *279*, 236–238.

Henrich, V. C. (1995). Comparison of ecdysteroid production in Drosophila and Manduca: Pharmacology and cross-species neural reactivity. *Archives of Insect Biochemistry and Physiology*, *30*, 239–254.

Huang, J., Tian, L., Peng, C., Abdou, M., Wen, D., Wang, Y., et al. (2011). DPP-mediated TGFbeta signaling regulates juvenile hormone biosynthesis by activating the expression of juvenile hormone acid methyltransferase. *Development (Cambridge, England)*, *138*, 2283–2291.

Huang, X., Warren, J. T., & Gilbert, L. I. (2008). New players in the regulation of ecdysone biosynthesis. *Journal of Genetics and Genomics = Yi chuan xue bao*, *35*, 1–10.

Ishikawa, A., Ogawa, K., Gotoh, H., Walsh, T. K., Tagu, D., Brisson, J. A., et al. (2012). Juvenile hormone titre and related gene expression during the change of reproductive modes in the pea aphid. *Insect Molecular Biology*, *21*, 49–60.

Jackson, F. R., Schroeder, A. J., Roberts, M. A., McNeil, G. P., Kume, K., & Akten, B. (2001). Cellular and molecular mechanisms of circadian control in insects. *Journal of Insect Physiology*, *47*, 833–842.

Jiang, C., Lamblin, A. F., Steller, H., & Thummel, C. S. (2000). A steroid-triggered transcriptional hierarchy controls salivary gland cell death during Drosophila metamorphosis. *Molecular Cell*, *5*, 445–455.

Jones, D., Jones, G., Teal, P., Hammac, C., Messmer, L., Osborne, K., et al. (2010). Suppressed production of methyl farnesoid hormones yields developmental defects and lethality in Drosophila larvae. *General and Comparative Endocrinology*, *165*, 244–254.

Jones, G., Jones, D., Teal, P., Sapa, A., & Wozniak, M. (2006). The retinoid-X receptor ortholog, ultraspiracle, binds with nanomolar affinity to an endogenous morphogenetic ligand. *FEBS Journal*, *273*, 4983–4996.

Kadener, S., Menet, J. S., Schoer, R., & Rosbash, M. (2008). Circadian transcription contributes to core period determination in Drosophila. *PLoS Biology*, *6*, e119.

Kawakami, A., Kataoka, H., Oka, T., Mizoguchi, A., Kimura-Kawakami, M., Adachi, T., et al. (1990). Molecular cloning of the *Bombyx mori* prothoracicotropic hormone. *Science*, *247*, 1333–1335.

Kayukawa, T., Minakuchi, C., Namiki, T., Togawa, T., Yoshiyama, M., Kamimura, M., et al. (2012). Transcriptional regulation of juvenile hormone-mediated induction of Kruppel homolog 1, a repressor of insect metamorphosis. *Proceedings of the National Academy of Sciences of the United States of America*, *109*, 11729–11734.

Kennaway, D. J. (2005). The role of circadian rhythmicity in reproduction. *Human Reproduction Update, 11*, 91–101.
Kewley, R. J., Whitelaw, M. L., & Chapman-Smith, A. (2004). The mammalian basic helix-loop-helix/PAS family of transcriptional regulators. *The International Journal of Biochemistry & Cell Biology, 36*, 189–204.
Kim, E. Y., & Edery, I. (2006). Balance between DBT/CKIepsilon kinase and protein phosphatase activities regulate phosphorylation and stability ofDrosophila CLOCK protein. *Proceedings of the National Academy of Sciences of the United States of America, 103*, 6178–6183.
Kim, E. Y., Ko, H. W., Yu, W., Hardin, P. E., & Edery, I. (2007). A DOUBLETIME kinase binding domain on the Drosophila PERIOD protein is essential for its hyper-phosphorylation, transcriptional repression, and circadian clock function. *Molecular and Cellular Biology, 27*, 5014–5028.
Kim, Y. J., Zitnan, D., Cho, K. H., & Adams, M. E. (2006). Central peptidergic ensembles associated with organization of an innate behavior. *Proceedings of the National Academy of Sciences of the United States of America, 103*, 14211–14216.
King-Jones, K., & Thummel, C. S. (2005). Nuclear receptors—A perspective from Drosophila. *Nature Reviews Genetics, 6*, 311–323.
Kingan, T. G., & Adams, M. E. (2000). Ecdysteroids regulate secretory competence in Inka cells. *Journal of Experimental Biology, 203*, 3011–3018.
Kinjoh, T., Kaneko, Y., Itoyama, K., Mita, K., Hiruma, K., & Shinoda, T. (2007). Control of juvenile hormone biosynthesis in *Bombyx mori*: Cloning of the enzymes in the mevalonate pathway and assessment of their developmental expression in the corpora allata. *Insect Biochemistry and Molecular Biology, 37*, 808–818.
Kivimae, S., Saez, L., & Young, M. W. (2008). Activating PER repressor through a DBT-directed phosphorylation switch. *PLoS Biology, 6*, e183.
Koelle, M. R., Talbot, W. S., Segraves, W. A., Bender, M. T., Cherbas, P., & Hogness, D. S. (1991). The Drosophila EcR gene encodes an ecdysone receptor, a new member of the steroid receptor superfamily. *Cell, 67*, 59–77.
Konopova, B., & Jindra, M. (2007). Juvenile hormone resistance gene Methoprene-tolerant controls entry into metamorphosis in the beetle *Tribolium castaneum*. *Proceedings of the National Academy of Sciences of the United States of America, 104*, 10488–10493.
Krishnan, B., Levine, J. D., Lynch, M. K., Dowse, H. B., Funes, P., Hall, J. C., et al. (2001). A new role for cryptochrome in a Drosophila circadian oscillator. *Nature, 411*, 313–317.
Lee, R. C., Feinbaum, R. L., & Ambros, V. (1993). The *C. elegans* heterochronic gene lin-4 encodes small RNAs with antisense complementarity to lin-14. *Cell, 75*, 843–854.
Li, M., Mead, E. A., & Zhu, J. (2011). Heterodimer of two bHLH-PAS proteins mediates juvenile hormone-induced gene expression. *Proceedings of the National Academy of Sciences of the United States of America, 108*, 638–643.
Lightman, S. L., & Conway-Campbell, B. L. (2010). The crucial role of pulsatile activity of the HPA axis for continuous dynamic equilibration. *Nature Reviews Neuroscience, 11*, 710–718.
Liu, X., Lorenz, L., Yu, Q. N., Hall, J. C., & Rosbash, M. (1988). Spatial and temporal expression of the period gene in Drosophila melanogaster. *Genes & Development, 2*, 228–238.
Martinez-Padron, M., & Ferrus, A. (1997). Presynaptic recordings from Drosophila: Correlation of macroscopic and single-channel K+currents. *Journal of Neuroscience, 17*, 3412–3424.
McBrayer, Z., Ono, H., Shimell, M., Parvy, J. P., Beckstead, R. B., Warren, J. T., et al. (2007). Prothoracicotropic hormone regulates developmental timing and body size in Drosophila. *Developmental Cell, 13*, 857–871.

McNabb, S. L., Baker, J. D., Agapite, J., Steller, H., Riddiford, L. M., & Truman, J. W. (1997). Disruption of a behavioral sequence by targeted death of peptidergic neurons in Drosophila. *Neuron*, *19*, 813–823.
Meissner, R. A., Kilman, V. L., Lin, J. M., & Allada, R. (2008). TIMELESS is an important mediator of CK2 effects on circadian clock function in vivo. *Journal of Neuroscience*, *28*, 9732–9740.
Mesce, K. A., & Fahrbach, S. E. (2002). Integration of endocrine signals that regulate insect ecdysis. *Frontiers in Neuroendocrinology*, *23*, 179–199.
Meyer, P., Saez, L., & Young, M. W. (2006). PER–TIM interactions in living Drosophila cells: An interval timer for the circadian clock. *Science*, *311*, 226–229.
Minakuchi, C., Namiki, T., & Shinoda, T. (2009). Kruppel homolog 1, an early juvenile hormone-response gene downstream of Methoprene-tolerant, mediates its anti-metamorphic action in the red flour beetle *Tribolium castaneum*. *Developmental Biology*, *325*, 341–350.
Minakuchi, C., Namiki, T., Yoshiyama, M., & Shinoda, T. (2008). RNAi-mediated knock-down of juvenile hormone acid O-methyltransferase gene causes precocious metamorphosis in the red flour beetle *Tribolium castaneum*. *FEBS Journal*, *275*, 2919–2931.
Minakuchi, C., Zhou, X., & Riddiford, L. M. (2008). Kruppel homolog 1 (Kr-h1) mediates juvenile hormone action during metamorphosis of Drosophila melanogaster. *Mechanisms of Development*, *125*, 91–105.
Miura, K., Oda, M., Makita, S., & Chinzei, Y. (2005). Characterization of the Drosophila Methoprene—Tolerant gene product. Juvenile hormone binding and ligand-dependent gene regulation. *FEBS Journal*, *272*, 1169–1178.
Mizoguchi, A. (2001). Effects of juvenile hormone on the secretion of prothoracicotropic hormone in the last- and penultimate-instar larvae of the silkworm *Bombyx mori*. *Journal of Insect Physiology*, *47*, 767–775.
Mizoguchi, A., Dedos, S. G., Fugo, H., & Kataoka, H. (2002). Basic pattern of fluctuation in hemolymph PTTH titers during larval-pupal and pupal-adult development of the silkworm, *Bombyx mori*. *General and Comparative Endocrinology*, *127*, 181–189.
Mizoguchi, A., & Ishizaki, H. (1982). Prothoracic glands of the saturniid moth *Samia cynthia ricini* possess a circadian clock controlling gut purge timing. *Proceedings of the National Academy of Sciences of the United States of America*, *79*, 2726–2730.
Mizoguchi, A., Ohashi, Y., Hosoda, K., Ishibashi, J., & Kataoka, H. (2001). Developmental profile of the changes in the prothoracicotropic hormone titer in hemolymph of the silkworm *Bombyx mori*: Correlation with ecdysteroid secretion. *Insect Biochemistry and Molecular Biology*, *31*, 349–358.
Morioka, E., Matsumoto, A., & Ikeda, M. (2012). Neuronal influence on peripheral circadian oscillators in pupal Drosophila prothoracic glands. *Nature Communications*, *3*, 909.
Morton, D. B., & Truman, J. W. (1988). The EGPs: The eclosion hormone and cyclic GMP-regulated phosphoproteins. II. Regulation of appearance by the steroid hormone 20-hydroxyecdysone in *Manduca sexta*. *Journal of Neuroscience*, *8*, 1338–1345.
Myers, E. M., Yu, J., & Sehgal, A. (2003). Circadian control of eclosion: Interaction between a central and peripheral clock in Drosophila melanogaster. *Current Biology*, *13*, 526–533.
Nitabach, M. N., & Taghert, P. H. (2008). Organization of the Drosophila circadian control circuit. *Current Biology*, *18*, R84–R93.
Ou, Q., & King-Jones, K. (2013). What goes up must come down: Transcription factors have their say in making ecdysone pulses. *Current Topics in Developmental Biology*, *103*, 35–71.
Ou, Q., Magico, A., & King-Jones, K. (2011). Nuclear receptor DHR4 controls the timing of steroid hormone pulses during Drosophila development. *PLoS Biology*, *9*, e1001160.
Park, J. H., Schroeder, A. J., Helfrich-Forster, C., Jackson, F. R., & Ewer, J. (2003). Targeted ablation of CCAP neuropeptide-containing neurons of Drosophila causes specific defects

in execution and circadian timing of ecdysis behavior. *Development (Cambridge, England)*, *130*, 2645–2656.
Pecasse, F., Beck, Y., Ruiz, C., & Richards, G. (2000). Kruppel-homolog, a stage-specific modulator of the prepupal ecdysone response, is essential for Drosophila metamorphosis. *Developmental Biology*, *221*, 53–67.
Pilorz, V., & Steinlechner, S. (2008). Low reproductive success in Per1 and Per2 mutant mouse females due to accelerated ageing? *Reproduction*, *135*, 559–568.
Pittendrigh, C. S. (1964). The entrainment of circadian oscillations by skeleton photoperiods. *Science*, *144*, 565.
Plautz, J. D., Kaneko, M., Hall, J. C., & Kay, S. A. (1997). Independent photoreceptive circadian clocks throughout Drosophila. *Science*, *278*, 1632–1635.
Preitner, N., Damiola, F., Lopez-Molina, L., Zakany, J., Duboule, D., Albrecht, U., et al. (2002). The orphan nuclear receptor REV-ERBalpha controls circadian transcription within the positive limb of the mammalian circadian oscillator. *Cell*, *110*, 251–260.
Prokop, A. (2006). Organization of the efferent system and structure of neuromuscular junctions in Drosophila. *International Review of Neurobiology*, *75*, 71–90.
Rewitz, K. F., O'Connor, M. B., & Gilbert, L. I. (2007). Molecular evolution of the insect Halloween family of cytochrome P450s: Phylogeny, gene organization and functional conservation. *Insect Biochemistry and Molecular Biology*, *37*, 741–753.
Rewitz, K. F., Yamanaka, N., Gilbert, L. I., & O'Connor, M. B. (2009). The insect neuropeptide PTTH activates receptor tyrosine kinase torso to initiate metamorphosis. *Science*, *326*, 1403–1405.
Rewitz, K. F., Yamanaka, N., & O'Connor, M. B. (2013). Developmental checkpoints and feedback circuits time insect maturation. *Current Topics in Developmental Biology*, *103*, 1–33.
Richter, K. (2001). Daily changes in neuroendocrine control of moulting hormone secretion in the prothoracic gland of the cockroach *Periplaneta americana* (L.). *Journal of Insect Physiology*, *47*, 333–338.
Riddiford, L. M. (1970a). Effects of juvenile hormone on the programming of post-embryonic development in eggs of the silkworm, *Hyalophora cecropia*. *Developmental Biology*, *22*, 249–263.
Riddiford, L. M. (1970b). Prevention of metamorphosis by exposure of insect eggs to juvenile hormone analogs. *Science*, *167*, 287–288.
Riddiford, L. M. (1993). Hormone receptors and the regulation of insect metamorphosis. *Receptor*, *3*, 203–209.
Riddiford, L. M., Truman, J. W., Mirth, C. K., & Shen, Y. C. (2010). A role for juvenile hormone in the prepupal development of Drosophila melanogaster. *Development (Cambridge, England)*, *137*, 1117–1126.
Röller, H., Dahm, K. H., Sweeley, C. C., & Trost, B. M. (1967). Die Struktur des Juvenilhormons. *Angewandte Chemie*, *79*, 190–191.
Rountree, D. B., & Bollenbacher, W. E. (1986). The release of the prothoracicotropic hormone in the tobacco hornworm, *Manduca sexta*, is controlled intrinsically by juvenile hormone. *Journal of Experimental Biology*, *120*, 41–58.
Russel, S. a. A. M. (1996). *Ecdysone-regulated some puffing in Drosophila melanogaster*. New York: Academic Press.
Rybczynaki, R. (2005). Prothoracicotropic hormone. In L. I. Gilbert, K. Latrou, & S. S. Gill (Eds.), *Molecular insect science* (pp. 61–123). Oxford: Elsevier Pergamon.
Sakurai, S., Okuda, M., & Ohtaki, T. (1989). Juvenile hormone inhibits ecdysone secretion and responsiveness to prothoracicotropic hormone in prothoracic glands of *Bombyx mori*. *General and Comparative Endocrinology*, *75*, 222–230.
Satake, S., Kaya, M., & Sakurai, S. (1998). Hemolymph ecdysteroid titer and ecdysteroid-dependent developmental events in the last-larval stadium of the silkworm, *Bombyx mori*:

Role of low ecdysteroid titer in larval-pupal metamorphosis and a reappraisal of the head critical period. *Journal of Insect Physiology*, *44*, 867–881.
Sato, T. K., Panda, S., Miraglia, L. J., Reyes, T. M., Rudic, R. D., McNamara, P., et al. (2004). A functional genomics strategy reveals Rora as a component of the mammalian circadian clock. *Neuron*, *43*, 527–537.
Saunders, D. S. (1982). *Insect clocks* (2nd ed.). Oxford: Pergamon Pess.
Saunders, D. S., Richard, D. S., Applebaum, S. W., Ma, M., & Gilbert, L. I. (1990). Photoperiodic diapause in Drosophila melanogaster involves a block to the juvenile hormone regulation of ovarian maturation. *General and Comparative Endocrinology*, *79*, 174–184.
Schwartz, L. M., & Truman, J. W. (1983). Hormonal control of rates of metamorphic development in the tobacco hornworm *Manduca sexta*. *Developmental Biology*, *99*, 103–114.
Shin, S. W., Zou, Z., Saha, T. T., & Raikhel, A. S. (2012). bHLH-PAS heterodimer of methoprene-tolerant and Cycle mediates circadian expression of juvenile hormone-induced mosquito genes. *Proceedings of the National Academy of Sciences of the United States of America*, *109*, 16576–16581.
Siegmund, T., & Korge, G. (2001). Innervation of the ring gland of Drosophila melanogaster. *The Journal of Comparative Neurology*, *431*, 481–491.
So, W. V., Sarov-Blat, L., Kotarski, C. K., McDonald, M. J., Allada, R., & Rosbash, M. (2000). Takeout, a novel Drosophila gene under circadian clock transcriptional regulation. *Molecular and Cellular Biology*, *20*, 6935–6944.
Solt, L. A., Wang, Y., Banerjee, S., Hughes, T., Kojetin, D. J., Lundasen, T., et al. (2012). Regulation of circadian behaviour and metabolism by synthetic REV-ERB agonists. *Nature*, *485*, 62–68.
Song, Q., & Gilbert, L. I. (1995). Multiple phosphorylation of ribosomal protein S6 and specific protein synthesis are required for prothoracicotropic hormone-stimulated ecdysteroid biosynthesis in the prothoracic glands of *Manduca sexta*. *Insect Biochemistry and Molecular Biology*, *25*, 591–602.
Song, Q., & Gilbert, L. I. (1997). Molecular cloning, developmental expression, and phosphorylation of ribosomal protein S6 in the endocrine gland responsible for insect molting. *Journal of Biological Chemistry*, *272*, 4429–4435.
Song, Q., & Gilbert, L. I. (1998). Alterations in ultraspiracle (USP) content and phosphorylation state accompany feedback regulation of ecdysone synthesis in the insect prothoracic gland. *Insect Biochemistry and Molecular Biology*, *28*, 849–860.
Sroka, P., & Gilbert, L. I. (1974). The timing of juvenile hormone release for ovarian maturation in *Manduca sexta*. *Journal of Insect Physiology*, *20*, 1173–1180.
Stanewsky, R. (2003). Genetic analysis of the circadian system in Drosophila melanogaster and mammals. *Journal of Neurobiology*, *54*, 111–147.
Steel, C. G., & Vafopoulou, X. (2006). Circadian orchestration of developmental hormones in the insect, *Rhodnius prolixus*. *Comparative Biochemistry and Physiology. Part A, Molecular & Integrative Physiology*, *144*, 351–364.
Sullivan, A. A., & Thummel, C. S. (2003). Temporal profiles of nuclear receptor gene expression reveal coordinate transcriptional responses during Drosophila development. *Molecular Endocrinology*, *17*, 2125–2137.
Tan, A., Tanaka, H., Tamura, T., & Shiotsuki, T. (2005). Precocious metamorphosis in transgenic silkworms overexpressing juvenile hormone esterase. *Proceedings of the National Academy of Sciences of the United States of America*, *102*, 11751–11756.
Teboul, M., Guillaumond, F., Grechez-Cassiau, A., & Delaunay, F. (2008). The nuclear hormone receptor family round the clock. *Molecular Endocrinology*, *22*, 2573–2582.
Thummel, C. S. (1995). From embryogenesis to metamorphosis: The regulation and function of Drosophila nuclear receptor superfamily members. *Cell*, *83*, 871–877.

Thummel, C. S. (1996). Flies on steroids—Drosophila metamorphosis and the mechanisms of steroid hormone action. *Trends in Genetics, 12*, 306–310.

Thummel, C. S. (2001). Molecular mechanisms of developmental timing in *C. elegans* and Drosophila. *Developmental Cell, 1*, 453–465.

Thummel, C. S. (2002). Ecdysone-regulated puff genes 2000. *Insect Biochemistry and Molecular Biology, 32*, 113–120.

Truman, J. (1972). Physiology of insect rhythms. I. Circadian organization of the endocrine events underlying the moulting cycle of larval tobacco hornworms. *Journal of Experimental Biology, 57*, 805–820.

Truman, J. W. (1981). Interaction between ecdysteroid, eclosion hormone, and bursicon titers in *Manduca sexta. American Zoologist, 21*, 655–661.

Truman, J. W., & Riddiford, L. M. (1974). Physiology of insect rhythms. 3. The temporal organization of the endocrine events underlying pupation of the tobacco hornworm. *Journal of Experimental Biology, 60*, 371–382.

Truman, J. W., Riddiford, L. M., & Safranek, L. (1974). Temporal patterns of response to ecdysone and juvenile hormone in the epidermis of the tobacco hornworm, *Manduca sexta. Developmental Biology, 39*, 247–262.

Truman, J. W., Rountree, D. B., Reiss, S. E., & Schwartz, L. M. (1983). Ecdysteroids regulate the release and action of eclosion hormone in the tobacco hornworm, *Manduca sexta* (L.). *Journal of Insect Physiology, 29*, 895–900.

Vafopoulou, X., & Steel, C. G. (1991). Circadian regulation of synthesis of ecdysteroids by prothoracic glands of the insect *Rhodnius prolixus*: Evidence of a dual oscillator system. *General and Comparative Endocrinology, 83*, 27–34.

Vafopoulou, X., & Steel, C. G. (1996). Circadian regulation of a daily rhythm of release of prothoracicotropic hormone from the brain retrocerebral complex of *Rhodnius prolixus* (hemiptera) during larval-adult development. *General and Comparative Endocrinology, 102*, 123–129.

Vafopoulou, X., & Steel, C. G. (2006). Ecdysteroid hormone nuclear receptor (EcR) exhibits circadian cycling in certain tissues, but not others, during development in *Rhodnius prolixus* (Hemiptera). *Cell and Tissue Research, 323*, 443–455.

Vafopoulou, X., & Steel, C. G. (2012a). Cytoplasmic travels of the ecdysteroid receptor in target cells: Pathways for both genomic and non-genomic actions. *Frontiers in Endocrinology, 3*, 43.

Vafopoulou, X., & Steel, C. G. (2012b). Metamorphosis of a clock: Remodeling of the circadian timing system in the brain of *Rhodnius prolixus* (Hemiptera) during larval-adult development. *The Journal of Comparative Neurology, 520*, 1146–1164.

Vafopoulou, X., Steel, C. G., & Terry, K. L. (2005). Edysteroid receptor (EcR) shows marked differences in temporal patterns between tissues during larval-adult development in *Rhodnius prolixus*: Correlations with haemolymph ecdysteroid titres. *Journal of Insect Physiology, 51*, 27–38.

Venkatesh, K., & Hasan, G. (1997). Disruption of the IP3 receptor gene of Drosophila affects larval metamorphosis and ecdysone release. *Current Biology*, 7, 500–509.

Veverytsa, L., & Allan, D. W. (2011). Retrograde BMP signaling controls Drosophila behavior through regulation of a peptide hormone battery. *Development (Cambridge, England), 138*, 3147–3157.

Veverytsa, L., & Allan, D. W. (2012). Temporally tuned neuronal differentiation supports the functional remodeling of a neuronal network in Drosophila. *Proceedings of the National Academy of Sciences of the United States of America, 109*, E748–E756.

Wang, H. B., Ali, S. M., Moriyama, M., Iwanaga, M., & Kawasaki, H. (2012). 20-Hydroxyecdysone and juvenile hormone analog prevent precocious metamorphosis in recessive trimolter mutants of *Bombyx mori. Insect Biochemistry and Molecular Biology, 42*, 102–108.

Westbrook, A. L., & Bollenbacher, W. E. (1990). The development of identified neurosecretory cells in the tobacco hornworm, *Manduca sexta*. *Developmental Biology*, *140*, 291–299.
Wigglesworth, V. B. (1934). The physiology of ecdysis in *Rhodnius prolixus* (Hemiptera). Factors controlling moulting and 'metamorphosis'. *Quarterly Journal of Microscopical Science II*, *77*, 191–222.
Wigglesworth, V. B. (1936). The function of the corpus allatum in the growth and reproduction of *Rhodnius prolixus* (Hemiptera). *Quarterly Journal of Microscopical Science*, *79*, 91–121.
Williams, C. M. (1963). The juvenile hormone. III. Its accumulation and storage in the abdomens of certain male moths. *The Biological Bulletin*, *124*, 355–367.
Wilson, G. N. (1988). Heterochrony and human malformation. *American Journal of Medical Genetics*, *29*, 311–321.
Wilson, T. G., & Fabian, J. (1986). A Drosophila melanogaster mutant resistant to a chemical analog of juvenile hormone. *Developmental Biology*, *118*, 190–201.
Winfree, A. T. (1970). Integrated view of resetting a circadian clock. *Journal of Theoretical Biology*, *28*, 327–374.
Woodard, C. T., Baehrecke, E. H., & Thummel, C. S. (1994). A molecular mechanism for the stage specificity of the Drosophila prepupal genetic response to ecdysone. *Cell*, *79*, 607–615.
Xu, K., Zheng, X., & Sehgal, A. (2008). Regulation of feeding and metabolism by neuronal and peripheral clocks in Drosophila. *Cell Metabolism*, *8*, 289–300.
Yamada, M., Murata, T., Hirose, S., Lavorgna, G., Suzuki, E., & Ueda, H. (2000). Temporally restricted expression of transcription factor betaFTZ-F1: Significance for embryogenesis, molting and metamorphosis in Drosophila melanogaster. *Development (Cambridge, England)*, *127*, 5083–5092.
Yamanaka, N., Rewitz, K. F., & O'Connor, M. B. (2013). Ecdysone control of developmental transitions: Lessons from Drosophila research. *Annual Review of Entomology*, *58*, 497–516.
Yang, X., Downes, M., Yu, R. T., Bookout, A. L., He, W., Straume, M., et al. (2006). Nuclear receptor expression links the circadian clock to metabolism. *Cell*, *126*, 801–810.
Yao, T. P., Forman, B. M., Jiang, Z., Cherbas, L., Chen, J. D., McKeown, M., et al. (1993). Functional ecdysone receptor is the product of EcR and Ultraspiracle genes. *Nature*, *366*, 476–479.
Yu, W., & Hardin, P. E. (2006). Circadian oscillators of Drosophila and mammals. *Journal of Cell Science*, *119*, 4793–4795.
Yu, W., Zheng, H., Houl, J. H., Dauwalder, B., & Hardin, P. E. (2006). PER-dependent rhythms in CLK phosphorylation and E-box binding regulate circadian transcription. *Genes & Development*, *20*, 723–733.
Zhao, T., Gu, T., Rice, H. C., McAdams, K. L., Roark, K. M., Lawson, K., et al. (2008). A Drosophila gain-of-function screen for candidate genes involved in steroid-dependent neuroendocrine cell remodeling. *Genetics*, *178*, 883–901.
Zheng, X., & Sehgal, A. (2008). Probing the relative importance of molecular oscillations in the circadian clock. *Genetics*, *178*, 1147–1155.
Zhu, J., Busche, J. M., & Zhang, X. (2010). Identification of juvenile hormone target genes in the adult female mosquitoes. *Insect Biochemistry and Molecular Biology*, *40*, 23–29.

CHAPTER TWO

Nutrient Signaling and Developmental Timing of Maturation

E. Thomas Danielsen[2], Morten E. Moeller[2], Kim F. Rewitz[1]
Department of Biology, Cell and Neurobiology, University of Copenhagen, Copenhagen, Denmark
[1]Corresponding author: e-mail address: Kim.Rewitz@bio.ku.dk
[2]Equal contributions and order chosen randomly.

Contents

1. Introduction 38
2. Interplay Between Nutrients, Systemic Growth, and Developmental Transitions 41
   2.1 Switching to a nutrient-independent developmental timing program 41
   2.2 Insulin/target of rapamycin (TOR) couples nutrient-dependent growth to timing of maturation 42
   2.3 Influence of dietary amino acids, sugar, and fat on growth and onset of maturation 44
3. A Nutritional Checkpoint Based on Crosstalk Between Different Tissues 48
   3.1 Coordinating nutrient uptake and systemic growth 48
   3.2 Interactions between the gut microbiota, nutrition, and developmental progression 52
   3.3 Crosstalk of nutrient signaling and neuroendocrine control of maturation 53
4. Fine-Tuning Systemic Insulin Signaling 54
   4.1 Antagonistic interactions between ecdysone and insulin 54
   4.2 Adjusting circulating DILP activity 55
5. A Developmental Checkpoint Couples Organ Growth and Tissue Regeneration to Timing 57
6. Conclusions and Perspectives 60
Acknowledgments 61
References 61

Abstract

In animals, developmental timing of sexual maturation is tightly linked to nutrition and growth. Maturation only occurs once the juvenile has acquired sufficient nutrients and completed enough growth to produce a reproductively mature adult with a genetically predefined body size. Animals therefore adjust the duration of juvenile development to the dietary conditions. When nutrients are scarce the juvenile growth phase is extended to compensate for slow growth. Conversely, development is accelerated in nutrient rich environments where animals rapidly reach their genetic target size. To achieve such

Current Topics in Developmental Biology, Volume 105
ISSN 0070-2153
http://dx.doi.org/10.1016/B978-0-12-396968-2.00002-6

flexibility, nutrient-dependent growth regulators must feed into the endocrine system that controls the timing of maturation. Work on the fruit fly *Drosophila* has revealed a central role of secreted signal molecules with similarity to the conserved insulin-like growth factors (IGFs) in the decision making process. These molecules are involved in checkpoints that allow the endocrine system to decide whether to release the steroid hormone, ecdysone, that triggers maturation or extent development, depending on nutrient levels and growth status. Importantly, different dietary components influence timing of maturation in *Drosophila*, with proteins having the greatest impact; fat and sugar play a minor role, at least within the limits of what can be considered a balanced diet. Remarkably, excess dietary sugar concentrations that mimic physiological conditions associated with diabetes, negatively affect growth and delays maturation. Altogether, this shows that the source of energy in the diet is important for timing and may provide a paradigm for understanding the emerging links between diet, obesity and diabetes, and the onset of puberty. Here, we provide an overview of the system underlying developmental timing of maturation in *Drosophila* and review recent success in understanding its coupling to nutrition and growth.

1. INTRODUCTION

Animals grow towards a genetically determined adult size primarily defined by the amount of growth during the juvenile stages. Since the growth rate depends on the nutritional conditions, animals must be able to adjust the duration of the growth period to reach a similar final size under different environmental conditions. This ensures that the transition to adulthood only occurs after a certain amount of growth has been completed during development (Mirth & Riddiford, 2007; Rewitz & O'Connor, 2011; Tennessen & Thummel, 2013, 2011; Yamanaka, Rewitz, & O'Connor, 2013). Based on recent insight, it has become clear that the system coordinating growth and developmental timing of maturation is remarkably conserved from flies and humans. Work on *Drosophila* has unveiled a central role for conserved insulin-like peptides in determining this transition. These signaling molecules are regulators of nutrient-dependent growth and are involved in crosstalk between different organs that eventually converge on the endocrine system to time the release of ecdysone, the temporal cue that triggers maturation (Colombani, Andersen, & Leopold, 2012; Colombani et al., 2005, 2003; Garelli, Gontijo, Miguela, Caparros, & Dominguez, 2012). Consequently, the *Drosophila* larva is able to slow its developmental progression to adulthood under poor nutrient conditions and accelerate development in nutrient rich environments. Insulin-like peptides also affect the decision between delaying or continuing development to the adult stage in both *Caenorhabditis elegans* (*C. elegans*) and mammals (Kjaer, Hagen, Sando, & Eshoj, 1992; Liu & LeRoith, 1999; Tennessen & Thummel, 2011).

Drosophila has three larval instars where it rapidly accumulates mass and nutrients. At the end of the third instar (L3), when the larva has accumulated enough mass and stored sufficient nutrients to survive the non-feeding metamorphic development to the adult stage, it exits the food and initiates a characteristic wandering behavior before it pupariates. This marks the onset of metamorphosis, a process that transforms the juvenile larva into a sexually mature adult by the destruction of obsolete larval tissues and the morphogenesis and differentiation of adult precursor tissues (Baehrecke, 2000; Siaussat, Porcheron, & Debernard, 2009; Thummel, 2001). Pulses of ecdysone that causes the larva to shed its old cuticle and form a new and bigger one that allows continued growth trigger developmental transitions during the larval stages. In *Drosophila*, a single ecdysone pulse triggers the transition through the first two larval stages, while pupariation (onset of maturation) is initiated in response to three low-level pulses followed by a high-level pulse during L3 (Rewitz et al., 2013; Rewitz, Yamanaka, & O'Connor, 2010; Yamanaka, Rewitz, & O'Connor, 2013; Warren et al., 2006).

Ecdysone is produced and released from the prothoracic gland (PG), the major endocrine tissues of insects, in response to a brain-derived neuropeptide called prothoracicotropic hormone (PTTH; McBrayer et al., 2007; Smith & Rybczynski, 2012). PTTH activates its receptor Torso, a receptor tyrosine kinase that signals through the mitogen-activated protein kinase (MAPK) pathway to stimulate the production and release of ecdysone from the PG (Rewitz, Yamanaka, Gilbert, & O'Connor, 2009). Stimulation of the PG by PTTH activates the ecdysone biosynthetic pathway that converts dietary cholesterol into ecdysone. The biosynthesis of ecdysone is mediated by several cytochrome P450 enzymes, a rieske-domain protein and a short-chain dehydrogenase/reductase (Niwa et al., 2010; Petryk et al., 2003; Rewitz, Rybczynski, Warren, & Gilbert, 2006a, 2006b; Warren et al., 2002, 2004; Yoshiyama, Namiki, Mita, Kataoka, & Niwa, 2006; Yoshiyama-Yanagawa et al., 2011). Lack of either PTTH or Torso attenuates ecdysone synthesis, which delays development by prolonging primarily the L3 stage. This extends the duration of the growth period and increases adult body size. Therefore, PTTH triggers the onset of maturation once the animal has completed enough growth during larval development. Moreover, the release of PTTH is regulated by the photoperiod (Nijhout, 1981)

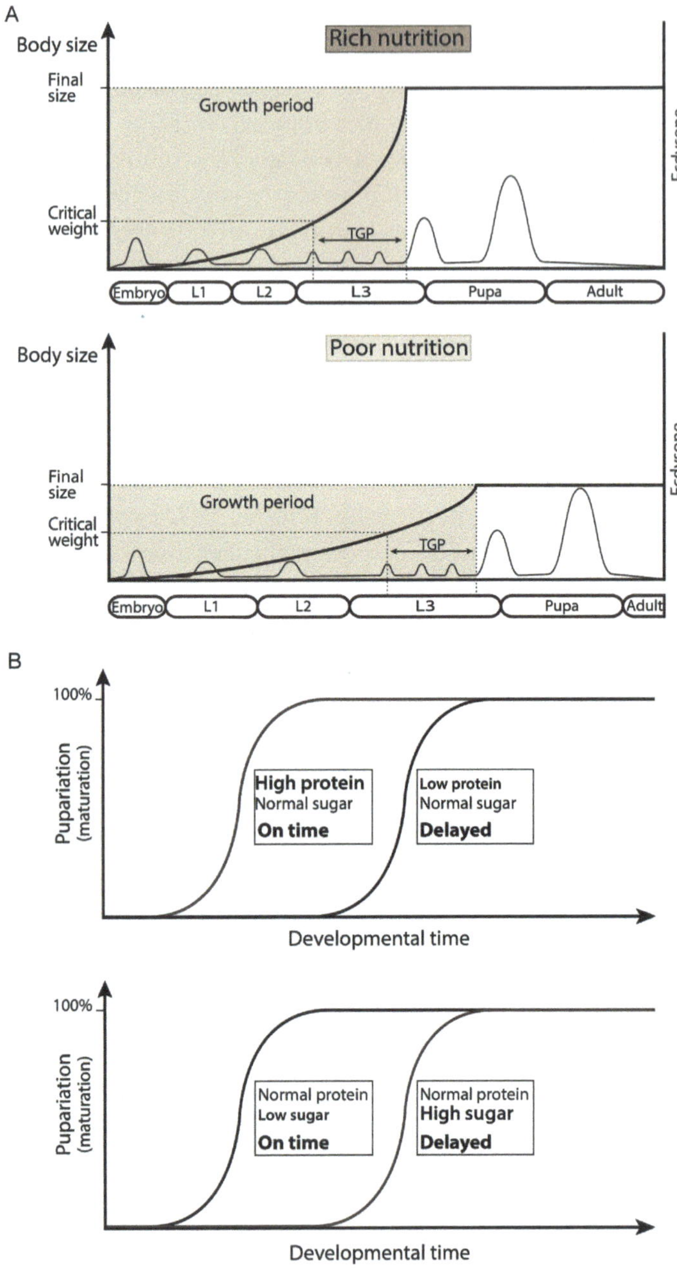

Figure 2.1 Influence of nutritional conditions on developmental timing and body size. (A) The developmental program is dictated by pulses of ecdysone which triggers molting during the three larval instars (L1, L2, and L3) and the larval–pupal transition at the onset of metamorphosis. Nutrient rich conditions promote rapid larval growth,

and in *Drosophila* the clock neurons, producing the pigment dispersing factor (PDF), impinge on the PTTH-producing neurons (McBrayer et al., 2007). The photoperiodic mechanism gating PTTH release therefore imposes an additional checkpoint control on the release of ecdysone.

Here, we review the system responsible for developmental timing of maturation based on research on *Drosophila* with focus on the role of conserved molecules related to insulin in connecting onset of maturation with the consumption of calories and growth.

2. INTERPLAY BETWEEN NUTRIENTS, SYSTEMIC GROWTH, AND DEVELOPMENTAL TRANSITIONS

2.1. Switching to a nutrient-independent developmental timing program

Ecdysone is produced and released from the PG in response to cues that serve as nutritional and developmental checkpoints. This means that the neuroendocrine system monitors these cues to ensure that the larva has accumulated sufficient mass and completed enough growth before the release of ecdysone that triggers metamorphosis (Mirth & Riddiford, 2007; Rewitz et al., 2013; Yamanaka et al., 2013). In early L3, *Drosophila* larvae reach critical weight, a nutritional checkpoint that mediates a switch in the developmental response to starvation (Fig. 2.1A). Critical weight is defined as the time when the larva is committing itself to undergo metamorphosis even in the absence of any further nutritional uptake. If starved before critical weight, larvae extend the growth period by delaying pupariation, whereas starvation after the critical weight point will result in cessation of growth but without affecting the timing of pupariation (Mirth & Riddiford, 2007; Rewitz et al., 2013). Much of the larval growth takes place during the

---

while nutrient poor environments reduce the growth rate. Critical weight in the beginning of L3 marks a nutritional checkpoint that switches the larva to a nutrient-independent schedule for the release of ecdysone and initiation of metamorphosis. This fixes the period between the critical weight and pupariation, called the terminal growth period (TGP). Critical weight is related to size, and not time, which means that the larva adjusts the duration of the growth depending on nutrient conditions pre-critical weight. Conversely, post-critical weight, the TGP is fixed and cannot be adjusted according to nutrition supply. The amount of growth during this period therefore determines the final body size of the adult. (B) Opposing effects of high protein and high sugar concentrations in the diet on development timing. (For color version of this figure, the reader is referred to the online version of this chapter.)

period between critical weight and the cessation of feeding, known as the terminal growth period (TGP). During this period the larva can quadruple its size in nutrient rich environments (Beadle, Tatum, & Clancy, 1938; Edgar, 2006). Since nutrition affects growth, but not developmental time during the TGP, nutrient restriction during this period has a large impact on final body size.

As critical weight relates to body size and not developmental time, there must be a system that continuously assesses nutrient levels and growth status to determine if critical weight has been attained. The insulin-like peptides play a key role in determining critical weight, a genetically determined size, and the outcome of surpassing this checkpoint is the production of a low-level ecdysone pulse in early L3 (Mirth & Riddiford, 2007; Mirth & Shingleton, 2012; Rewitz et al., 2013). Insulin-like peptides are ancient signal molecules that appear to play a conserved role in coordinating growth and progression through life stages in animals. In both *Drosophila* and *C. elegans* these factors adjust systemic growth in response to environmental conditions and relay information to the endocrine system to control timing of steroid release and transition to the adult stage (Tennessen & Thummel, 2011). In *Drosophila*, disrupting insulin signaling before critical weight delays puparation, but after surpassing this checkpoint reduced insulin signaling affects growth and final size but not timing of metamorphosis (Shingleton, Das, Vinicius, & Stern, 2005). This demonstrates that insulin signaling is essential in setting the critical weight parameter.

2.2. Insulin/target of rapamycin (TOR) couples nutrient-dependent growth to timing of maturation

In *Drosophila*, seven insulin-like peptides called DILP1–7 are believed to signal through a single insulin receptor (InR) of the receptor tyrosine kinase family (Ikeya, Galic, Belawat, Nairz, & Hafen, 2002; Rulifson, Kim, & Nusse, 2002). These *DILPs* are expressed in different tissues including the gut, imaginal discs, fat body, glia cells, and neurosecretory cells of the brain, known as the insulin producing cells (IPCs). *DILP2*, *DILP3*, and *DILP5* are expressed in the IPCs which are the main source of circulating DILPs during development and thought to be functionally equivalent to the mammalian pancreatic islet β-cells. Genetic ablation of these cells causes severe reduction in systemic growth and final adult size. Binding of the ligand to InR activates a conserved intracellular network of signaling molecules including phosphoinositide 3 kinase (PI3K) and AKT (Fig. 2.2). The activation of AKT (also known as protein kinase B; PKB) promotes growth by

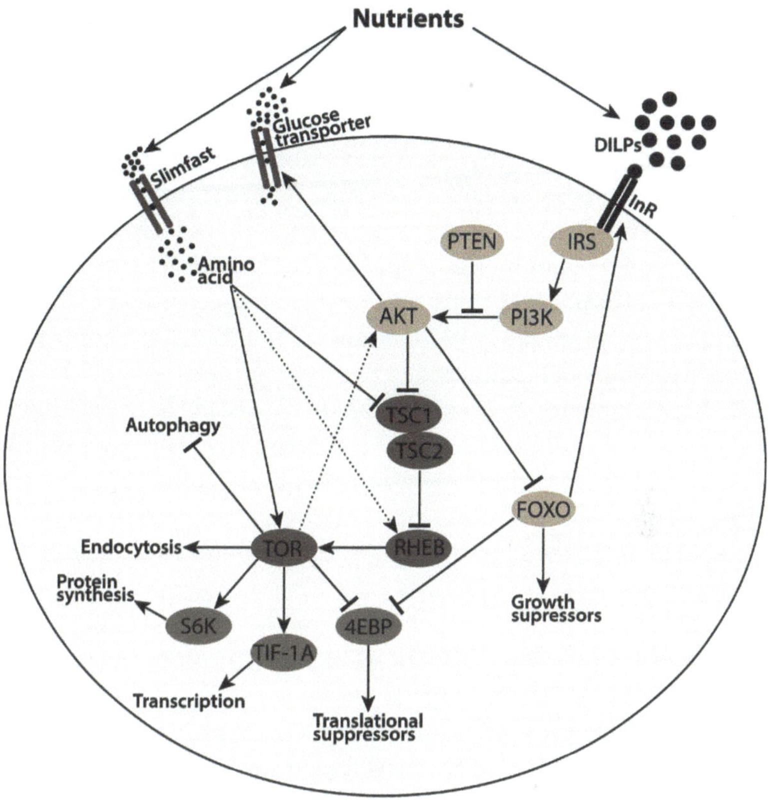

Figure 2.2 A schematic illustration of downstream effectors of insulin/TOR in *Drosophila*. Circulating *Drosophila* insulin-like peptides (DILPs) activate the insulin receptor (InR) and promote cell growth through effectors (grey) that include PI3K/AKT which affects the transcription factor FOXO. The TOR branch (green effectors) is activated by intracellular amino acids imported through Slimfast, a membrane transporter. The insulin pathway through AKT can also regulate TOR. Both pathways affect growth by regulating common downstream effectors such as S6K, TIF-1A, and 4E-BP (orange). (For interpretation of the references to color in this figure legend, the reader is referred to the online version of this chapter.)

suppressing Forkhead box Class O (FOXO) activity, a negative growth regulator, and also increases cellular energy levels through an upregulation of glucose transporters in the cell membrane (Junger et al., 2003; Kramer, Davidge, Lockyer, & Staveley, 2003; Puig, Marr, Ruhf, & Tjian, 2003; Puig & Tjian, 2005). Phosphorylation of FOXO by AKT results in the translocation of FOXO from the nucleus to the cytosol which suppresses its transcriptional activity. During nutritional deprivation, reduced insulin signaling lowers PI3K/AKT activity, which promotes nuclear localization of FOXO and suppresses cell growth.

In concert with systemic growth regulators such as insulin, the TOR pathway acts cell-autonomously to coordinate growth according to the internal nutrient status (Oldham & Hafen, 2003; Wullschleger, Loewith, & Hall, 2006). The TOR complex is regulated by cellular levels of amino acids and ATP and promotes cell growth by activating ribosome biogenesis and protein synthesis through ribosomal S6 kinase (S6K). Cellular influx of amino acids during feeding stimulates TOR activity and promotes growth. On the other end, amino acid starvation causes developmental arrest and phenotypes associated with disruption of TOR activity phenocopies amino acid deprived animals (Colombani et al., 2003; Geminard, Rulifson, & Leopold, 2009; Oldham, Montagne, Radimerski, Thomas, & Hafen, 2000; Zhang, Stallock, Ng, Reinhard, & Neufeld, 2000). Crosstalk between the TOR and insulin pathways exists through the tuberous sclerosis proteins (TSC1 and TSC2) that are regulated by PI3K/AKT and inhibits TOR. Thus, TOR has a distinct role in sensing local amino acid concentrations, although it also responds to systemic nutritional signals through its regulation by the insulin pathway.

2.3. Influence of dietary amino acids, sugar, and fat on growth and onset of maturation

The developmental time of *Drosophila* from embryo to adulthood takes approximately nine days on a nutritionally rich balanced diet but can be prolonged depending on the quality and composition of the food. Amino acids are a fundamental part of the diet required to drive developmental progression. Newly hatched larvae fed only on a sucrose diet die as first instar larvae (L1) as they fail to grow (Britton & Edgar, 1998; Colombani et al., 2003). The L1 growth arrest of sucrose fed larvae is completely reversible by adding protein to media which allows continued growth, while adding fat or nucleotides precursors only allows the larva to develop to the second larval instar (L2) or early L3 before arresting their development (Britton & Edgar, 1998). Larvae fed on a low protein diet develop to adulthood, but with a negative correlation between the amount of protein added to the food and the developmental time (Fig. 2.1B; Layalle, Arquier, & Leopold, 2008). This suggests a tight connection between amino acid status, growth, and ecdysone release.

Developmental time and growth is, however, not only dependent on sufficient protein but can also be affected by other dietary components such as sugar. In a study by Matzkin, Johnson, Paight, Bozinovic, and Markow (2011) *Drosophila* larvae were fed on different diets with the same amount of calories, but containing either a high sugar to protein ratio or a low sugar to high protein ratio. Development to adulthood was delayed by one day on the high sugar to protein diet as a result of a prolonged larval stage. This was accompanied with elevated levels of triglycerides and glycogen and with reduced total protein levels compared to the animals fed on a low sugar to protein diet (Matzkin et al., 2011). Studies of larvae raised on nutritional caloric excess from sugar, protein, or fat demonstrate that excess sugar has the most severe impact on growth and developmental timing. Remarkably, increasing calories from sugar (high sugar diet, HSD), but not fat and protein, in medium that is otherwise sufficient for a maximal developmental rate, decreases growth and delays pupariation by three to four days (Fig. 2.1B; Musselman et al., 2011; Pasco & Leopold, 2012). Despite the prolonged developmental time on HSD, animals are smaller, but with an increased hemolymph concentration of glucose as well as stored fat. This is combined with a twofold increase in the expression of *DILP2* and *DILP5* which may be a compensatory mechanism to counteract the high levels of glucose in circulation. The impaired growth on HSD is a result of insulin resistance which leads to a reduction of insulin signaling in peripheral tissues that suppresses growth and not a deficiency of circulating DILPs (Musselman et al., 2011; Pasco & Leopold, 2012). Moreover, fat body specific knockdown of neural lazarillo (Nlaz), a lipocalin family member and homolog of the mammalian apolipoprotein D (ApoD), and retinol binding protein 4 (RBP4), is sufficient to protect against the development of insulin resistance on a HSD (Pasco & Leopold, 2012). This illustrates the physiological consequences of HSD-induced hyperglycemia and disruption of insulin homeostasis on developmental timing. Similar consequences on developmental rate and growth are seen when the IPCs are ablated genetically and in *InR* deficient larvae (Brogiolo et al., 2001; Rulifson et al., 2002). The situation with elevated levels of glucose, stored fat and insulin combined with insulin resistance resembles that of type 2 diabetes in mammals (Muoio & Newgard, 2008). The developmental delay in larvae raised on a HSD can be explained by the reduced growth rate, which means that these animals reach critical weight later. One interesting possibility is that under these nutrient conditions, the PG may develop resistance to insulin, which regulates ecdysone synthesis in the gland (Caldwell, Walkiewicz, & Stern, 2005; Colombani et al., 2005; Mirth, Truman, & Riddiford, 2005).

Recently, studies have also coupled the function of insulin/TOR signaling to lipid metabolism (Birse et al., 2010; Oldham, 2011). Like excess sugar, the consequences of feeding *Drosophila* on a diet containing high amounts of saturated fats have also been shown to affect insulin/TOR signaling. Animals fed on a high fat diet become obese in an insulin/TOR dependent manner with an increase in free glucose accompanied with a drop in *DILP2* transcript levels (Birse et al., 2010). Unlike fatty acids, cholesterol is a neutral lipid that serves as a substrate for steroid synthesis (Huang, Suyama, Buchanan, Zhu, & Scott, 2005; Huang, Warren, & Gilbert, 2008). Since insects are sterol auxotrophs, they require a dietary source of cholesterol or other sterols that can be converted into ecdysteroids for their development. Although many steps in the uptake and trafficking of cholesterol remain unclear, it has been demonstrated that the *niemann-pick type C* (*npc*) genes are essential for delivery of cholesterol for ecdysone synthesis (Huang et al., 2005; Huang, Warren, Buchanan, Gilbert, & Scott, 2007). These genes are believed to be important for intestinal absorption of cholesterol and are required specifically in the PG to deliver cholesterol for the first step in the ecdysone biosynthetic pathway, the conversion of cholesterol to 7-dehydrocholesterol (7dC). Larvae lacking *npc1a* arrest development in L1, a phenotype that can be at least partially rescued by providing high levels of cholesterol in the diet (Huang et al., 2005). Interestingly, a recent study suggests that dietary sterols from hempseed and cholesterol may increase growth and accelerate pupariation in *Drosophila* (Lee et al., 2010). While cholesterol is a precursor of ecdysone, which might explain the acceleration of developmental progression, it is less clear how elevated cholesterol levels increases growth, although cholesterol has essential roles in cell membranes.

Although poor nutrition delays development, nutrition past an upper threshold limited by the maximal growth rate cannot further accelerate pupariation. Beyond this point only genetic perturbations that result in premature release of ecdysone are known to accelerate pupariation (Ou, Chwalla, Landgraf, & van Meyel, 2008; Rewitz et al., 2009). However, addition of royalactin to the diet was recently found to accelerate development (Kamakura, 2011; Yamanaka & O'Connor, 2011). Royalactin is a protein isolated from the food source compound Royal Jelly, that bees feed their larvae to promote them to become queens instead of workers. This changes the epigenetic program of the larvae (Kucharski, Maleszka, Foret, & Maleszka, 2008; Maleszka, 2008) and accelerates growth, giving rise to animals with a larger body size and 10 times the lifespan of the workers. Surprisingly, *Drosophila* feed with royalactin show a similar phenotype of increased body size, extended lifespan and shortened developmental time (Kamakura, 2011; Kayashima, Yamanashi, Sato, Kumazawa, & Yamakawa-Kobayashi, 2012). The effect of royalactin is not nutritional, as increasing protein levels in food do not have a similar effect. Instead, dietary royalactin activates either directly or indirectly the epidermal growth factor receptor (EGFR) in the fat body, which activates pathways that include PI3K/TOR and Ras/Erk. This increases systemic growth and upregulates the synthesis of ecdysone in the PG which accelerates pupariation (Kamakura, 2011; Yamanaka & O'Connor, 2011).

Although major progress has been made in understanding how nutrients regulate systemic growth and developmental timing, some major questions still remain open. It is clear that insulin signaling is affected by sugar, protein, and fat, but the outcome seems to be different. While high protein levels increases growth and developmental rate, high sugar has the opposite effect. However, it is important to remember that *Drosophila* maintained on a HSD develop hallmarks of type 2 diabetes, including perturbation of insulin homeostasis. Thus, the developmental delay observed on a HSD may not be associated with normal physiology, but pathological conditions relating to diabetes which have also been found to delay puberty in humans (Kjaer et al., 1992). It will be important to investigate how the insulin system of *Drosophila* responds differentially to sugar, protein, and fat. This may involve different responses of DILPs to specific nutrient components, like observed for expression of *DILP3* and *DILP5* that are affected by dietary sugar, but not protein restriction (Colombani et al., 2003; Ikeya et al., 2002), or changes in the cellular response as in the case of peripheral insulin resistance (Musselman et al., 2011; Pasco & Leopold, 2012).

Intriguingly, the genetic program that controls developmental timing also influences organismal life span. A number of studies have shown that dietary restrictions, genetic reduction of insulin or ecdysone signaling increase average life span in *Drosophila* (Broughton et al., 2010; Galikova, Klepsatel, Senti, & Flatt, 2011; Simon, Shih, Mack, & Benzer, 2003; van Heemst, 2010; Wessells, Fitzgerald, Cypser, Tatar, & Bodmer, 2004). Therefore, the genetic mechanism controlling growth and developmental timing in response to nutritional conditions during development is also used to prioritize reproduction versus maintenance and survival during adulthood by assessments of environmental conditions. However, there are exceptions for the trade-off between reproduction and longevity where increased life expectancy does not come at a reproductive cost, as observed in insects fed royalactin (Kamakura, 2011). Like the regulation of growth and timing during development, dietary proteins play a major role in determining longevity versus reproduction in adults. Increasing amounts of proteins in the diet decrease adult lifespan and increase fecundity, while sugar and lipids have little influence (Grandison et al., 2009). This shows that dietary amino acids, and not the calories *per se*, regulate these processes. Amino acids that reduce average life span generally increase reproduction through activation of insulin signaling. Of all the amino acids, addition of methionine (Met) alone to a protein restricted diet increases fecundity as much as a non-restricted diet, indicating that this essential amino acid plays a key role. Strikingly, an extra copy of the initiator methionine tRNA ($tRNA_i^{Met}$) also accelerates pupariation (Rideout, Marshall, & Grewal, 2012). Met may therefore be a limiting factor for a nutritional checkpoint that couples dietary uptake of amino acids with growth, developmental timing and reproduction. Another interesting observation is that leucine and isoleucine, two other essential amino acids, are the only single amino acids that promotes DILP secretion from the IPCs (Geminard et al., 2009).

Studies unraveling the nutritional cues that affect the endocrine system and onset of maturation in *Drosophila* may provide important insight that will serve as a paradigm to understand the correlations between diet, obesity, diabetes, and timing of puberty in mammals. One important aspect is that imbalanced diets such as carbohydrate overload affect developmental timing in *Drosophila*, which may provide the framework for understanding how obesity and diabetes influence pubertal timing. Studies on *Drosophila* have established a key role for dietary proteins in affecting developmental timing. Consistent with these observations, it has recently been found that dietary intake of protein during early childhood is associated with timing of puberty in humans (Gunther, Karaolis-Danckert, Kroke, Remer, & Buyken, 2010). How are amino acids sensed at the organismal level? In *Drosophila* the fat body is a central relay that communicates dietary protein intake to peripheral tissues and the endocrine system.

3. A NUTRITIONAL CHECKPOINT BASED ON CROSSTALK BETWEEN DIFFERENT TISSUES

3.1. Coordinating nutrient uptake and systemic growth

Amino acids are more intimately linked to growth than sugar in *Drosophila*. Although TOR cell-autonomously senses nutrients and regulates growth according to amino acid levels, evidence suggests that growth is regulated at the cellular level in response to circulating DILPs rather than levels of amino acids (Britton, Lockwood, Li, Cohen, & Edgar, 2002). Thus, when larvae are deprived of dietary proteins, growth is suppressed due to a reduction of systemic insulin and not as a direct consequence of amino acid starvation at the cellular level. This implies the existence of a mechanism that monitors amino acid levels and controls the secretion of DILPs. In *Drosophila*, the fat body plays a central role in a nutritional checkpoint that requires dietary protein and controls growth and developmental progression by regulation of systemic insulin signaling (Britton & Edgar, 1998; Colombani et al., 2003; Geminard et al., 2009). The fat body is the functional equivalent of the mammalian adipose tissue and liver and also has endocrine functions. In order to regulate systemic growth, the fat body harbors a nutrient sensor that monitors energy status and conveys that information to the IPCs in the brain to control the secretion of DILPs (Fig. 2.3). Evidence for the existence of a fat body amino acid sensor comes from several studies showing that fat body

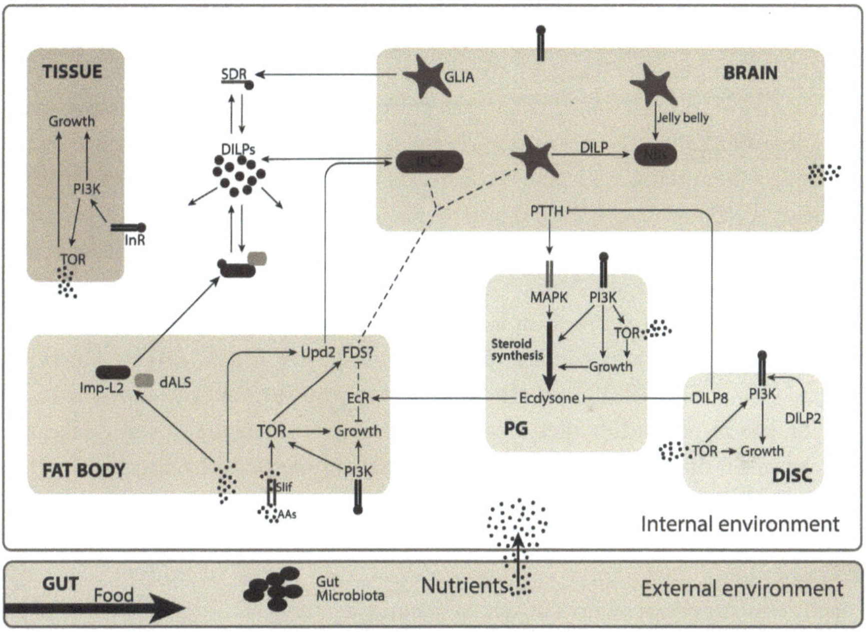

Figure 2.3 Inter-organ signaling coordinates growth and developmental timing. Uptake of nutrients in the gut from the external environment is influenced by the gut microbiota. Nutrient sensing in the fat body leads to the production of secreted factors that regulate the release of local acting and systemic DILPs from the brain affecting growth in response to nutrition. Activity of circulating DILPs is regulated by the formation of binding complexes with SDR, Imp-L2, and dALS. The PG controls developmental timing based on the production of ecdysone which is regulated by developmental cues (PTTH signaling and disc development; DILP8) and by nutritional inputs via InR/PI3K and TOR. IPCs, insulin-producing cells; NBs, neuroblasts; PG, prothoracic gland; DILPs, *Drosophila* insulin-like peptides; SDR, secreted decoy receptor; Imp-L2, Imaginal morphogenesis protein-Late 2; dALS, *Drosophila* acid-labile unit; FDS, fat body-derived signal; Upd2, Unpaired 2; PTTH, prothoracicotropic hormone. (For color version of this figure, the reader is referred to the online version of this chapter.)

specific suppression of amino acid import or TOR activity, non-autonomously affects insulin signaling and growth (Britton & Edgar, 1998; Colombani et al., 2003; Geminard et al., 2009; Rideout et al., 2012; Storelli et al., 2011). Reduced expression of the amino acid importer, *slimfast* (*slif*), in the larval fat body leads to inhibition of organismal growth by repression of insulin mediated PI3K signaling at the systemic level (Colombani et al., 2003).

This inter-organ communication allows the organism to monitor amino acid levels centrally via TOR in the fat body and globally adjust growth and developmental progression. A main target of the fat body nutrient sensor is the IPCs of the brain that secrete DILP2, DILP3, and DILP5 into the hemolymph and promote systemic growth (Colombani et al., 2003; Geminard et al., 2009). During starvation, expression of *DILPs* is decreased in the IPCs and *ex vivo* experiments with organ cultures have shown that isolated fat body from fed animals promotes secretion of DILPs from these neurosecretory cells (Geminard et al., 2009). This suggests that amino acid import into fat body cells activates TOR and leads to the production of a secreted signal(s) from the fat body that acts remotely on the IPCs (Britton et al., 2002; Colombani et al., 2003; Geminard et al., 2009). Recently, it was discovered that the cytokine, Unpaired2 (Upd2), is a diet-dependent fat body-derived signal that acts on the IPCs (Rajan & Perrimon, 2012). Fat body specific loss of *upd2* affects global growth resulting in the formation of smaller larvae and adult flies with altered metabolism including decreased storage of fat and increased levels of circulating sugar. As expected for the fat body derived signal, expression of *upd2* is nutrient-dependent, but the expression is only regulated by dietary fat and sugar, not protein. Starvation decreases *upd2* levels and overexpression of *upd2* in the fat body under such conditions suppresses breakdown of fat and lowers DILP levels in the IPCs suggesting that Udp2 is involved in signaling the fed state. However, Upd2 does not act directly on the IPCs to regulate the release of DILPs. Instead, Upd2 affects JAK/STAT signaling in GABAergic neurons which impinge on the IPCs and regulate their activity (Enell, Kapan, Soderberg, Kahsai, & Nassel, 2010). Remarkably, the phenotype of *upd2* mutants is rescued by supplying human leptin, which suggest that Udp2 and leptin are related signal molecules. Since dietary proteins do not induce Upd2, the fat body most likely secretes other yet unidentified nutrient-specific signals in response to amino acid levels. Although *upd2* expression is not directly affected by dietary protein, fat body specific knock down of the amino acid transporter *slif*, which mimics protein deprivation, results in an upregulation of *upd2* levels. This may be explained by remodeling of the metabolic program by using fat or carbohydrates as an energy source. Importantly, this finding highlights that not only proteins, but also dietary sugar and fat affects systemic insulin signaling and may be important for the regulation of ecdysone production and developmental timing.

Stimulation of systemic growth by the fat body also seems to depend on the protein synthesis rate within the fat body. It was recently shown that Maf1, a repressor of RNA polymerase III mediated tRNA synthesis, is involved in nutrient dependent regulation of growth and developmental timing (Marshall, Rideout, & Grewal, 2012; Rideout et al., 2012). In nutrient rich conditions, TOR-dependent inhibition of Maf1 in the fat body promotes tRNA synthesis in this tissue, which increases growth at the organismal level by stimulating brain-derived *DILP2* and *DILP5* transcription. This provides additional evidence for the fat-to-brain axis and the importance of fat body-derived secreted factors in regulating insulin secretion from the IPCs. Interestingly, suppressing *Maf1* specifically in the fat body, but not muscle or brain, is sufficient to accelerate pupariation. $tRNA_i^{Met}$ is an important target of Maf1 and increasing the dose of $tRNA_i^{Met}$ phenocopies the inhibition of Maf1 in the fat body. This suggests a model in which this single tRNA, important for initiation of translation, is a limiting factor for protein synthesis and plays a key role in fine-tuning growth and timing of ecdysone release in response to diet.

The fat body itself has also been shown to be a source of insulin as it is the predominant site of *DILP6* expression, which is increased by ecdysone and during starvation (Okamoto et al., 2009; Slaidina, Delanoue, Gronke, Partridge, & Leopold, 2009). In contrast to the DILPs derived from the brain, DILP6 sustains growth during the non-feeding pupal stages. As the major source of stored energy, the fat body is able to sustain levels of hemolymph nutrients from initiation of autophagy during nutrient deprivation which is otherwise suppressed by TOR signaling (Oldham, 2011). Furthermore, nutrient dependent fat body signaling also induces *DILP* expression in glia cells which promotes local proliferation and growth of neural stem cells (Chell & Brand, 2010). Activation of TOR in the fat body is required to produce a secreted signal that reactivates quiescent neuroblasts. This reactivation relies on insulin/PI3K/TOR signaling in the neuroblast, but is independent of the IPC-derived circulating DILPs that regulate systemic growth. Instead, reactivation is a result of diet-dependent *DILP2* and *DILP6* expression in glia cells that specifically reactivates neuroblast without affecting systemic growth. The fact that the secreted fat body signal that stimulates glia *DILP2* and *DILP6* expression requires dietary amino acids and fat body TOR activation suggests that it is not Upd2, but may be the same signal that regulates the insulin pool in the IPCs in response to dietary proteins. Together, this shows that the fat body with its nutrient sensing and endocrine functions plays a key role in coordinating nutrient uptake with systemic growth required for developmental progression. Future research should focus on the identification of additional nutrient-dependent secreted signals from the fat body and investigate their potential redundancy, interactions, and targets. It is also possible that signals from the fat body may not only include secreted proteins. A recent study, for example, indicates that TCA cycle intermediates and upstream metabolites from the fat body crosstalk with the neuroendocrine system of the brain to control ecdysone release (Xu, Lu, & Denlinger, 2012).

3.2. Interactions between the gut microbiota, nutrition, and developmental progression

Although extensive studies have focused on the fat body, recent reports have highlighted the important role of intestinal homeostasis in coordinating nutrient uptake with growth and development timing (Akman Gunduz & Douglas, 2009; Ridley, Wong, Westmiller, & Douglas, 2012; Shin et al., 2011; Storelli et al., 2011). One of the challenges facing animals in nature is to find a nutritionally adequate food source. It has been shown that by engaging in symbiotic relationship with microorganisms some insects can meet their dietary requirements even on low protein diets (Akman et al., 2009). The host-intestinal microbiota affects the organisms nutritional supply, and hence, larval growth rate and developmental time (Ridley et al., 2012). Interestingly, two recent reports show that the microenvironment of the gut seems to have a critical influence on systemic insulin signaling in *Drosophila* (Shin et al., 2011; Storelli et al., 2011). These studies have investigated the impact of commensal bacteria on *Drosophila* growth and development and show that larvae reared under germ-free conditions attenuate ecdysone release and delay development as nutrients become limited. The gut bacteria *Acetobacter pomorum* and *Lactobacillus plantarum* modulates insulin signaling of the host, demonstrating a host-microbe crosstalk that influences nutrient uptake and developmental progression in *Drosophila*. The role of the host microbiota in regulating the internal metabolic milieu and insulin homeostasis has also received increasing attention in mammals (Boroni Moreira, Fiche Salles Teixeira, do C Gouveia Peluzio, & de Cassia Goncalves Alfenas, 2012; De Bandt, Waligora-Dupriet, & Butel, 2011). Considering that diet affects the host gut microbiota which modulate the metabolism of the host, a deeper understanding of crosstalk between microbes, metabolism, and growth will be important to elucidate how these interactions may affect timing of maturation.

3.3. Crosstalk of nutrient signaling and neuroendocrine control of maturation

The findings that insulin/TOR signaling in the PG regulates ecdysone synthesis (Boulan, Martín, & Milán, 2013; Caldwell et al., 2005; Colombani et al., 2005; Layalle et al., 2008; Mirth et al., 2005) have provided a connection between ecdysone synthesis and nutrition. These studies demonstrate that both pathways converge on the regulation of ecdysone synthesis in the PG and suggest that the PG itself acts as a nutrient sensor to some extent. Since insulin signaling only affects developmental time before critical weight (Shingleton et al., 2005), it presumably plays a role in producing the first low-level ecdysone pulse in the early L3, but not the subsequent peaks. If insulin is directly responsible for the production of the low-level ecdysone pulse associated with the critical weight switch, it requires that insulin is a regulator of the ecdysone biosynthetic pathway, a view supported by the fact that bombyxin, an insulin-like molecule, was first identified for its ability to stimulate ecdysone biosynthesis in the silkworm *Bombyx mori* (Nagasawa et al., 1986). One possible mechanism is that a certain threshold level of insulin signaling switches the PG to a nutrient independent program for the production of ecdysone. An alternative explanation for the role of insulin in the PG is that when insulin signaling passes a certain threshold it provides the PG with competence to respond to other signals like PTTH that are then directly responsible for producing the ecdysone pulses. Potential crosstalk between the insulin and PTTH signaling pathways in the PG via MAPK/Erk also opens the possibility that both signals are required simultaneously to produce the critical weight ecdysone pulse. Another scenario is that insulin signaling promotes PG cell growth and is permissive for ecdysone synthesis in the gland, but that its direct coupling to the production of the ecdysone pulse is upstream of PTTH. Some evidence suggests that PTTH is responsible for producing the ecdysone peak associated with critical weight (Ou, Magico, & King-Jones, 2011; Rewitz & O'Connor, 2011). Insulin may therefore potentially act directly or indirectly via the brain to control the timing of PTTH release.

In addition to its response to insulin, the activity of TOR in the PG also regulates ecdysone synthesis (Layalle et al., 2008). This suggests that local nutrient sensing occurs at the level of the PG, although it is unlikely that this is the primary mechanism for coordinating nutrient uptake with ecdysone synthesis. Nevertheless, inhibition of TOR signaling in the PG reduces ecdysone synthesis and delays pupariation. This implies that cellular levels of amino acids regulate the steroidogenic activity of the PG and/or that TOR is cross-activated by the insulin pathway. The insulin/TOR pathway may also crosstalk with the MAPK pathway downstream of PTTH or they may converge on common targets to promote ecdysone synthesis as reported in other systems (Mendoza, Er, & Blenis, 2011). The nutrition-dependent checkpoint for ecdysone biosynthesis may potentially also be controlled by a direct response of the PG to a fat body derived signal.

4. FINE-TUNING SYSTEMIC INSULIN SIGNALING

4.1. Antagonistic interactions between ecdysone and insulin

While peaks of ecdysone promote transitions between developmental stages and eventually the onset of pupariation, basal levels of ecdysone produced and released from the PG negatively regulates insulin mediated growth (Boulan et al., 2013; Colombani et al., 2005; Delanoue, Slaidina, & Leopold, 2010; Francis, Zorzano, & Teleman, 2010). Stimulation of insulin signaling in most cases promotes growth, however, PG-specific activation of insulin signaling reduces final body size and *vice versa* (Caldwell et al., 2005; Colombani et al., 2005; Mirth et al., 2005). A recent report shows that insulin signaling modulates the basal level of ecdysone synthesis in the PG by repressing *bantam*, a microRNA that suppresses ecdysone synthesis (Boulan et al., 2013). These studies suggest that insulin regulates basal ecdysone levels, which makes it possible that insulin signaling is not involved in the production of the ecdysone pulses during L3. Since ecdysone inhibits growth by influencing the ecdysone receptor (EcR) activity in the fat body leading to a systemic reduction of insulin signaling (Fig. 2.3) this may be a way to adjust growth to the internal nutritional state. Removal of EcR function in this tissue, but not other organs, increases growth and final pupal size without affecting developmental time (Colombani et al., 2005; Delanoue et al., 2010). The transcription factor dMyc appears to be a specific target of EcR in the fat body, since downregulation of dMyc in the fat body is sufficient to mediate the systemic inhibitory effect of ecdysone. Ecdysone acts by inhibiting dMyc in the fat body which leads to nonautonomous suppression of growth (Delanoue et al., 2010; Grewal, 2009). This further supports the notion that the fat body is a central relay secreting factors into the hemolymph in response to both intrinsic and extrinsic cues that control systemic growth by regulating the pool of insulin in the IPCs. This also highlights the importance of basal ecdysone concentrations, lower than those required to trigger developmental transitions, in regulation of growth.

In addition to the effect on systemic growth, ecdysone signaling in the fat body cell-autonomously suppresses insulin signaling at the level of PI3K activity, dFOXO localization and 4E-BP expression (Colombani et al., 2005; Delanoue et al., 2010). The EcR coactivator, dDOR, is an important factor for the relationship between insulin and ecdysone signaling (Francis et al., 2010). Expression of dDOR is suppressed by insulin, which means that insulin signaling attenuates the ecdysone response. Conversely, ecdysone inhibits insulin signaling which stimulates dDOR expression demonstrating a positive feedback where ecdysone potentiates its own genetic response. The antagonistic relationship between insulin and ecdysone illustrates a mechanism that fine-tunes their action to coordinate growth and developmental transitions and may also prevent run-away synthesis of ecdysone.

4.2. Adjusting circulating DILP activity

Nutrient-dependent systemic insulin signaling is regulated at the level of transcriptional expression and secretion. However, circulating DILP levels are also fine-tuned by other mechanisms. In the mammalian system, IGF-binding proteins (IGFBPs) form binary complexes with circulating IGFs and ternary complexes including the acid-labile subunit (ALS) that regulates the availability of circulating IGF (Hall, Hilding, & Thoren, 1999; Holly & Perks, 2012). Formation of such complexes stabilizes the IGFs and extends their half-life as well as prevents ligand-receptor interaction (Fig. 2.3). The *Drosophila* Imp-L2 is a secreted protein, related to the mammalian IGFBPs, that modulates insulin homeostasis by binding to DILP2 and DILP5 (Alic, Hoddinott, Vinti, & Partridge, 2011; Honegger et al., 2008; Sloth Andersen, Hertz Hansen, Schaffer, & Kristensen, 2000). Expression of *Imp-L2* is nutrient-dependent and genetic loss of *Imp-L2* is associated with an increased body size without affecting developmental timing. On the other end, increased levels of Imp-L2 lead to reduction of growth and delays pupariation, implying that Imp-L2 is a negative regulator of insulin signaling (Honegger et al., 2008). Imp-L2 is secreted from the fat body under adverse nutrient conditions to reduce insulin signaling (Honegger et al., 2008). In parallel to the situation in mammals, the *Drosophila* ortholog of ALS (dALS) forms a trimeric complex with Imp-L2 and DILP2 that suppresses insulin-mediated growth (Arquier et al., 2008). Like

Imp-L2, expression of dALS is nutrient-dependent, however, its expression is suppressed when amino acid restriction is induced genetically in the fat body (Colombani et al., 2003). During conditions with restricted nutrients, loss of *dALS* further reduces the body size suggesting that dALS is required for adjusting insulin signaling under nutritional stress (Arquier et al., 2008). The physiological function of dALS may therefore be to reduce systemic insulin signaling under conditions where dietary protein is limiting for growth. Manipulating dALS expression affects the larval growth rate, but not developmental timing (Arquier et al., 2008).

Recently a new key player in the regulation of circulating DILPs activity was discovered, a secreted protein with similarity to the extracellular domain of InR named *secreted decoy receptor* (SDR; Okamoto et al., 2013). Like Imp-L2 and dALS, SDR is a negative regulator that functions by forming a complex with circulating DILPs to antagonize their activity during development. Larvae with loss of *SDR* have increased growth rate and final size, while overexpression of *SDR* results in animals with a reduced size. This shows that SDR affects physiological functions by limiting insulin-mediated growth. Although *SDR* is expressed in various tissues including the midgut and the imaginal discs, its non-autonomous effect on growth is a result of glia cell-derived expression that regulates insulin signaling at the systemic level. Expression levels of the prominent *DILPs* are unaffected by *SDR* loss of function, supporting the idea that SDR modulates insulin signaling in the hemolymph downstream of secretion. Like Imp-L2, SDR has the ability to bind with DILPs but possibly independent of Imp-L2 and dALS. It is interesting to note that SDR and Imp-L2 exhibit differential affinities for different DILPs.

Nutrients have opposing effect on the expression of the different factors regulating circulating DILP activity. While Imp-L2 levels are increased in fat body during starvation, dALS expression is decreased under such conditions (Colombani et al., 2003; Geminard et al., 2009; Honegger et al., 2008; Slaidina et al., 2009). In contrast to these more dynamic regulators, expression of *SDR* and its secretion into the hemolymph is nutrient independent (Okamoto et al., 2013). Indeed, *SDR* mutants have increased body size independent of nutrient supply. However, animals lacking *SDR* have increased lethality under nutrient restrictions, suggesting that SDR is required to limit insulin-mediated growth in such conditions. The difference in the action of SDR, Imp-L2 and dALS may reflect a robustness of the system that is important under fluctuating nutrient conditions to fine-tuning the activity of the ligands in circulation. This may also be important for downstream responses that have been shown to selectively decode sustained or pulse-like insulin stimulation on the basis of distinctions in the kinetics of their signal response networks (Kubota et al., 2012). Adjusting the downstream response to insulin also occurs by regulating InR levels via a feedback loop involving the transcription factor FOXO and the MAPK/Erk pathway (Puig & Tjian, 2005; Zhang, Thompson, Hietakangas, & Cohen, 2011), either sensitizing or desensitizing the cellular response.

5. A DEVELOPMENTAL CHECKPOINT COUPLES ORGAN GROWTH AND TISSUE REGENERATION TO TIMING

An important physiological aspect of growth is to produce an adult body with the correct size and dimensions. This is determined by the mechanisms that keep individual organ size proportions and time the termination of their growth at the correct size. Compared to the abundant larval-specific cells, organs that are destined to form the mature adult respond differently to nutrient deprivation and are to some degree "spared" under starvation (Britton & Edgar, 1998; Cheng et al., 2011). While most endoreplicative larval-specific cells become quiescent, adult precursor tissues continue to grow under nutrient restrictions, at least during post critical weight development. The protection of these critical organs against starvation reflects their important role in the formation of the adult. The brain and discs are organs that have the ability to sustain growth even during nutrient deprivation by mechanisms that are independent of DILPs secreted from the IPCs. These organs do not scale up or down with the overall body size in response to manipulation of DILP secretion from the IPCs. In the brain, activation of the anaplastic lymphoma kinase (Alk) receptor by its ligand Jelly belly makes the growth of neuroblasts independent of both the growth promoting DILP/InR and TOR pathways (Cheng et al., 2011). Remarkably, PI3K is activated in neuroblasts by Alk under nutrient restriction rather than by InR. Alk signaling also bypasses neuroblast requirements for TOR mediated nutrient sensing by interacting with downstream effectors including 4E-BP (Cheng et al., 2011). Jelly belly is constitutively expressed in glia cells showing that these cells play a central role in sustaining brain development under limiting nutrient conditions. Growth of the imaginal discs is also more resistant to nutrient deprivation than larval-specific tissues that shut down all growth under such conditions. In contrast to neuroblasts, disc growth is TOR dependent but may be decoupled from nutrition due to the expression of *DILP2* in the discs (Brogiolo et al., 2001). Together this demonstrates that the growth of the central nervous system and the imaginal discs is buffered against fluctuations in the diet at the expense of other larval-specific organs, including the fat body that under such conditions mobilize stored energy to maintain hemolymph nutrients.

An intriguing observation in insects is the synchronization of organ growth and the developmental timing program. Growth perturbation and damage to the imaginal disc delay pupariation, which allows extra time for tissues to regenerate or grow (Halme, Cheng, & Hariharan, 2010; Poodry & Woods, 1990; Simpson, Berreur, & Berreur-Bonnenfant, 1980; Stieper, Kupershtok, Driscoll, & Shingleton, 2008). During the extended growth period, non-affected discs reduce their growth rate to prevent overgrowth in order to maintain their final size relationship with the damaged or growth-perturbed discs (Parker & Shingleton, 2011). Discs therefore sense growth perturbation and damage, and signal to other tissues to slow their growth as well as to the neuroendocrine system to attenuate the production of ecdysone and extent the growth period. The signal from damaged disc may include a role for retinoids (Halme et al., 2010). However, recently two independent studies demonstrated that growth-retarded discs secrete a peptide exhibiting structural similarity with insulin, named *Drosophila insulin-like peptide 8* (*DILP8*), that delays pupariation and suppresses systemic growth (Colombani et al., 2012; Garelli et al., 2012). This peptide has a conserved cysteine motif that is characteristic for members of the insulin-relaxin peptide family. *DILP8* is strongly induced in discs when their growth is retarded or if they are damaged, indicating that it might be a downstream signal of the JNK pathway. Knock down of *DILP8* is sufficient to suppress the delay of pupariation in larvae with damaged discs. Furthermore, *DILP8* overexpression delays pupariation and also prevents overgrowth during the prolonged growth period (Garelli et al., 2012). Overexpression of *DILP8* and discs damage mostly prolong the duration of L3 (Colombani et al., 2012; Garelli et al., 2012), like lack of PTTH signaling (McBrayer et al., 2007; Rewitz et al., 2009). The effect of DILP8 on ecdysone synthesis may therefore be mediated through its suppressive effects on PTTH-producing neurons as well as on the level of the PG response (Colombani et al., 2012; Garelli et al., 2012). DILP8 has no influence on patterning and morphology of the discs, indicating that it acts downstream of a disc growth or a damage sensing system but upstream of the neuroendocrine system (Colombani et al., 2012). Moreover, *ex vivo* organ co-culture experiments demonstrate that DILP8 is a secreted factor that suppresses ecdysone production of brains with the PG attached (Colombani et al., 2012).

The discs growth program is therefore a checkpoint for activation of the neuroendocrine system and for the general growth control of other tissues.

This ensures a certain amount of disc growth before the larvae commits to a fixed schedule for metamorphosis. Since imaginal discs continue to grow even when nutrients are restricted, this checkpoint is, at least partly, nutrient independent and different from the nutritional checkpoint associated with the attainment of critical weight. This view is supported by a recent study showing that disc damage can delay the ecdysone-mediated development program until the mid-L3 (Hackney, Zolali-Meybodi, & Cherbas, 2012), after the critical weight switch in the early L3 where limiting nutrition has no influence on timing. Altogether, these studies show that at least two different growth-related checkpoints control the developmental transition to adulthood, and that both are based on crosstalk between tissues and the ecdysone system via secreted insulin-like molecules (Fig. 2.4): (I) a nutritional checkpoint during early L3 that depends on systemic nutrient-dependent insulin signaling regulated by dietary factors and primarily amino acids via the fat body; (II) a mid-L3 checkpoint that allows developmental progression depending on growth status of adult precursor tissues. Interestingly, imaginal discs grow to a final size independently of the developmental stage of their host (Hackney et al., 2012). This suggests that their target size is controlled by a

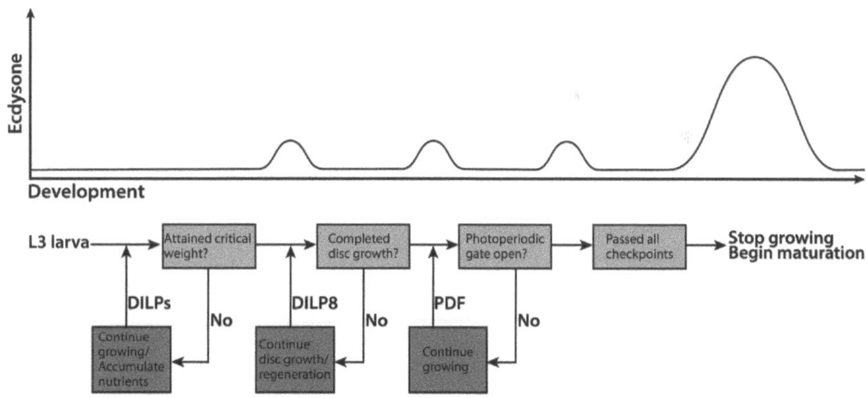

Figure 2.4 Control of developmental progression to adulthood by checkpoints. Developmental progression is controlled by a surveillance system monitoring the completion of events associated with the accumulation of nutrients/mass and discs growth. A circadian checkpoint is superimposed on this system, to gate the release of PTTH according to the photoperiod. The time-line indicates the third larval instar (L3). The predicted translation of passing these checkpoints into ecdysone pulses that coordinates and control developmental progression is shown. DILPs, *Drosophila* insulin-like peptides; DILP8, *Drosophila* insulin-like peptide 8; PDF, pigment dispersing factor. (For color version of this figure, the reader is referred to the online version of this chapter.)

tissue-autonomous program and provides a mechanism for how genetic programs determining organ size allows growth assessments by the developmental timing system.

Despite the recent progress, some important questions remain: What are the targets of DILP8 and how does this signal converge on ecdysone synthesis? Identifying its receptor will help clarifying these issues. Although it is possible that DILP8 acts antagonistically on InR, the fact that the disc checkpoint can delay pupariation post critical weight (Hackney et al., 2012) suggests otherwise, since timing is then independent of InR (Shingleton et al., 2005). DILP8 is structurally related to the relaxin family and may act through one of the *Drosophila* relaxin family receptors, Lgr3 and Lgr4, for which the ligand and their function remain unknown (Hauser et al., 2008; Van Hiel, Vandersmissen, Van Loy, & Vanden, 2012).

6. CONCLUSIONS AND PERSPECTIVES

In summary, developmental timing of maturation in *Drosophila* is controlled by the endocrine system that makes assessments of nutrition and growth status of the developing organism. This requires crosstalk between the endocrine system and organs that communicate nutritional levels and their growth status. Recent advances in *Drosophila* have shown that the overall architecture underlying timing of maturation is remarkably conserved and has demonstrated that secreted insulin-like molecules establish an important connection between nutrients, growth sensing, and the endocrine system. Like in mammals, maturation in *Drosophila* is controlled by a neuroendocrine axis that ultimately produces a steroid signal triggering developmental and behavioral changes associated with this transition. The release of ecdysone only occurs once the larva has passed certain checkpoints ensuring accumulation of enough nutrients and completion of sufficient growth during development to reach sexual maturity at a genetically determined final adult size. Consistent with the overall conservation of the system, the insulin-like molecules that are key to adjust the duration of juvenile development (the developmental timer) to dietary conditions have been conserved during evolution.

The recent identification of the secreted signal molecules Upd2 and DILP8 has provided important new information, and also raised several questions. Upd2 is functionally related to the human leptin which is linked to obesity and is expressed in adipose tissue in response to fat and affects timing of maturation, like *upd2* in *Drosophila* (Roa et al., 2009). Considering the link between childhood obesity and early puberty, it has become increasingly important to better understand the ways nutrition affects timing of maturation. Together with the influence of gut microbes on metabolism and obesity in humans, the link between the gut microbiota, dietary amino acids, and timing in *Drosophila* opens avenues for future research. Genetic studies on model organisms like *Drosophila* may provide valuable insight into how pathological conditions, such as obesity and diabetes, may affect onset of puberty.

Although systemic insulin signaling affects the steroidogenic activity of the PG, it might not be the primary means by which the timing of ecdysone pulses is determined. Both systemic insulin and DILP8 may act via PTTH to regulate ecdysone synthesis and future studies may shed light on how these signals converge on the PTTH-ecdysone axis. Another important question is how the developmental timing program decodes inputs from different nutrients, like sugar and protein, that both affect systemic insulin signaling. Perhaps DILP1–7 have different roles, which may be related to the type of diet. Interestingly, *upd2* is regulated in response to fat and sugar, but not protein which is the primary nutritional cue that promotes growth and maturation via systemic insulin signaling. This implies the existence of other amino acid-dependent secreted factors from the fat body. Further investigations of the "secretome" therefore undoubtedly will reveal diffusible molecules that are important for timing of maturation in environments with fluctuating nutrient availability.

ACKNOWLEDGMENTS

This work was supported by the Danish Council for Independent Research, Natural Sciences grant 11-105446 to K.F.R.

REFERENCES

Akman Gunduz, E., & Douglas, A. E. (2009). Symbiotic bacteria enable insect to use a nutritionally inadequate diet. *Proceedings of the Royal Society B: Biological Sciences*, *276*, 987–991.

Alic, N., Hoddinott, M. P., Vinti, G., & Partridge, L. (2011). Lifespan extension by increased expression of the Drosophila homologue of the IGFBP7 tumour suppressor. *Aging Cell*, *10*, 137–147.

Arquier, N., Geminard, C., Bourouis, M., Jarretou, G., Honegger, B., Paix, A., et al. (2008). Drosophila ALS regulates growth and metabolism through functional interaction with insulin-like peptides. *Cell Metabolism*, *7*, 333–338.

Baehrecke, E. H. (2000). Steroid regulation of programmed cell death during Drosophila development. *Cell Death and Differentiation*, *7*, 1057–1062.

Beadle, G. W., Tatum, E. L., & Clancy, C. W. (1938). Food level in relation to rate of development and eye pigmentation in Drosophila melanogaster. *The Biological Bulletin*, *75*, 447–462.

Birse, R. T., Choi, J., Reardon, K., Rodriguez, J., Graham, S., Diop, S., et al. (2010). High-fat-diet-induced obesity and heart dysfunction are regulated by the TOR pathway in Drosophila. *Cell Metabolism, 12*, 533–544.

Boroni Moreira, A. P., Fiche Salles Teixeira, T., do C Gouveia Peluzio, M., & de Cassia Goncalves Alfenas, R. (2012). Gut microbiota and the development of obesity. *Nutrición Hospitalaria, 27*, 1408–1414.

Boulan, L., Martín, D., & Milán, M. (2013). Bantam miRNA promotes systemic growth by connecting insulin signaling and ecdysone production. *Current Biology, 23*, 473–478.

Britton, J. S., & Edgar, B. A. (1998). Environmental control of the cell cycle in Drosophila: Nutrition activates mitotic and endoreplicative cells by distinct mechanisms. *Development, 125*, 2149–2158.

Britton, J. S., Lockwood, W. K., Li, L., Cohen, S. M., & Edgar, B. A. (2002). Drosophila's insulin/PI3-kinase pathway coordinates cellular metabolism with nutritional conditions. *Developmental Cell, 2*, 239–249.

Brogiolo, W., Stocker, H., Ikeya, T., Rintelen, F., Fernandez, R., & Hafen, E. (2001). An evolutionarily conserved function of the Drosophila insulin receptor and insulin-like peptides in growth control. *Current Biology, 11*, 213–221.

Broughton, S. J., Slack, C., Alic, N., Metaxakis, A., Bass, T. M., Driege, Y., et al. (2010). DILP-producing median neurosecretory cells in the Drosophila brain mediate the response of lifespan to nutrition. *Aging Cell, 9*, 336–346.

Caldwell, P. E., Walkiewicz, M., & Stern, M. (2005). Ras activity in the Drosophila prothoracic gland regulates body size and developmental rate via ecdysone release. *Current Biology, 15*, 1785–1795.

Chell, J. M., & Brand, A. H. (2010). Nutrition-responsive glia control exit of neural stem cells from quiescence. *Cell, 143*, 1161–1173.

Cheng, L. Y., Bailey, A. P., Leevers, S. J., Ragan, T. J., Driscoll, P. C., & Gould, A. P. (2011). Anaplastic lymphoma kinase spares organ growth during nutrient restriction in Drosophila. *Cell, 146*, 435–447.

Colombani, J., Andersen, D. S., & Leopold, P. (2012). Secreted peptide Dilp8 coordinates Drosophila tissue growth with developmental timing. *Science, 336*, 582–585.

Colombani, J., Bianchini, L., Layalle, S., Pondeville, E., Dauphin-Villemant, C., Antoniewski, C., et al. (2005). Antagonistic actions of ecdysone and insulins determine final size in Drosophila. *Science, 310*, 667–670.

Colombani, J., Raisin, S., Pantalacci, S., Radimerski, T., Montagne, J., & Leopold, P. (2003). A nutrient sensor mechanism controls Drosophila growth. *Cell, 114*, 739–749.

De Bandt, J. P., Waligora-Dupriet, A. J., & Butel, M. J. (2011). Intestinal microbiota in inflammation and insulin resistance: Relevance to humans. *Current Opinion in Clinical Nutrition and Metabolic Care, 14*, 334–340.

Delanoue, R., Slaidina, M., & Leopold, P. (2010). The steroid hormone ecdysone controls systemic growth by repressing dMyc function in Drosophila fat cells. *Developmental Cell, 18*, 1012–1021.

Edgar, B. A. (2006). How flies get their size: Genetics meets physiology. *Nature Reviews Genetics, 7*, 907–916.

Enell, L. E., Kapan, N., Soderberg, J. A., Kahsai, L., & Nassel, D. R. (2010). Insulin signaling, lifespan and stress resistance are modulated by metabotropic GABA receptors on insulin producing cells in the brain of Drosophila. *PLoS One, 5*, e15780.

Francis, V. A., Zorzano, A., & Teleman, A. A. (2010). dDOR is an EcR coactivator that forms a feed-forward loop connecting insulin and ecdysone signaling. *Current Biology, 20*, 1799–1808.

Galikova, M., Klepsatel, P., Senti, G., & Flatt, T. (2011). Steroid hormone regulation of *C. elegans* and Drosophila aging and life history. *Experimental Gerontology, 46*, 141–147.

Garelli, A., Gontijo, A. M., Miguela, V., Caparros, E., & Dominguez, M. (2012). Imaginal discs secrete insulin-like peptide 8 to mediate plasticity of growth and maturation. *Science, 336*, 579–582.

Geminard, C., Rulifson, E. J., & Leopold, P. (2009). Remote control of insulin secretion by fat cells in Drosophila. *Cell Metabolism, 10*, 199–207.

Grandison, R. C., Piper, M. D. W., & Partridge, L. (2009). Amino-acid imbalance explains extension of lifespan by dietary restriction in *Drosophila. Nature, 462*, 1061–1064.

Grewal, S. S. (2009). Insulin/TOR signaling in growth and homeostasis: A view from the fly world. *The International Journal of Biochemistry & Cell Biology, 41*, 1006–1010.

Gunther, A. L., Karaolis-Danckert, N., Kroke, A., Remer, T., & Buyken, A. E. (2010). Dietary protein intake throughout childhood is associated with the timing of puberty. *Journal of Nutrition, 140*, 565–571.

Hackney, J. F., Zolali-Meybodi, O., & Cherbas, P. (2012). Tissue damage disrupts developmental progression and ecdysteroid biosynthesis in Drosophila. *PLoS One*, 7, e49105.

Hall, K., Hilding, A., & Thoren, M. (1999). Determinants of circulating insulin-like growth factor-I. *Journal of Endocrinological Investigation, 22*, 48–57.

Halme, A., Cheng, M., & Hariharan, I. K. (2010). Retinoids regulate a developmental checkpoint for tissue regeneration in Drosophila. *Current Biology, 20*, 458–463.

Hauser, F., Cazzamali, G., Williamson, M., Park, Y., Li, B., Tanaka, Y., et al. (2008). A genome-wide inventory of neurohormone GPCRs in the red flour beetle *Tribolium castaneum. Frontiers in Neuroendocrinology, 29*, 142–165.

Holly, J. M., & Perks, C. M. (2012). Insulin-like growth factor physiology: What we have learned from human studies. *Endocrinology and Metabolism Clinics of North America, 41*, 249–263.

Honegger, B., Galic, M., Kohler, K., Wittwer, F., Brogiolo, W., Hafen, E., et al. (2008). Imp-L2, a putative homolog of vertebrate IGF-binding protein 7, counteracts insulin signaling in Drosophila and is essential for starvation resistance. *Journal of Biology*, 7, 10.

Huang, X., Suyama, K., Buchanan, J., Zhu, A. J., & Scott, M. P. (2005). A Drosophila model of the Niemann–Pick type C lysosome storage disease: dnpc1a is required for molting and sterol homeostasis. *Development, 132*, 5115–5124.

Huang, X., Warren, J. T., Buchanan, J., Gilbert, L. I., & Scott, M. P. (2007). Drosophila Niemann–Pick type C-2 genes control sterol homeostasis and steroid biosynthesis: A model of human neurodegenerative disease. *Development, 134*, 3733–3742.

Huang, X., Warren, J. T., & Gilbert, L. I. (2008). New players in the regulation of ecdysone biosynthesis. *Journal of Genetics and Genomics, 35*, 1–10.

Ikeya, T., Galic, M., Belawat, P., Nairz, K., & Hafen, E. (2002). Nutrient-dependent expression of insulin-like peptides from neuroendocrine cells in the CNS contributes to growth regulation in Drosophila. *Current Biology, 12*, 1293–1300.

Junger, M. A., Rintelen, F., Stocker, H., Wasserman, J. D., Vegh, M., Radimerski, T., et al. (2003). The Drosophila forkhead transcription factor FOXO mediates the reduction in cell number associated with reduced insulin signaling. *Journal of Biology, 2*, 20.

Kamakura, M. (2011). Royalactin induces queen differentiation in honeybees. *Nature, 473*, 478–483.

Kayashima, Y., Yamanashi, K., Sato, A., Kumazawa, S., & Yamakawa-Kobayashi, K. (2012). Freeze-dried royal jelly maintains its developmental and physiological bioactivity in Drosophila melanogaster. *Bioscience, Biotechnology, and Biochemistry, 76*, 2107–2111.

Kjaer, K., Hagen, C., Sando, S. H., & Eshoj, O. (1992). Epidemiology of menarche and menstrual disturbances in an unselected group of women with insulin-dependent diabetes mellitus compared to controls. *Journal of Clinical Endocrinology and Metabolism, 75*, 524–529.

Kramer, J. M., Davidge, J. T., Lockyer, J. M., & Staveley, B. E. (2003). Expression of Drosophila FOXO regulates growth and can phenocopy starvation. *BMC Developmental Biology, 3*, 5.

Kubota, H., Noguchi, R., Toyoshima, Y., Ozaki, Y., Uda, S., Watanabe, K., et al. (2012). Temporal coding of insulin action through multiplexing of the AKT pathway. *Molecular Cell, 46*, 820–832.

Kucharski, R., Maleszka, J., Foret, S., & Maleszka, R. (2008). Nutritional control of reproductive status in honeybees via DNA methylation. *Science, 319*, 1827–1830.

Layalle, S., Arquier, N., & Leopold, P. (2008). The TOR pathway couples nutrition and developmental timing in Drosophila. *Developmental Cell, 15*, 568–577.

Lee, M. J., Park, M. S., Hwang, S., Hong, Y. K., Choi, G., Suh, Y. S., et al. (2010). Dietary hempseed meal intake increases body growth and shortens the larval stage via the upregulation of cell growth and sterol levels in Drosophila melanogaster. *Molecules and Cells, 30*, 29–36.

Liu, J. L., & LeRoith, D. (1999). Insulin-like growth factor I is essential for postnatal growth in response to growth hormone. *Endocrinology, 140*, 5178–5184.

Maleszka, R. (2008). Epigenetic integration of environmental and genomic signals in honey bees: The critical interplay of nutritional, brain and reproductive networks. *Epigenetics, 3*, 188–192.

Marshall, L., Rideout, E. J., & Grewal, S. S. (2012). Nutrient/TOR-dependent regulation of RNA polymerase III controls tissue and organismal growth in Drosophila. *EMBO Journal, 31*, 1916–1930.

Matzkin, L. M., Johnson, S., Paight, C., Bozinovic, G., & Markow, T. A. (2011). Dietary protein and sugar differentially affect development and metabolic pools in ecologically diverse Drosophila. *Journal of Nutrition, 141*, 1127–1133.

McBrayer, Z., Ono, H., Shimell, M., Parvy, J. P., Beckstead, R. B., Warren, J. T., et al. (2007). Prothoracicotropic hormone regulates developmental timing and body size in Drosophila. *Developmental Cell, 13*, 857–871.

Mendoza, M. C., Er, E. E., & Blenis, J. (2011). The Ras-ERK and PI3K-mTOR pathways: Cross-talk and compensation. *Trends in Biochemical Sciences, 36*, 320–328.

Mirth, C. K., & Riddiford, L. M. (2007). Size assessment and growth control: How adult size is determined in insects. *BioEssays, 29*, 344–355.

Mirth, C. K., & Shingleton, A. W. (2012). Integrating body and organ size in Drosophila: Recent advances and outstanding problems. *Frontiers in Endocrinology, 3*, 49.

Mirth, C., Truman, J. W., & Riddiford, L. M. (2005). The role of the prothoracic gland in determining critical weight for metamorphosis in Drosophila melanogaster. *Current Biology, 15*, 1796–1807.

Muoio, D. M., & Newgard, C. B. (2008). Mechanisms of disease: Molecular and metabolic mechanisms of insulin resistance and beta-cell failure in type 2 diabetes. *Nature Reviews Molecular Cell Biology, 9*, 193–205.

Musselman, L. P., Fink, J. L., Narzinski, K., Ramachandran, P. V., Sukumar Hathiramani, S., Cagan, R. L., et al. (2011). A high-sugar diet produces obesity and insulin resistance in wild-type Drosophila. *Disease Models & Mechanisms, 4*, 842–849.

Nagasawa, H., Kataoka, H., Isogai, A., Tamura, S., Suzuki, A., Mizoguchi, A., et al. (1986). Amino acid sequence of a prothoracicotropic hormone of the silkworm *Bombyx mori*. *Proceedings of the National Academy of Sciences of the United States of America, 83*, 5840–5843.

Nijhout, H. F. (1981). Physiological control of molting in insects. *American Zoologist, 21*, 631–640.

Niwa, R., Namiki, T., Ito, K., Shimada-Niwa, Y., Kiuchi, M., Kawaoka, S., et al. (2010). Non-molting glossy/shroud encodes a short-chain dehydrogenase/reductase that functions in the 'Black Box' of the ecdysteroid biosynthesis pathway. *Development, 137*, 1991–1999.

Okamoto, N., Nakamori, R., Murai, T., Yamauchi, Y., Masuda, A., & Nishimura, T. (2013). A secreted decoy of InR antagonizes insulin/IGF signaling to restrict body growth in Drosophila. *Genes & Development, 27*, 87–97.

Okamoto, N., Yamanaka, N., Yagi, Y., Nishida, Y., Kataoka, H., O'Connor, M. B., et al. (2009). A fat body-derived IGF-like peptide regulates postfeeding growth in Drosophila. *Developmental Cell, 17*, 885–891.

Oldham, S. (2011). Obesity and nutrient sensing TOR pathway in flies and vertebrates: Functional conservation of genetic mechanisms. *Trends in Endocrinology and Metabolism, 22*, 45–52.

Oldham, S., & Hafen, E. (2003). Insulin/IGF and target of rapamycin signaling: A TOR de force in growth control. *Trends in Cell Biology, 13*, 79–85.

Oldham, S., Montagne, J., Radimerski, T., Thomas, G., & Hafen, E. (2000). Genetic and biochemical characterization of dTOR, the Drosophila homolog of the target of rapamycin. *Genes & Development, 14*, 2689–2694.

Ou, Q., Magico, A., & King-Jones, K. (2011). Nuclear receptor DHR4 controls the timing of steroid hormone pulses during Drosophila development. *PLoS Biology, 9*, e1001160.

Ou, Y., Chwalla, B., Landgraf, M., & van Meyel, D. J. (2008). Identification of genes influencing dendrite morphogenesis in developing peripheral sensory and central motor neurons. *Neural Development, 3*, 16.

Parker, N. F., & Shingleton, A. W. (2011). The coordination of growth among Drosophila organs in response to localized growth-perturbation. *Developmental Biology, 357*, 318–325.

Pasco, M. Y., & Leopold, P. (2012). High sugar-induced insulin resistance in Drosophila relies on the lipocalin Neural Lazarillo. *PLoS One*, 7, e36583.

Petryk, A., Warren, J. T., Marques, G., Jarcho, M. P., Gilbert, L. I., Kahler, J., et al. (2003). Shade is the Drosophila P450 enzyme that mediates the hydroxylation of ecdysone to the steroid insect molting hormone 20-hydroxyecdysone. *Proceedings of the National Academy of Sciences of the United States of America, 100*, 13773–13778.

Poodry, C. A., & Woods, D. F. (1990). Control of the developmental timer for Drosophila pupariation. *Roux's Archives of Developmental Biology, 199*, 219–227.

Puig, O., Marr, M. T., Ruhf, M. L., & Tjian, R. (2003). Control of cell number by Drosophila FOXO: Downstream and feedback regulation of the insulin receptor pathway. *Genes & Development, 17*, 2006–2020.

Puig, O., & Tjian, R. (2005). Transcriptional feedback control of insulin receptor by dFOXO/FOXO1. *Genes & Development, 19*, 2435–2446.

Rajan, A., & Perrimon, N. (2012). Drosophila cytokine unpaired 2 regulates physiological homeostasis by remotely controlling insulin secretion. *Cell, 151*, 123–137.

Rewitz, K., & O'Connor, M. B. (2011). Timing is everything: PTTH mediated DHR4 nucleocytoplasmic trafficking sets the tempo of Drosophila steroid production. *Frontiers in Endocrinology, 2*, 108.

Rewitz, K. F., Rybczynski, R., Warren, J. T., & Gilbert, L. I. (2006a). Developmental expression of Manduca shade, the P450 mediating the final step in molting hormone synthesis. *Molecular and Cellular Endocrinology, 247*, 166–174.

Rewitz, K. F., Rybczynski, R., Warren, J. T., & Gilbert, L. I. (2006b). Identification, characterization and developmental expression of Halloween genes encoding P450 enzymes mediating ecdysone biosynthesis in the tobacco hornworm, *Manduca sexta. Insect Biochemistry and Molecular Biology, 36*, 188–199.

Rewitz, K. F., Yamanaka, N., Gilbert, L. I., & O'Connor, M. B. (2009). The insect neuropeptide PTTH activates receptor tyrosine kinase torso to initiate metamorphosis. *Science, 326*, 1403–1405.

Rewitz, K. F., Yamanaka, N., & OConnor, M. B. (2013). Developmental checkpoints and feedback circuits time insect maturation. In S. Yun-Bo (Ed.), *Current topics in developmental biology* (pp. 1–33). Academic Press.

Rewitz, K. F., Yamanaka, N., & O'Connor, M. B. (2010). Steroid hormone inactivation is required during the juvenile-adult transition in Drosophila. *Developmental Cell, 19*, 895–902.

Rideout, E. J., Marshall, L., & Grewal, S. S. (2012). Drosophila RNA polymerase III repressor Maf1 controls body size and developmental timing by modulating tRNAiMet synthesis and systemic insulin signaling. *Proceedings of the National Academy of Sciences of the United States of America, 109*, 1139–1144.

Ridley, E. V., Wong, A. C., Westmiller, S., & Douglas, A. E. (2012). Impact of the resident microbiota on the nutritional phenotype of Drosophila melanogaster. *PLoS One, 7*, e36765.

Roa, J., Garcia-Galiano, D., Castellano, J. M., Gaytan, F., Pinilla, L., & Tena-Sempere, M. (2010). Metabolic control of puberty onset: New players, new mechanisms. *Molecular and Cellular Endocrinology, 324*, 87–94.

Rulifson, E. J., Kim, S. K., & Nusse, R. (2002). Ablation of insulin-producing neurons in flies: Growth and diabetic phenotypes. *Science, 296*, 1118–1120.

Shin, S. C., Kim, S. H., You, H., Kim, B., Kim, A. C., Lee, K. A., et al. (2011). Drosophila microbiome modulates host developmental and metabolic homeostasis via insulin signaling. *Science, 334*, 670–674.

Shingleton, A. W., Das, J., Vinicius, L., & Stern, D. L. (2005). The temporal requirements for insulin signaling during development in Drosophila. *PLoS Biology, 3*, e289.

Siaussat, D., Porcheron, P., & Debernard, S. (2009). The ecdysteroids' effects in the control of cell proliferation and differentiation. In G. Smagghe (Ed.), *Ecdysone: Structures and functions* (pp. 185–204). Springer.

Simon, A. F., Shih, C., Mack, A., & Benzer, S. (2003). Steroid control of longevity in Drosophila melanogaster. *Science, 299*, 1407–1410.

Simpson, P., Berreur, P., & Berreur-Bonnenfant, J. (1980). The initiation of pupariation in Drosophila: Dependence on growth of the imaginal discs. *Journal of Embryology and Experimental Morphology, 57*, 155–165.

Slaidina, M., Delanoue, R., Gronke, S., Partridge, L., & Leopold, P. (2009). A Drosophila insulin-like peptide promotes growth during nonfeeding states. *Developmental Cell, 17*, 874–884.

Sloth Andersen, A., Hertz Hansen, P., Schaffer, L., & Kristensen, C. (2000). A new secreted insect protein belonging to the immunoglobulin superfamily binds insulin and related peptides and inhibits their activities. *Journal of Biological Chemistry, 275*, 16948–16953.

Smith, W., & Rybczynski, R. (2012). Prothoracicotropic hormone. In L. I. Gilbert (Ed.), *Insect endocrinology* (pp. 1–62). San Diego: Academic Press.

Stieper, B. C., Kupershtok, M., Driscoll, M. V., & Shingleton, A. W. (2008). Imaginal discs regulate developmental timing in Drosophila melanogaster. *Developmental Biology, 321*, 18–26.

Storelli, G., Defaye, A., Erkosar, B., Hols, P., Royet, J., & Leulier, F. (2011). *Lactobacillus plantarum* promotes Drosophila systemic growth by modulating hormonal signals through TOR-dependent nutrient sensing. *Cell Metabolism, 14*, 403–414.

Tennessen, J. M., & Thummel, C. S. (2011). Coordinating growth and maturation—Insights from Drosophila. *Current Biology, 21*, R750–R757.

Thummel, C. S. (2001). Molecular mechanisms of developmental timing in *C. elegans* and Drosophila. *Developmental Cell, 1*, 453–465.

van Heemst, D. (2010). Insulin, IGF-1 and longevity. *Aging and Disease, 1*, 147–157.

Van Hiel, M. B., Vandersmissen, H. P., Van Loy, T., & Vanden, J. (2012). An evolutionary comparison of leucine-rich repeat containing G protein-coupled receptors reveals a novel LGR subtype. *Peptides, 34*, 193–200.

Warren, J. T., Petryk, A., Marques, G., Jarcho, M., Parvy, J. P., Dauphin-Villemant, C., et al. (2002). Molecular and biochemical characterization of two P450 enzymes in the ecdysteroidogenic pathway of Drosophila melanogaster. *Proceedings of the National Academy of Sciences of the United States of America, 99*, 11043–11048.

Warren, J. T., Petryk, A., Marques, G., Parvy, J. P., Shinoda, T., Itoyama, K., et al. (2004). Phantom encodes the 25-hydroxylase of Drosophila melanogaster and Bombyx mori: A P450 enzyme critical in ecdysone biosynthesis. *Insect Biochemistry and Molecular Biology*, *34*, 991–1010.

Warren, J. T., Yerushalmi, Y., Shimell, M. J., O'Connor, M. B., Restifo, L. L., & Gilbert, L. I. (2006). Discrete pulses of molting hormone, 20-hydroxyecdysone, during late larval development of Drosophila melanogaster: Correlations with changes in gene activity. *Developmental Dynamics*, *235*, 315–326.

Wessells, R. J., Fitzgerald, E., Cypser, J. R., Tatar, M., & Bodmer, R. (2004). Insulin regulation of heart function in aging fruit flies. *Nature Genetics*, *36*, 1275–1281.

Wullschleger, S., Loewith, R., & Hall, M. N. (2006). TOR signaling in growth and metabolism. *Cell*, *124*, 471–484.

Xu, W. H., Lu, Y. X., & Denlinger, D. L. (2012). Cross-talk between the fat body and brain regulates insect developmental arrest. *Proceedings of the National Academy of Sciences of the United States of America*, *109*, 14687–14692.

Yamanaka, N., & O'Connor, M. B. (2011). Apiology: Royal secrets in the queen's fat body. *Current Biology*, *21*, R510–R512.

Yamanaka, N., Rewitz, K. F., & O'Connor, M. B. (2013). Ecdysone control of developmental transitions: Lessons from Drosophila research. *Annual Review of Entomology*, *58*, 497–516.

Yoshiyama, T., Namiki, T., Mita, K., Kataoka, H., & Niwa, R. (2006). Neverland is an evolutionally conserved Rieske-domain protein that is essential for ecdysone synthesis and insect growth. *Development*, *133*, 2565–2574.

Yoshiyama-Yanagawa, T., Enya, S., Shimada-Niwa, Y., Yaguchi, S., Haramoto, Y., Matsuya, T., et al. (2011). The conserved Rieske oxygenase DAF-36/Neverland is a novel cholesterol-metabolizing enzyme. *Journal of Biological Chemistry*, *286*, 25756–25762.

Zhang, H., Stallock, J. P., Ng, J. C., Reinhard, C., & Neufeld, T. P. (2000). Regulation of cellular growth by the Drosophila target of rapamycin dTOR. *Genes & Development*, *14*, 2712–2724.

Zhang, W., Thompson, B. J., Hietakangas, V., & Cohen, S. M. (2011). MAPK/ERK signaling regulates insulin sensitivity to control glucose metabolism in Drosophila. *PLoS Genetics*, 7, e1002429.

CHAPTER THREE

Temporal Patterning of Neural Progenitors in *Drosophila*

Xin Li, Zhenqing Chen, Claude Desplan[1]
Department of Biology, New York University, New York, New York, USA
[1]Corresponding author: e-mail address: cd38@nyu.edu

Contents

1. Introduction 70
   1.1 Embryonic and larval ventral nerve cord 71
   1.2 Olfactory system 72
   1.3 Medulla in the optic lobes 72
   1.4 Type II NBs in the central brain 73
2. Sequentially Expressed Transcription Factors in NBs Control Birth-Order-Dependent Neuron Fates 73
   2.1 Temporal sequence of transcription factors in *Drosophila* embryonic VNC NBs 73
   2.2 Similar or different TF sequence in other systems? 75
   2.3 Different TF sequence in medulla NBs 76
   2.4 Combinatorial temporal patterning of INPs and type II NBs 77
3. How are Temporal Transitions in NBs Controlled? 78
   3.1 Cross-regulations between the TFs play important roles in the transitions 78
   3.2 Differential requirements for cell cycle and asymmetric division 80
   3.3 Switching factors involved in TF transitions 80
4. Relationship Between Temporal Sequence and NB Competence 81
   4.1 Restriction of NB competence 81
   4.2 TF sequence and regulation of NB competence 81
   4.3 Epigenetic mechanism for loss of NB competence 82
5. Feedforward Loops Downstream of the Temporal Sequence Control Neuron Fates 84
6. Progression of the TF Sequence Required for the End of Neurogenesis 85
   6.1 Apoptosis of abdominal pNBs 85
   6.2 Pros-dependent cell-cycle exit of thoracic and central brain pNBs 86
   6.3 Studies in other systems about the end of neurogenesis 87
7. Integration of Temporal and Spatial Information Determines Lineages 87
8. Integration of Temporal Identity with Binary Fate Choice 88
9. Conclusions and Future Questions 90
References 91

Current Topics in Developmental Biology, Volume 105
ISSN 0070-2153
http://dx.doi.org/10.1016/B978-0-12-396968-2.00003-8

Abstract

Drosophila has recently become a powerful model system to understand the mechanisms of temporal patterning of neural progenitors called neuroblasts (NBs). Two different temporal sequences of transcription factors (TFs) have been found to be sequentially expressed in NBs of two different systems: the Hunchback, Krüppel, Pdm1/Pdm2, Castor, and Grainyhead sequence in the *Drosophila* ventral nerve cord; and the Homothorax, Klumpfuss, Eyeless, Sloppy-paired, Dichaete, and Tailless sequence that patterns medulla NBs. In addition, the intermediate neural progenitors of type II NB lineages are patterned by a different sequence: Dichaete, Grainyhead, and Eyeless. These three examples suggest that temporal patterning of neural precursors by sequences of TFs is a common theme to generate neural diversity. Cross-regulations, including negative feedback regulation and positive feedforward regulation among the temporal factors, can facilitate the progression of the sequence. However, there are many remaining questions to understand the mechanism of temporal transitions. The temporal sequence progression is intimately linked to the progressive restriction of NB competence, and eventually determines the end of neurogenesis. Temporal identity has to be integrated with spatial identity information, as well as with the Notch-dependent binary fate choices, in order to generate specific neuron fates.

1. INTRODUCTION

One fundamental question in developmental neurobiology is to understand how to generate the remarkable diversity of neurons and glia present in adult brains, from a small number of seemingly homogenous neural stem cells in the embryo. Spatial patterning of neural stem cells that is achieved by various morphogens and their signaling cascades contributes to the generation of neural diversity (Bhat, 1999a; Dessaud, McMahon, & Briscoe, 2008). Furthermore, single neural stem cells can generate different neural types in a stereotyped order: This is achieved by temporal patterning of neural stem cells. Extensive studies in the vertebrate central nervous system (CNS), especially cerebral cortex and retina, have revealed that birth order correlates with distinct neuronal/glial identity (reviewed in Jacob, Maurange, & Gould, 2008; Livesey & Cepko, 2001; Molyneaux, Arlotta, Menezes, & Macklis, 2007; Okano & Temple, 2009; Pearson & Doe, 2004). Although both extrinsic and intrinsic factors are required for the correct specification of temporal identify, isolated neural stem cells cultured *in vitro* can recapitulate the sequential generation of different neuron types, thus underscoring a cell-intrinsic temporal control of stem cells (Gaspard et al., 2008; Naka, Nakamura, Shimazaki, & Okano, 2008; Shen et al., 2006).

Neural stem cells are called neuroblasts (NBs) in *Drosophila*. *Drosophila* NBs usually undergo several rounds of Notch-dependent asymmetric divisions to produce a self-renewed NB, and a smaller daughter cell called ganglion mother cell (GMC). GMCs generally divide asymmetrically once to produce two postmitotic progeny with different fates (reviewed in Doe, 2008; Karcavich, 2005; Knoblich, 2010; Lin & Lee, 2012). Several *Drosophila* NB lineages in the CNS have been partially or completely characterized. They generate multiple distinct progeny in a birth-order-dependent manner (Akiyama-Oda, Hosoya, & Hotta, 1999; Baek & Mann, 2009; Bossing, Udolph, Doe, & Technau, 1996; Isshiki, Pearson, Holbrook, & Doe, 2001; Jefferis, Marin, Stocker, & Luo, 2001; Karcavich & Doe, 2005; Lee, Lee, & Luo, 1999; Lundell & Hirsh, 1998; Pearson & Doe, 2003; Schmid, Chiba, & Doe, 1999; Schmidt et al., 1997; Skeath & Thor, 2003; Yu, Chen, Shi, Huang, & Lee, 2009; Yu et al., 2010). *Drosophila* has recently become a powerful model system to understand the mechanisms of temporal specification of neurons and there have been several extensive and excellent reviews on this topic (Brand & Livesey, 2011; Kao & Lee, 2010; Maurange, 2012; Pearson & Doe, 2004). This review will therefore focus on the most recent advances in temporal patterning of NBs in several systems of *Drosophila* that are now well characterized. Below is a brief introduction of each system to be described.

1.1. Embryonic and larval ventral nerve cord

Classic work on the temporal patterning of NBs in the *Drosophila* embryonic ventral nerve cord (VNC) has made it undoubtedly the best-characterized system. The VNC of *Drosophila*, spanning three thoracic and eight abdominal segments, is the counterpart of the vertebrate spinal cord. During early embryonic development, NBs delaminate from a neuroectoderm at stereotypical positions. Each hemi-segment contains approximately 30 NBs arranged in seven rows, and each NB is identifiable based on its position (e.g., NB1-1 is in row1, column1) and molecular markers (Bhat, 1999b; Doe, 1992; Skeath & Thor, 2003; Technau, Berger, & Urbach, 2006). The distinction between these NBs is achieved by spatial patterning that will not be discussed in detail here. The neuronal and glial progeny composition of each embryonic NB lineage have been determined by early DiI NB labeling experiments: A single NB can generate multiple neuron types and glia (Bossing et al., 1996; Schmid et al., 1999, 1997). At the embryo to larval transition, some of these NBs commit to apoptosis, while others enter into a quiescence phase. They are then reactivated during the larval stage and continue to proliferate during larval and pupal stages. These postembryonic NBs generate ~90% of neurons that constitute the adult CNS (Prokop & Technau, 1991; Truman & Bate, 1988). Relatively less is known about the postembryonic lineages as compared to embryonic lineages, but several recent studies have shed some light on the temporal control of larval lineages (see below).

1.2. Olfactory system

Another well-characterized system is the olfactory system, including the antennal lobe, which receives olfactory inputs from olfactory receptor neurons (ORNs), and the mushroom body, which receives inputs from projection neurons (PNs) in the antenna lobe. Systematic clonal analysis of mushroom body neurons using the mosaic analysis with a repressible cell marker (MARCM) technique has demonstrated that each of the four mushroom body NBs sequentially generates at least three types of neurons, γ neurons first, followed by $\alpha'\beta'$, and then $\alpha\beta$ neurons (Lee et al., 1999), suggesting birth-order-dependent neuronal specification. Similarly, the antennal lobe PNs that are derived from three NBs (an anterodorsal, a lateral, and a ventral NB) are specified by lineage and birth order to connect with specific classes of ORN axons (Jefferis et al., 2001). These three lineages have been extensively characterized (Das, Reichert, & Rodrigues, 2010; Das et al., 2008; Lai, Awasaki, Ito, & Lee, 2008; Lin, Kao, Yu, Huang, & Lee, 2012; Lin et al., 2010; Yu et al., 2010). For example, the anterodorsal lineage produces 40 types of PNs in a stereotypic order, while the lateral lineage sequentially generates 48 pairs of local interneurons paired with distinct PNs.

1.3. Medulla in the optic lobes

The *Drosophila* optic lobes, composed of lamina, medulla, and lobula complex, are the processing centers of visual information in the brain. Among them, the medulla is the largest neuropil. It contains approximately 40,000 neurons, belonging to more than 70 different cell types (Fischbach & Dittrich, 1989; Morante & Desplan, 2008). These neurons are generated by NBs derived from a crescent-shaped single-layered neuroepithelium (NE) in the larval brain called the outer proliferation center (Green, Hartenstein, & Hartenstein, 1993; White & Kankel, 1978). During development, a wave of neurogenesis progresses from the edge of the crescent toward its center, and sequentially converts NE cells into medulla NBs.

Therefore, NBs of different ages can be observed together in one snapshot (Egger, Boone, Stevens, Brand, & Doe, 2007; Yasugi, Umetsu, Murakami, Sato, & Tabata, 2008). These NBs generate GMCs, which then produce medulla neurons that form chains below each NB, with the first-born neurons located in the deepest position and latest-born neurons in the most superficial position, near the NB (Hasegawa et al., 2011; Morante, Erclik, & Desplan, 2011). Recently, the developing medulla has emerged as a new powerful system for studying temporal patterning of NBs (Li et al., 2013; Suzuki, Kaido, Takayama, & Sato, 2013).

1.4. Type II NBs in the central brain

In contrast to the regular type I NBs that divide asymmetrically to generate a series of GMCs that each divides once to produce two postmitotic cells, type II NBs divide asymmetrically to self-renew and to generate a series of transit amplifying GMCs called intermediate neural progenitors (INPs), with each INP dividing asymmetrically to generate several GMCs and eventually giving rise to 6–12 neurons and/or glia (Bayraktar, Boone, Drummond, & Doe, 2010; Boone & Doe, 2008; Viktorin, Riebli, Popkova, Giangrande, & Reichert, 2011; Weng, Golden, & Lee, 2010; Zhu, Barshow, Wildonger, Jan, & Jan, 2011). Type II lineages allow more neurons than type I lineages to be quickly generated.

2. SEQUENTIALLY EXPRESSED TRANSCRIPTION FACTORS IN NBs CONTROL BIRTH-ORDER-DEPENDENT NEURON FATES

2.1. Temporal sequence of transcription factors in *Drosophila* embryonic VNC NBs

How does a NB generate a stereotypical order of different neuron types as it ages? Extensive studies in the *Drosophila* embryonic VNC have led to the identification of a molecular mechanism of temporal specification, where a series of transcription factors (TFs) are sequentially and transiently expressed in NBs as they age. Postmitotic progeny born during each time window maintains the expression of the TF. This temporal sequence of TFs includes Hunchback (Hb), Krüppel (Kr), Pdm1/Pdm2 (Pdm), Castor (Cas), and Grainyhead (Grh) (Baumgardt, Karlsson, Terriente, Diaz-Benjumea, & Thor, 2009; Brody & Odenwald, 2000; Cui & Doe, 1992; Grosskortenhaus, Pearson, Marusich, & Doe, 2005; Isshiki et al., 2001; Kambadur et al., 1998; Mellerick, Kassis, Zhang, & Odenwald, 1992;

Romani et al., 1996; Yang, Yeo, Dick, & Chia, 1993). Although different NBs in the VNC delaminate and start differentiation at different times, most of them follow the same TF sequence. These temporally expressed TFs have been shown to be both required and sufficient for birth-order-dependent neuronal specification in several different NB lineages (Baumgardt et al., 2009; Benito-Sipos et al., 2010; Cleary & Doe, 2006; Grosskortenhaus, Robinson, & Doe, 2006; Isshiki et al., 2001; Kambadur et al., 1998; Novotny, Eiselt, & Urban, 2002; Pearson & Doe, 2003; Tran & Doe, 2008). For example, each of the first five GMCs in the NB7-1 lineage gives rise to a motor neuron (U1–U5) and a sibling cell. The unique identity of each of the U1–U5 motor neurons is specified by the sequentially expressed TFs: Hb is necessary and sufficient to specify U1 and U2 fates (high Hb: U1, low Hb: U2); Kr is necessary and sufficient to specify the U3 fate; Pdm alone specifies U4; while Pdm and Cas together specify U5 (Cleary & Doe, 2006; Grosskortenhaus et al., 2005; Isshiki et al., 2001; Pearson & Doe, 2003; Fig. 3.1). In the NB7-3 lineage, there are only three GMCs generated: GMC-1 is Hb^+,Kr^+, and generates the EW1

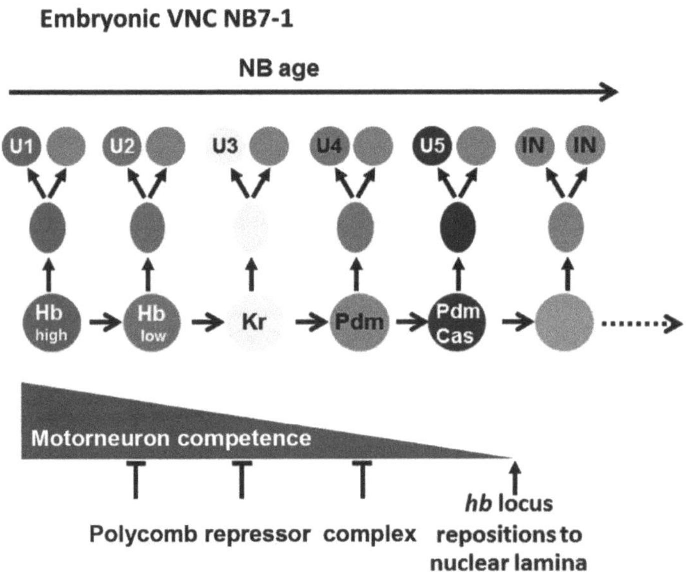

Figure 3.1 Schematic model showing the sequential expression of Hb, Kr, Pdm, and Cas in the embryonic VNC NB7-1. These TFs control the sequential generation of different U motor neurons. The Polycomb repressor complex is required for the progressive loss of NB competence to generate motor neurons. At the end of the first five divisions, the *hb* locus relocates to the nuclear periphery, and the NB completely loses its competence to generate motor neurons. (For color version of this figure, the reader is referred to the online version of this chapter.)

interneuron and the GW motor neuron; GMC-2 is Kr^{+}, and generates the EW2 interneuron and a sibling that undergoes apoptosis; GMC-3 is Pdm^{+}, and differentiates directly into the EW3 interneuron (Isshiki et al., 2001; Karcavich & Doe, 2005; Lundell & Hirsh, 1998). In this case, Hb is necessary and sufficient for the first-born neuron fates (EW1/GW). Loss of *hb* results in a lack of the EW1/GW, while EW2 and EW3 are still present. Prolonged expression of Hb in the NB leads to additional neurons with EW1/GW fates at the expense of EW2 and EW3 (Isshiki et al., 2001; Novotny et al., 2002). Kr is necessary and sufficient for the second-born neuron fate EW2 (Isshiki et al., 2001).

There are variations in the function of the temporal TFs: In the NB3-1 lineage, Pdm is not required for specification of the third temporal identity, but instead to close the preceding Kr^{+} temporal identity window (Tran & Doe, 2008). The first four GMCs of NB3-1 lineage produce $HB9^{+}$,$Islet^{+}$ RP motor neurons with a birth order of RP1 → RP4 → RP3 → RP5 (and their non-RP siblings): Hb specifies RP1 and RP4 (high Hb: RP1; low Hb: RP4); Kr specifies RP3; while Pdm is not required for specifying RP5, because there is only a modest expansion of Kr expression in NBs in *pdm* mutants, with a few extra RP3 neurons produced, but RP5 neurons are still specified. Similarly, Cas is required for closing the third (RP5) temporal identity window: In *cas* mutants, there are ectopic RP5 neurons (Tran & Doe, 2008).

Thus, a series of TFs sequentially expressed in NBs control the sequential generation of different neural types in multiple NB lineages. Since different NBs generate different lineages, these TFs do not specify a certain neuron type, but control the birth-order-dependent neuronal identity. The birth-order-dependent temporal identity is integrated with the spatial identity of the NB within each segment or between different segments, and is translated into specific cell types. For example, *hb* controls the first-born cell fates in multiple lineages, which can be motor neurons, interneurons or glial cells, depending on the NB lineage (Isshiki et al., 2001).

2.2. Similar or different TF sequence in other systems?

Since birth-order-dependent neuronal specification has been widely observed in various systems, is the same or a similar temporal TF sequence utilized to pattern neural stem cells of other systems?

In the mushroom body, which is generated by four NBs, each NB sequentially generates at least three types of neurons. Thus far, no temporal sequence of TFs that controls the fate of these neurons has been identified in the NB. However, a novel bric-a-brac,tramtrack, broad (BTB)-zinc finger protein, named Chinmo (chronologically inappropriate morphogenesis), acts in GMCs or young neurons to control the temporal identity of mushroom body neurons (Zhu et al., 2006). Although Chinmo mRNA is equally expressed throughout the entire NB lineage, Chinmo protein shows a temporal gradient in the neuronal progeny. It is absent in NBs and is expressed at its highest levels in the early-born neurons, lower in the next-born neurons, and undetectable in the latest-born neurons. Reducing or increasing Chinmo levels causes transformation of neurons toward later or earlier fates, respectively (Zhu et al., 2006). Recently, microRNAs of the Let-7-complex, the heterochronic miRNAs originally identified in *Caenorhabditis elegans*, were found to target *chinmo* to regulate the temporal identity of *Drosophila* mushroom body neurons (Wu, Chen, Mercer, & Sokol, 2012). Since Chinmo protein or Let-7 miRNAs are not detected in NBs, how the Chinmo gradient in neurons is regulated as the NB ages is not understood.

Postembryonic neuroblasts (pNBs) in the VNC generate 90% of neurons that constitute the adult CNS. Although a complete temporal TF sequence has not been identified, recent studies identified two members of the postembryonic TF sequence, Cas and Seven-up (Svp) (Maurange, Cheng, & Gould, 2008; Tsuji, Hasegawa, & Isshiki, 2008). They are required for a temporal switch of pNBs from generating small Chinmo$^+$ neurons to instead producing large Br-C$^+$ (Broad Complex) neurons in many different lineages. Since the switch from Chinmo$^+$ to Br-C$^+$ happens later than the transient expression of Cas or Svp, the switch must be directly controlled by an unknown member of the TF sequence whose expression is promoted by *cas* and *svp* by feedforward regulation (Maurange et al., 2008). Cas and Svp are also required for the NBs to eventually end neurogenesis. This is discussed in detail below.

In the anterodorsal lineage of the antennal lobe, Kr was shown to act in the NB to define one out of 40 temporal fates of PNs (Kao, Yu, He, Kao, & Lee, 2012). Loss of Kr from the NB causes a single PN fate to be skipped. However, loss of Hb, Pdm or Cas does not produce detectable phenotypes (Kao et al., 2012). It will be interesting to identify more TFs that are temporally expressed in antennal lobe NBs and control the temporal specification of a large number of distinct PNs and interneurons.

2.3. Different TF sequence in medulla NBs

In the developing medulla, NBs of different ages can be visualized in one snapshot, and thus the medulla provides another powerful system to study temporal patterning of NBs. Recent studies identified a different series of TFs expressed in medulla NBs: Homothorax (Hth), Klumpfuss (Klu), Eyeless (Ey), Sloppy paired (Slp), Dichaete (D), and Tailless (Tll) are sequentially expressed in NBs of increasing ages, with Hth expressed in newly differentiated NBs, Klu, Ey, Slp, and D expressed in increasingly older NBs, and Tll in the oldest NBs. Hth, Ey, and Slp were shown to control the generation of specific neuron types born during each time window (Li et al., 2013; Suzuki et al., 2013; Fig. 3.2). This identification of a second TF sequence that is different from the one in the embryonic VNC suggests that TF sequence-dependent temporal patterning of NBs is likely to be broadly utilized, and that different TF sequences can be recruited in different systems.

2.4. Combinatorial temporal patterning of INPs and type II NBs

A recent elegant paper showed that, in type II NB lineages, both NBs and INPs are temporally patterned by different temporal series of TFs (Bayraktar & Doe, 2013). Type II NBs sequentially express D/Cas and Svp as well as other not yet identified members of a TF sequence. Along the second temporal axis, INPs sequentially express D, Grh, and Ey as they age: These TFs are required for the sequential production of distinct neural subtypes. Thus, TF sequences on the two temporal axes act combinatorially to generate larger neural diversity (Bayraktar & Doe, 2013; Fig. 3.3).

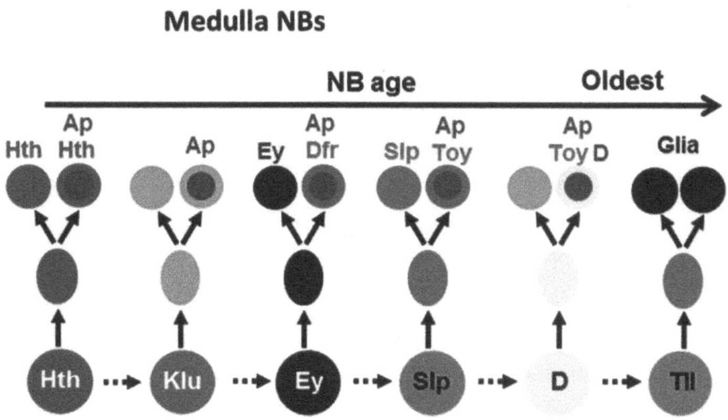

Figure 3.2 Schematic model showing the sequential expression of Hth, Klu, Ey, Slp, D, and Tll in medulla NBs. These TFs control the sequential generation of different neuronal types marked by specific combinations of neuronal markers. Ap is expressed in the N-on daughters of ganglion mother cells. Tll^{+} NBs generate glia. (For color version of this figure, the reader is referred to the online version of this chapter.)

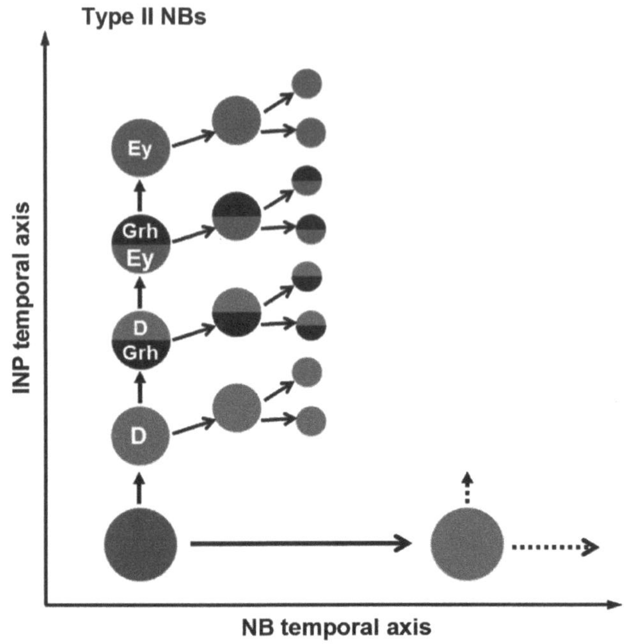

Figure 3.3 Schematic model showing combinatorial temporal patterning in type II NB lineages in the central brain. Both NBs and INPs are patterned by distinct temporal series of TFs. INPs sequentially express D, Grh, and Ey as they age. These TFs are required for the sequential production of distinct neuronal subtypes. (For color version of this figure, the reader is referred to the online version of this chapter.)

3. HOW ARE TEMPORAL TRANSITIONS IN NBs CONTROLLED?

3.1. Cross-regulations between the TFs play important roles in the transitions

In the embryonic VNC, the Hb → Kr → Pdm → Cas → Grh sequence can be recapitulated in isolated NBs cultured *in vitro*, suggesting a lineage-intrinsic mechanism for the TF transitions (Brody & Odenwald, 2000; Grosskortenhaus et al., 2005). Gain and loss of function studies suggest that multiple cross-regulations between these temporal TFs exist: one TF activates the next TF, while repressing the previous TF and the next-plus-one TF (Fig. 3.4). This, in theory, could allow the TF transitions (Baumgardt et al., 2009; Brody & Odenwald, 2000; Grosskortenhaus et al., 2006; Isshiki et al., 2001; Kambadur et al., 1998; Nakajima, Isshiki, Kaneko, & Ishihara, 2010; Tran & Doe, 2008). However, in most lineages, loss of

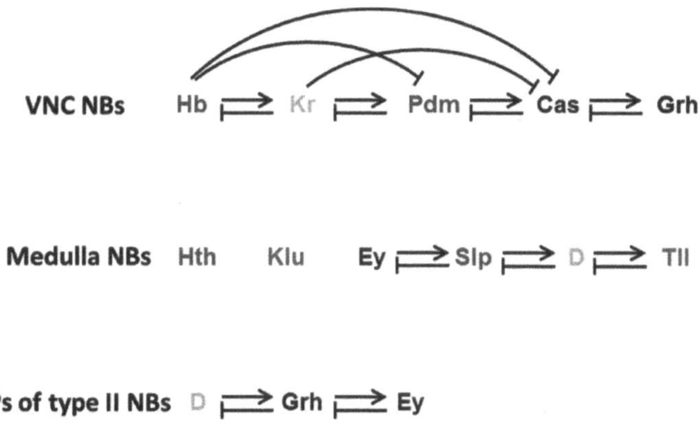

Figure 3.4 Cross-regulations between TFs in the three temporal sequences that pattern NBs or INPs. Each TF activates the next TF and represses the previous TF (with some exceptions). (For color version of this figure, the reader is referred to the online version of this chapter.)

Hb, Kr, or Pdm simply causes the corresponding temporal identity to be skipped, rather than a blockage of the temporal progression, suggesting that the cross-regulations are not essential and that other factors are involved in the switches. However, loss of Cas does lead to a block in the temporal progression and to persistent Pdm expression (Brody & Odenwald, 2000; Grosskortenhaus et al., 2006; Isshiki et al., 2001; Kambadur et al., 1998; Maurange et al., 2008; Tran & Doe, 2008).

In the Hth → Klu → Ey → Slp → D → Tll sequence that patterns medulla NBs, loss of Ey, Slp, or D does disrupt essential cross-regulations, and the NBs keep dividing but do not progress to the next TF stage. Slp and D are also required to repress the previous TF (Li et al., 2013; Suzuki et al., 2013). Thus, Ey, Slp, and D are required for the sequential TF switching, and, in this regard, they are similar to Cas in the embryonic sequence. However, there are a few exceptions to these modes of regulation: No cross-regulation is observed among Hth, Klu, and Ey; and Tll is not required, but it is sufficient to repress D (Fig. 3.4). This suggests that there are more TFs temporally expressed in medulla NBs that work together with the Hth → Klu → Ey → Slp → D → Tll sequence to ensure temporal transitions (Li et al., 2013; Suzuki et al., 2013).

In the D → Grh → Ey sequence that patterns INPs of type II NB lineages, D is necessary but not sufficient to activate the next TF, Grh. In turn, Grh is required to repress D and activate Ey. Ey is required and sufficient to repress Grh. Thus, a "feedforward activation/feedback repression" model for D → Grh → Ey cross-regulation appears to control the transitions of TFs (Bayraktar & Doe, 2013; Fig. 3.4).

Considering the three know sequences of TFs in the VNC, medulla and type II NBs, it is clear that a cross-regulatory network among the temporally expressed TFs does have important roles in the transition between TFs. However, more factors or timing mechanisms must be involved and may vary case by case. Redundant mechanisms might also be present to ensure the robustness of transitions.

3.2. Differential requirements for cell cycle and asymmetric division

Since the NB progresses through the TF sequence as it undergoes asymmetric divisions to produce neuronal progeny, cell cycle and cell division might play a role in these transitions, either by cell-cycle dependent signaling event, asymmetric localization of certain mRNA or protein, or by feedback signal from the progeny. Indeed, the first Hb to Kr transition requires cell cycle and cytokinesis, but it does not require asymmetric localization of *hb* mRNA or Hb protein. However, the Kr → Pdm → Cas transitions are regulated in a cell-cycle independent manner, and progress normally in cell-cycle arrested NBs (Grosskortenhaus et al., 2005).

3.3. Switching factors involved in TF transitions

As discussed above, Hb expression is downregulated at the transcription level in a cell-division dependent manner (Grosskortenhaus et al., 2005; Kanai, Okabe, & Hiromi, 2005; Kohwi, Hiebert, & Doe, 2011; Mettler, Vogler, & Urban, 2006). The orphan nuclear receptor TF Svp is required for the Hb to Kr switch in several NB lineages of the *Drosophila* embryonic CNS (Kanai et al., 2005). Loss of *svp* causes prolonged Hb expression in NBs as well as ectopic early-born neurons. The nuclear export of *svp* mRNA (and thus its efficient translation) is dependent on the cell cycle (Mettler et al., 2006), providing an explanation for the requirement of cell division for the Hb to Kr transition.

In addition to Svp, the nuclear proteins distal antenna and distal antenna-related proteins (Dan and Dan-r) function in parallel pathways with Svp to downregulate Hb expression, and ensure the timely transition of the NBs from making early-born neurons to generating late-born neurons (Kohwi et al., 2011).

In the medulla NB TF sequence, both Slp and D are required to turn off the preceding TF and turn on the next TF, therefore, acting as switching factors for themselves. Additional transition factors remain to be identified for other transitions (Li et al., 2013; Suzuki et al., 2013).

4. RELATIONSHIP BETWEEN TEMPORAL SEQUENCE AND NB COMPETENCE

4.1. Restriction of NB competence

Early studies in mammalian neurogenesis have demonstrated that neural stem cells undergo progressive restriction in their competence to generate different types of progeny in response to extrinsic signals (Desai & McConnell, 2000; Livesey & Cepko, 2001). For example, during development of the cerebral cortex, neural stem cells generate neurons in the six cortical layers in an "inside-out" defined order, in which layer 6 and 5 neurons are generated first, followed by layer 4, 3 and 2. The fate of the progeny also depends on environmental cues to which the cells respond prior to cell division. Transplantation experiments showed that mid-stage neural stem cells that are producing layer 4 neurons have the competence to generate layer 2/3 (later-born) neurons when transplanted into older brains. However, when these neural stem cells are transplanted into younger brains in which layer 6 neurons are being generated, they have lost the competence to generate layer 6 neurons. Instead, they produce layer 4 and 5 neurons. Thus, the competence of neural stem cells to generate a given cell type is progressively lost, although this loss of competence lags behinds the completion of generation of this neural type (Desai & McConnell, 2000).

4.2. TF sequence and regulation of NB competence

Loss of competence is also observed in *Drosophila* NBs, although, in this case, competence responds to intrinsic TFs rather than extrinsic cues. In the VNC, providing ectopic Hb to the NB after Hb has been downregulated is sufficient to generate additional early-born neurons. However, this competence to respond to Hb progressively decreases over time. In the NB7-1 lineage, a pulse of ectopic Hb at the time U1 and U2 are normally born leads to the generation of several ectopic U1 and U2 neurons. If the pulse of ectopic Hb is provided later, when U3 and U4 are normally born, only one ectopic U2 neuron can be generated; and if Hb is given after the fifth division when the NB switches to generate interneurons, the NB fails to generate ectopic U neurons (Pearson & Doe, 2003). Further studies showed that NB7-1 has a single early competence window spanning the first five divisions for responding to Hb, Kr, Pdm, or Cas to make the corresponding U motor neurons. Then, NBs completely lose the competence to produce U motor neurons in response to these TFs (Cleary & Doe, 2006; Fig. 3.1). Similarly in the NB3-1 lineage, when the NB switches to generate interneurons at the fifth division, the competence to generate RP3 motor neurons in response to Kr is lost (Tran & Doe, 2008). This suggests that NBs can go sequentially through several competence windows in which they respond differently to the same temporal TF, allowing the repeated use of the same TF within one lineage to specify different fates (Cleary & Doe, 2006; Pearson & Doe, 2003).

The NB competence window can be extended if ectopic Hb is continuously provided before endogenous Hb is downregulated. In this case, NBs generate many U1/U2 neurons. They also retain their competence to generate later-born U motor neurons, and the lineage is extended. Thus, downregulation of Hb expression is required for the gradual loss of NB competence to respond to Hb (Cleary & Doe, 2006; Grosskortenhaus et al., 2005; Isshiki et al., 2001). Repression of multiple target genes (including Pdm2) by sustained Hb is necessary and sufficient for the maintenance of NB competence. Indeed, sustained expression of a form of Hb that functions solely as a transcription activator cannot significantly extend the competence window (Tran, Miller, & Doe, 2010).

Therefore, loss of NB competence to generate certain neuron fates lags behind the progression of the TF sequence (e.g., for cell fate specified by Hb, it is 3 cell divisions later) (Kohwi, Lupton, Lai, Miller, & Doe, 2013; Pearson & Doe, 2003). However, progression of the TF sequence, that is, downregulation of Hb relieving its transcription repression on multiple target genes, including a later temporal TF (Pdm), is required for the NB to close the current competence window and transit to the next.

4.3. Epigenetic mechanism for loss of NB competence

Recent studies uncovered a close relationship between epigenetic mechanism and loss of NB competence. Mammalian cortical neural stem cells generate cortical neurons that populate the different layers, and then lose the competence to generate neurons and switch to producing glia. This is due to the Polycomb repressor complexes (PRCs) that are required to suppress transcription of the neural fate TF neurogenin 1 in late-stage progenitors (Hirabayashi et al., 2009). PRCs induce repressive chromatin marks at the neurogenin 1 locus that gradually accumulate over time, and thus may provide a timer for the loss of competence to make neurons (Hirabayashi et al., 2009).

In *Drosophila*, PRCs are also involved in progressively restricting competence for generating motor neurons in NB7-1 and NB3-1: PRC loss of function extends the competence window to generate the corresponding motor neuron fate in response to Kr, while PRC gain of function precociously restricts this competence. In contrast, PRC activity does not affect the production of interneurons in multiple lineages (Touma, Weckerle, & Cleary, 2012), suggesting that it is involved specifically in the repression of multiple target genes involved in motor neuron specification (Fig. 3.1).

Chromosomal architecture has also recently been shown to be involved in the loss of competence of NB7-1 to respond to Hb, three NB divisions after *hb* transcription have stopped. At this stage, the *hb* gene locus relocates to the nuclear periphery of the NB, a repressive subnuclear compartment, preventing further activation of the *hb* gene by ectopic Hb, which is a requirement for the specification of early-born neurons (Kohwi et al., 2013; Fig. 3.1). The timing of the relocation correlates with downregulation of the Pipsqueak domain nuclear protein, Dan. Prolonging the expression of Dan can extend the NB competence by preventing relocalization of the *hb* locus at the periphery. This study proposed that Dan might competitively inhibit other Pipsqueak-domain factors, for example, Pipsqueak, a GAGA-binding factor essential for sequence-specific targeting of PRCs (Huang, Chang, Yang, Pan, & King, 2002), from binding to and recruiting *hb* and other loci to the nuclear lamina (Kohwi et al., 2013). This is consistent with the model that PRCs are involved in restricting NB competence (Touma et al., 2012).

In the medulla temporal sequence, NB competence has not yet been well characterized. Although mis-expressing the first NB TF Hth in all NBs leads to ectopic neurons with early-born cell fate, the phenotype becomes less obvious in later part of the lineage, suggesting that the competence of NBs to generate early-born cell fate in response to the first NB TF decreases with time. Whether the progressive loss of NB competence involves similar epigenetic mechanism remains to be studied. Interestingly, similar to the mammalian cortical neural stem cells, the medulla NBs switch from neurogenic to gliogenic at their final temporal stage when they express Tll (Li et al., 2013). Whether and how they lose the competence to generate neurons at this stage awaits further study.

5. FEEDFORWARD LOOPS DOWNSTREAM OF THE TEMPORAL SEQUENCE CONTROL NEURON FATES

Some temporal genes are expressed in a broad temporal window, during which multiple cell types are specified. How does a given temporal gene control multiple cell fates? An elegant timing mechanism subdivides the broad castor window in the NB5-6 lineage (Baumgardt et al., 2009). This NB lineage (producing in total ~20 progeny cells) ends with a broad castor window, during which ~10 cells are born, including the four last-born neurons that express the LIM-homeodomain TF Apterous (Ap), designated the Ap cluster (Ap1-4). The broad Cas window actually consists of two phases of Cas expression with a brief interruption of Cas expression when the Ap1 neuron is born. Ap1 and Ap4 are peptidergic neurons expressing the Nplp1 and FMRFamide neuropeptides, respectively. Ap2 and Ap3 are interneurons (Baumgardt, Miguel-Aliaga, Karlsson, Ekman, & Thor, 2007; Benveniste, Thor, Thomas, & Taghert, 1998; Park, Han, Kim, Han, & Taghert, 2004). The Ap cluster neurons are each born directly from a NB without a GMC intermediate (Baumgardt et al., 2009). Cell fate diversification of the four Ap neurons is achieved through three regulatory events triggered by Cas: (1) Cas activates the COE class TF Collier/Knot (Col) in the NB throughout the time Ap cluster neurons are born. It is initially inherited in all four Ap neurons, but it is only maintained in Ap1 where it acts in a feedforward loop involving sequential activation of Ap/Eyes absent, Dimmed, and Nplp1 to specify this cell fate (Baumgardt et al., 2007). (2) Cas also activates the zinc-finger protein squeeze (Sqz) in NBs, which is maintained in all four Ap neurons. After Sqz had time to accumulate in slightly later NBs, it acts with Cas to activate Nab, which is thus only inherited in Ap2-4. Sqz and Nab together rapidly downregulate Col in Ap2-4 neurons, and allow them to adopt fates different from Ap1 (Baumgardt et al., 2009). (3) Cas activates Grh expression, leading to gradual increase of Grh levels. Grh is required for the specification of Ap4 (Baumgardt et al., 2009). Thus, the broad Cas window is subdivided by these opposing feedforward loops that ensure the generation of neurons with different identities (Fig. 3.5).

Similar regulatory logic might be widely used in other NB lineages to increase neural diversity. In the NB3-3 lineage, Nab and Sqz also function to specify temporal identities in the Cas window (Tsuji et al., 2008). In the medulla, NBs are estimated to generate 2–6 GMCs at each temporal TF

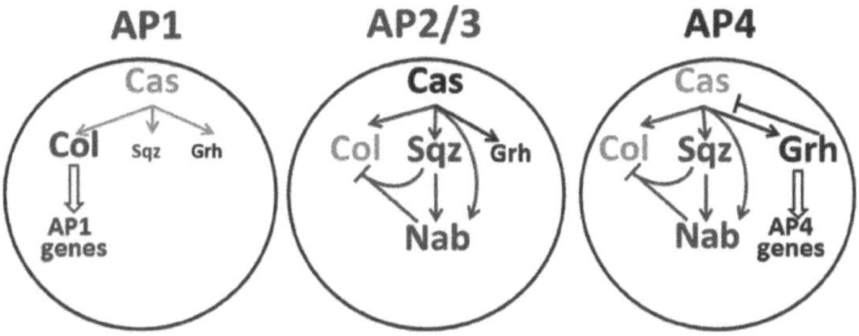

Figure 3.5 Model showing how three opposing feedforward loops activated by Cas subdivide the broad TF window. Ap1 loop (red): Transient Cas expression activates Col, which is initially expressed in all four Ap neurons, but is only maintained in Ap1, where it activates a feedforward loop that determines Ap1 fate. AP2-4 loop (green): Cas expression activates Sqz, and together they activate Nab in Ap2-4 neurons. Sqz and Nab together inactivate Col. AP4 loop (blue): Sustained Cas expression activates Grh, which is required for Ap4 neuron specification. Then, Grh represses Cas expression. (For interpretation of the references to color in this figure legend, the reader is referred to the online version of this chapter.)

stage, and there is evidence that the progeny from each GMC generated sequentially at the same TF stage express different TF combinations, and thus adopt different fates (Li et al., 2013). The temporal genes might act through multiple feedforward loops similar to the one described for the Ap cluster to subdivide each broad temporal window.

6. PROGRESSION OF THE TF SEQUENCE REQUIRED FOR THE END OF NEUROGENESIS

In order to generate the right number of neurons and avoid overproliferation, NBs must end neurogenesis after completion of their lineages. Studies in pNBs in the VNC and central brain show that the temporal progression of TF sequence is also required for a timely ending of neurogenesis.

6.1. Apoptosis of abdominal pNBs

In the VNC, pNBs have lineages of different lengths: Abdominal pNBs have much smaller lineages (4–12 neurons) than thoracic and central brain pNBs (~100 progeny each) (Bello, Hirth, & Gould, 2003; Truman & Bate, 1988). This is due to Reaper, Grim, and Head involution defective (hid)-dependent apoptosis of abdominal pNBs at larval stages, which is induced by a burst of expression of the Hox protein Abdominal-A (Bello et al., 2003; Peterson, Carney, Taylor, & White, 2002). This apoptosis also requires inputs from the temporal sequence. At late stages of embryogenesis, Cas turns on the expression of Grh, and turns off the expression of D in NBs. Thus, Grh is expressed in NBs that enter quiescence and is subsequently maintained in pNBs through larval stages. The state of $Grh^{+} D^{-}$ is required for the abdominal pNBs to undergo apoptosis in response to the burst of Abdominal-A (Almeida & Bray, 2005; Cenci & Gould, 2005; Maurange et al., 2008). Furthermore, completion of the larval temporal sequence is required for the pNBs to undergo apoptosis: Either loss of postembryonic Svp or persistent expression of Cas can block the larval temporal sequence progression, and these NBs do not die in spite of normal Abdominal-A expression (Maurange et al., 2008; Fig. 3.6).

6.2. Pros-dependent cell-cycle exit of thoracic and central brain pNBs

Most thoracic and central brain NBs (with the exception of mushroom body NBs) stop dividing at 120 h after larva hatching (~1 day into pupation). In contrast to the abdominal pNBs that undergo apoptosis, thoracic and central brain NBs end neurogenesis through cell-cycle exit that is dependent on a nuclear burst of expression of the homeodomain protein Prospero (Pros) (Maurange et al., 2008). Pros is tethered to cell membrane in NBs, and is asymmetrically localized to GMCs during asymmetric cell division. In GMCs, Pros is localized to nuclei where it promotes cell-cycle exit after one cell division (reviewed in Yu, Kuo, & Jan, 2006). Thoracic and central

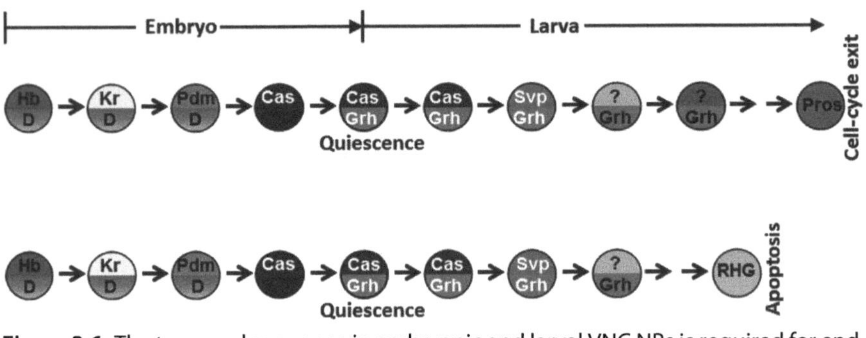

Figure 3.6 The temporal sequence in embryonic and larval VNC NBs is required for ending neurogenesis. At late stages of embryogenesis, Cas turns on Grh and turns off D in NBs. Grh, as well as two members of the larval temporal sequence, Cas and Svp, are required for the NBs to end neurogenesis at the correct time. Thoracic NB lineages are longer, and undergo pros-dependent cell-cycle exit. Abdominal NB lineages are shorter, and undergo apoptosis that depends on Reaper, Grim, and Hid. (For color version of this figure, the reader is referred to the online version of this chapter.)

brain NBs exhibit a burst of expression of nuclear Pros at 120 h. *pros* mutant clones contain multiple NB-like cells that do not die at 120 h, and even continue dividing in the adult brain (Maurange et al., 2008).

Pros-dependent cell-cycle exit of thoracic and central brain NBs also requires inputs from both embryonic and larval temporal sequences: Sustained Grh expression in pNBs induced by embryonic Cas is required to prevent premature cell-cycle exit of NBs; while completion of the larval temporal sequence is required for the NBs to undergo Pros-dependent cell-cycle exit at 120 h. Remarkably, stalled temporal sequence caused by either loss of postembryonic Svp expression or persistent Cas expression prevents cell-cycle exit and NBs continue to divide even in 7-day-old adults (Maurange et al., 2008; Fig. 3.6).

6.3. Studies in other systems about the end of neurogenesis

In contrast to VNC and the majority of central brain NBs, mushroom body NBs that are born at embryonic stages do not have a quiescent stage and continue proliferating until the end of pupal stage. Tll is required for the uninterrupted and prolonged proliferation of mushroom body NBs: Loss of *tll* causes premature loss of mushroom body NBs in early pupal stage (Kurusu et al., 2009). Surprisingly, in the medulla NB temporal sequence, Tll appears to have the opposite function: Tll is the last TF that is expressed in the oldest NBs. These Tll^{+} NBs show nuclear localization of Pros, indicating that they undergo Pros-dependent cell-cycle exit at the end of their life, similar to the thoracic and central brain NBs (Li et al., 2013). Whether Tll is required or sufficient for ending medulla neurogenesis is currently not known. If so, it will be interesting to understand how Tll plays completely opposite roles in the mushroom body and in medulla NBs.

7. INTEGRATION OF TEMPORAL AND SPATIAL INFORMATION DETERMINES LINEAGES

Although almost all NBs in the embryonic VNC follow the same temporal sequence, they generate different lineages depending on their spatial identity. There are approximately 30 NBs in each hemisegment, and each NB has an individual fate based on its position and the expression of specific molecular markers. Intrasegmental specification is achieved by superimposed activities of segment polarity and dorso-ventral patterning genes (Technau et al., 2006). In addition to the intrasegmental spatial patterning, homologous NBs in different segments along the antero-posterior axis have slightly different lineages, although they also share significant similarities in their lineages and gene expression patterns. Hox genes are critical players in determining these differences. For example, in Section 6 we discussed how the *abdA* Hox gene, together with temporal factors, determines the length of lineages and the end of neurogenesis. Hox genes also work with temporal genes to control the generation of specific cell types only found in certain segments. For example, NB5-6 generates the Ap cluster neurons only in thoracic segments. Late temporal genes *cas* and *grh*, and the thoracic Hox gene *Antennapedia* (*Antp*), are required to specify the Ap cluster neurons (Karlsson, Baumgardt, & Thor, 2010). One of the target genes for integration of temporal and spatial information is Collier, which plays important roles in the feedforward loops that specify the Ap cluster neurons as discussed in Section 5. It will be interesting to further characterize the exact molecular mechanism of the integration (Karlsson et al., 2010).

Spatial identity also modulates the progression of the temporal sequence of TFs in NBs. For example, young NB3-3 never expresses Hb, but sequentially expresses Kr, Pdm, Cas, Cas/Grh, and Svp/Grh (Tsuji et al., 2008). The absence of Hb in the sequence is common to both thoracic and abdominal NB3-3, suggesting that this is modulated by the intrasegmental spatial patterning genes common to NB3-3 in each segment. Although the temporal sequence is the same in both thoracic and abdominal NB3-3, the speed of transitions between TFs is faster in abdominal NB3-3s than in thoracic NB3-3s. Particularly, the switch from Cas to Svp occurs in late embryo in abdominal NB3-3, while it occurs in the larval stage for thoracic NB3-3. This spatial difference in the timing of switching to Svp is also true for other NB lineages and is regulated by Hox genes. Loss of *Antp* or misexpression of *abd-A* in thoracic NB3-3 causes precocious Svp expression during embryogenesis (Tsuji et al., 2008). How the Hox genes modulate the speed of the temporal progression and how the spatial–temporal information is integrated to regulate the NB lineage will be interesting questions for the future.

8. INTEGRATION OF TEMPORAL IDENTITY WITH BINARY FATE CHOICE

In embryonic NB lineages, GMCs usually divide asymmetrically to give rise to two progeny with different fates (either two different neuron types, one neuron and one glia, or one neuron and the other undergoing programmed cell death, etc.), and this depends on Notch signaling between the two GMC daughter cells (Buescher et al., 1998; Karcavich & Doe, 2005; Lundell, Lee, Perez, & Chadwell, 2003; Novotny et al., 2002; Schuldt & Brand, 1999; Skeath & Doe, 1998; Spana & Doe, 1996; Udolph, Rath, & Chia, 2001). During the asymmetric division of GMCs, Numb, a repressor of Notch signaling, is asymmetrically localized to one daughter cell, and this cell adopts a Notch-off (N-off) fate, while the other cell adopts a Notch-on (N-on) fate (Buescher et al., 1998; Spana & Doe, 1996). All the N-on (or N-off) cells within a lineage are collectively called a hemilineage. In postembryonic NB lineages, Notch/Numb also function in binary-fate choices of GMC progeny (Kumar, Bello, & Reichert, 2009; Li et al., 2013; Lin et al., 2010; Truman, Moats, Altman, Marin, & Williams, 2010).

Notch signaling only differentiates between two alternative fates, but the actual fates depend on the spatial and temporal identity of the NB. For example, Notch has distinct lineage-specific effects in the three antennal lobe NB lineages: antero-dorsal lineage (N-off: PN fates, N-on: apoptosis); ventral lineage (N-off: apoptosis, N-on: PN fates); and lateral lineage (N-off: PN fates, N-on: local interneuron fates) (Lin et al., 2010).

How does the Notch-dependent binary fate choice integrate with the temporal identity of NBs? In the developing medulla, the Notch pathway regulates the maintenance of the temporal TFs in postmitotic neurons and the expression of Ap. In the Ey and Slp NB stages, Ey or Slp is only maintained in the N-off daughter, while the N-on daughter turns on Ap. At the D NB stage, D is only maintained in the N-on daughter together with Ap. Although most if not all N-on daughters of medulla GMCs express Ap, they express different combinations of other TFs and adopt different fates depending on which NB stage they are born from. For example, the N-on progeny of Ey^+ GMCs express Ap and Drifter (Dfr), and this is dependent on both Notch signaling and the expression of Ey in NBs. The N-on progeny of Slp^+ GMCs express Ap and twin of eyeless (Toy), and similarly, this depends on both Notch signaling and Slp expression in NBs. Thus, the temporal TFs and the Notch pathway together control the expression of downstream TFs like Drifter and Toy to control neuron fates (Li et al., 2013; Fig. 3.2).

Studies in the antennal lobe lateral lineage illustrate another interesting question. In this lineage, PNs and local interneurons are produced as siblings of each GMC division (N-off: PN, N-on: local interneuron). However, as there is more diversity of PNs than of local interneurons, the same local interneuron can be the sibling of different PNs. Thus, it appears that the tempo of birth-order-dependent fate changes is different between the two N-off and N-on hemilineages (Lin et al., 2012). How the same temporal identity information regulates the independent pace of temporal switching of neuronal fates in the two hemilineages is an interesting question for the future.

9. CONCLUSIONS AND FUTURE QUESTIONS

Currently, two different temporal sequences have been identified in NBs of two different systems: the Hb → Kr → Pdm → Cas → Grh sequence in the *Drosophila* VNC, and the Hth → Klu → Ey → Slp → D → Tll sequence that patterns medulla NBs. In addition, the INPs of type II NB lineages are patterned by a TF sequence D → Grh → Ey. These three examples suggest that TF-sequence-dependent temporal patterning of neural precursors is a common theme to generate neural diversity. This suggests that more temporal sequences will be identified in other systems. For example, in the antennal lobe antero-dorsal lineage, Kr specifies one out of 40 temporal identities in the NB. Further identification of a complete temporal sequence will rely on candidate gene approaches, and/or screening based on either gene expression or mutant phenotypes. Does TF-sequence-dependent temporal patterning of neural precursors also function in vertebrate systems? There is some evidence suggesting that this might be the case. In the vertebrate retina, one ortholog of *hb*, *Ikaros*, specifies early-born cell fates (Elliott, Jolicoeur, Ramamurthy, & Cayouette, 2008). In mammalian cortical neurogenesis, Foxg1, an ortholog of Slp, functions in cortical progenitors to suppress early-born cortical cell fates (Hanashima, Li, Shen, Lai, & Fishell, 2004). Although the vertebrate systems are much more complex than *Drosophila*, studies in flies have provided important concepts that might be applicable to vertebrates.

Cross-regulations between temporal TFs are important for temporal transitions: Feedback negative regulation and feedforward positive regulation among the temporal TFs can facilitate the progression of the sequence. There are many remaining questions to understand the mechanism of temporal transitions. Even in the best-known systems, there can be missing factors or timing mechanisms in addition to the identified TF sequence. When all factors involved and their regulatory relationships are characterized by genetic analysis, theoretical modeling will provide insights into how the genetic network precisely times the NB temporal progression.

Loss of NB competence is related to epigenetic changes, such as chromatin modifications and chromosome architecture. More studies are needed to examine how the expression dynamics of the temporal genes leads to epigenetic changes and thus regulates NB competence, and eventually determines the end of neurogenesis. Such studies are relevant for the development of cell replacement therapies using stem cells to treat various diseases.

Another big challenge is to elucidate how the temporal and spatial identity of NBs, as well as the Notch-dependent binary fate choices, are integrated to generate specific neuron fates.

Finally, how the temporal sequence evolved is a great question for evolutionary developmental neurobiologists. An interesting observation is that the Hb → Kr → Pdm → Cas temporal sequence in VNC NBs mimics the anterior to posterior spatial distribution of Hb → Kr → Pdm → Cas expression domains at cellular blastoderm in the embryo (Isshiki et al., 2001). Whether they use the same regulatory logic is not known. The Hth → Klu → Ey → Slp → D → Tll or D → Grh → Ey regulatory cascades have not been described in other contexts. Evolutionary studies might provide clues as to how this powerful mechanism has evolved to pattern neural precursors to generate neural diversity.

REFERENCES

Akiyama-Oda, Y., Hosoya, T., & Hotta, Y. (1999). Asymmetric cell division of thoracic neuroblast 6-4 to bifurcate glial and neuronal lineage in Drosophila. *Development, 126*, 1967–1974.

Almeida, M. S., & Bray, S. J. (2005). Regulation of post-embryonic neuroblasts by Drosophila Grainyhead. *Mechanisms of Development, 122*, 1282–1293.

Baek, M., & Mann, R. S. (2009). Lineage and birth date specify motor neuron targeting and dendritic architecture in adult Drosophila. *The Journal of Neuroscience, 29*, 6904–6916.

Baumgardt, M., Karlsson, D., Terriente, J., Diaz-Benjumea, F. J., & Thor, S. (2009). Neuronal subtype specification within a lineage by opposing temporal feed-forward loops. *Cell, 139*, 969–982.

Baumgardt, M., Miguel-Aliaga, I., Karlsson, D., Ekman, H., & Thor, S. (2007). Specification of neuronal identities by feedforward combinatorial coding. *PLoS Biology, 5*, e37.

Bayraktar, O. A., Boone, J. Q., Drummond, M. L., & Doe, C. Q. (2010). Drosophila type II neuroblast lineages keep Prospero levels low to generate large clones that contribute to the adult brain central complex. *Neural Development, 5*, 26.

Bayraktar, O. A., & Doe, C. Q. (2013). Temporal patterning in intermediate progenitors increases neural diversity. *Nature, 498*, 449–455.

Bello, B. C., Hirth, F., & Gould, A. P. (2003). A pulse of the Drosophila Hox protein Abdominal-A schedules the end of neural proliferation via neuroblast apoptosis. *Neuron, 37*, 209–219.

Benito-Sipos, J., Estacio-Gomez, A., Moris-Sanz, M., Baumgardt, M., Thor, S., & Diaz-Benjumea, F. J. (2010). A genetic cascade involving klumpfuss, nab and castor specifies the abdominal leucokinergic neurons in the Drosophila CNS. *Development, 137*, 3327–3336.

Benveniste, R. J., Thor, S., Thomas, J. B., & Taghert, P. H. (1998). Cell type-specific regulation of the Drosophila FMRF-NH2 neuropeptide gene by Apterous, a LIM homeodomain transcription factor. *Development, 125*, 4757–4765.

Bhat, K. M. (1999a). Segment polarity genes in neuroblast formation and identity specification during Drosophila neurogenesis. *Bioessays, 21*, 472–485.

Bhat, K. M. (1999b). Segment polarity genes in neuroblast formation and identity specification during Drosophila neurogenesis. *BioEssays: News and Reviews in Molecular, Cellular and Developmental Biology, 21*, 472–485.

Boone, J. Q., & Doe, C. Q. (2008). Identification of Drosophila type II neuroblast lineages containing transit amplifying ganglion mother cells. *Developmental Neurobiology, 68*, 1185–1195.

Bossing, T., Udolph, G., Doe, C. Q., & Technau, G. M. (1996). The embryonic central nervous system lineages of Drosophila melanogaster. I. Neuroblast lineages derived from the ventral half of the neuroectoderm. *Developmental Biology, 179*, 41–64.

Brand, A. H., & Livesey, F. J. (2011). Neural stem cell biology in vertebrates and invertebrates: More alike than different? *Neuron, 70*, 719–729.

Brody, T., & Odenwald, W. F. (2000). Programmed transformations in neuroblast gene expression during Drosophila CNS lineage development. *Developmental Biology, 226*, 34–44.

Buescher, M., Yeo, S. L., Udolph, G., Zavortink, M., Yang, X., Tear, G., et al. (1998). Binary sibling neuronal cell fate decisions in the Drosophila embryonic central nervous system are nonstochastic and require inscuteable-mediated asymmetry of ganglion mother cells. *Genes & Development, 12*, 1858–1870.

Cenci, C., & Gould, A. P. (2005). Drosophila Grainyhead specifies late programmes of neural proliferation by regulating the mitotic activity and Hox-dependent apoptosis of neuroblasts. *Development, 132*, 3835–3845.

Cleary, M. D., & Doe, C. Q. (2006). Regulation of neuroblast competence: Multiple temporal identity factors specify distinct neuronal fates within a single early competence window. *Genes & Development, 20*, 429–434.

Cui, X., & Doe, C. Q. (1992). ming is expressed in neuroblast sublineages and regulates gene expression in the Drosophila central nervous system. *Development, 116*, 943–952.

Das, A., Reichert, H., & Rodrigues, V. (2010). Notch regulates the generation of diverse cell types from the lateral lineage of Drosophila antennal lobe. *Journal of Neurogenetics, 24*, 42–53.

Das, A., Sen, S., Lichtneckert, R., Okada, R., Ito, K., Rodrigues, V., et al. (2008). Drosophila olfactory local interneurons and projection neurons derive from a common neuroblast lineage specified by the empty spiracles gene. *Neural Development, 3*, 33.

Desai, A. R., & McConnell, S. K. (2000). Progressive restriction in fate potential by neural progenitors during cerebral cortical development. *Development, 127*, 2863–2872.

Dessaud, E., McMahon, A. P., & Briscoe, J. (2008). Pattern formation in the vertebrate neural tube: A sonic hedgehog morphogen-regulated transcriptional network. *Development, 135*, 2489–2503.

Doe, C. Q. (1992). Molecular markers for identified neuroblasts and ganglion mother cells in the Drosophila central nervous system. *Development, 116*, 855–863.

Doe, C. Q. (2008). Neural stem cells: Balancing self-renewal with differentiation. *Development, 135*, 1575–1587.

Egger, B., Boone, J. Q., Stevens, N. R., Brand, A. H., & Doe, C. Q. (2007). Regulation of spindle orientation and neural stem cell fate in the Drosophila optic lobe. *Neural Development, 2*, 1.

Elliott, J., Jolicoeur, C., Ramamurthy, V., & Cayouette, M. (2008). Ikaros confers early temporal competence to mouse retinal progenitor cells. *Neuron, 60*, 26–39.

Fischbach, K. F., & Dittrich, A. P. M. (1989). The optic lobe of Drosophila melanogaster. I. A Golgi analysis of wild-type structure. *Cell and Tissue Research, 258*, 441–475.
Gaspard, N., Bouschet, T., Hourez, R., Dimidschstein, J., Naeije, G., van den Ameele, J., et al. (2008). An intrinsic mechanism of corticogenesis from embryonic stem cells. *Nature, 455*, 351–357.
Green, P., Hartenstein, A. Y., & Hartenstein, V. (1993). The embryonic development of the Drosophila visual system. *Cell and Tissue Research, 273*, 583–598.
Grosskortenhaus, R., Pearson, B. J., Marusich, A., & Doe, C. Q. (2005). Regulation of temporal identity transitions in Drosophila neuroblasts. *Developmental Cell, 8*, 193–202.
Grosskortenhaus, R., Robinson, K. J., & Doe, C. Q. (2006). Pdm and Castor specify late-born motor neuron identity in the NB7-1 lineage. *Genes & Development, 20*, 2618–2627.
Hanashima, C., Li, S. C., Shen, L., Lai, E., & Fishell, G. (2004). Foxg1 suppresses early cortical cell fate. *Science, 303*, 56–59.
Hasegawa, E., Kitada, Y., Kaido, M., Takayama, R., Awasaki, T., Tabata, T., et al. (2011). Concentric zones, cell migration and neuronal circuits in the Drosophila visual center. *Development, 138*, 983–993.
Hirabayashi, Y., Suzki, N., Tsuboi, M., Endo, T. A., Toyoda, T., Shinga, J., et al. (2009). Polycomb limits the neurogenic competence of neural precursor cells to promote astrogenic fate transition. *Neuron, 63*, 600–613.
Huang, D. H., Chang, Y. L., Yang, C. C., Pan, I. C., & King, B. (2002). Pipsqueak encodes a factor essential for sequence-specific targeting of a polycomb group protein complex. *Molecular Cell Biology, 22*, 6261–6271.
Isshiki, T., Pearson, B., Holbrook, S., & Doe, C. Q. (2001). Drosophila neuroblasts sequentially express transcription factors which specify the temporal identity of their neuronal progeny. *Cell, 106*, 511–521.
Jacob, J., Maurange, C., & Gould, A. P. (2008). Temporal control of neuronal diversity: Common regulatory principles in insects and vertebrates? *Development, 135*, 3481–3489.
Jefferis, G. S., Marin, E. C., Stocker, R. F., & Luo, L. (2001). Target neuron prespecification in the olfactory map of Drosophila. *Nature, 414*, 204–208.
Kambadur, R., Koizumi, K., Stivers, C., Nagle, J., Poole, S. J., & Odenwald, W. F. (1998). Regulation of POU genes by castor and hunchback establishes layered compartments in the Drosophila CNS. *Genes & Development, 12*, 246–260.
Kanai, M. I., Okabe, M., & Hiromi, Y. (2005). Seven-up controls switching of transcription factors that specify temporal identities of Drosophila neuroblasts. *Developmental Cell, 8*, 203–213.
Kao, C. F., & Lee, T. (2010). Birth time/order-dependent neuron type specification. *Current Opinion in Neurobiology, 20*, 14–21.
Kao, C. F., Yu, H. H., He, Y., Kao, J. C., & Lee, T. (2012). Hierarchical deployment of factors regulating temporal fate in a diverse neuronal lineage of the Drosophila central brain. *Neuron, 73*, 677–684.
Karcavich, R. E. (2005). Generating neuronal diversity in the Drosophila central nervous system: A view from the ganglion mother cells. *Developmental Dynamics, 232*, 609–616.
Karcavich, R., & Doe, C. Q. (2005). Drosophila neuroblast 7-3 cell lineage: A model system for studying programmed cell death, Notch/Numb signaling, and sequential specification of ganglion mother cell identity. *The Journal of Comparative Neurology, 481*, 240–251.
Karlsson, D., Baumgardt, M., & Thor, S. (2010). Segment-specific neuronal subtype specification by the integration of anteroposterior and temporal cues. *PLoS Biology, 8*, e1000368.
Knoblich, J. A. (2010). Asymmetric cell division: Recent developments and their implications for tumour biology. *Nature Reviews. Molecular Cell Biology, 11*, 849–860.

Kohwi, M., Hiebert, L. S., & Doe, C. Q. (2011). The pipsqueak-domain proteins Distal antenna and Distal antenna-related restrict Hunchback neuroblast expression and early-born neuronal identity. *Development*, *138*, 1727–1735.

Kohwi, M., Lupton, J. R., Lai, S. L., Miller, M. R., & Doe, C. Q. (2013). Developmentally regulated subnuclear genome reorganization restricts neural progenitor competence in Drosophila. *Cell*, *152*, 97–108.

Kumar, A., Bello, B., & Reichert, H. (2009). Lineage-specific cell death in postembryonic brain development of Drosophila. *Development*, *136*, 3433–3442.

Kurusu, M., Maruyama, Y., Adachi, Y., Okabe, M., Suzuki, E., & Furukubo-Tokunaga, K. (2009). A conserved nuclear receptor, Tailless, is required for efficient proliferation and prolonged maintenance of mushroom body progenitors in the Drosophila brain. *Developmental Biology*, *326*, 224–236.

Lai, S. L., Awasaki, T., Ito, K., & Lee, T. (2008). Clonal analysis of Drosophila antennal lobe neurons: Diverse neuronal architectures in the lateral neuroblast lineage. *Development*, *135*, 2883–2893.

Lee, T., Lee, A., & Luo, L. (1999). Development of the Drosophila mushroom bodies: Sequential generation of three distinct types of neurons from a neuroblast. *Development*, *126*, 4065–4076.

Li, X., Erclik, T., Bertet, C., Chen, Z., Voutev, R., Venkatesh, S., et al. (2013). Temporal patterning of Drosophila medulla neuroblasts controls neural fates. *Nature*, *498*, 456–462.

Lin, S., Kao, C. F., Yu, H. H., Huang, Y., & Lee, T. (2012). Lineage analysis of Drosophila lateral antennal lobe neurons reveals notch-dependent binary temporal fate decisions. *PLoS Biology*, *10*, e1001425.

Lin, S., Lai, S. L., Yu, H. H., Chihara, T., Luo, L., & Lee, T. (2010). Lineage-specific effects of Notch/Numb signaling in post-embryonic development of the Drosophila brain. *Development*, *137*, 43–51.

Lin, S., & Lee, T. (2012). Generating neuronal diversity in the Drosophila central nervous system. *Developmental Dynamics*, *241*, 57–68.

Livesey, F. J., & Cepko, C. L. (2001). Vertebrate neural cell-fate determination: Lessons from the retina. *Nature Reviews Neuroscience*, *2*, 109–118.

Lundell, M. J., & Hirsh, J. (1998). Eagle is required for the specification of serotonin neurons and other neuroblast 7-3 progeny in the Drosophila CNS. *Development*, *125*, 463–472.

Lundell, M. J., Lee, H. K., Perez, E., & Chadwell, L. (2003). The regulation of apoptosis by Numb/Notch signaling in the serotonin lineage of Drosophila. *Development*, *130*, 4109–4121.

Maurange, C. (2012). Temporal specification of neural stem cells: Insights from Drosophila neuroblasts. *Current Topics in Developmental Biology*, *98*, 199–228.

Maurange, C., Cheng, L., & Gould, A. P. (2008). Temporal transcription factors and their targets schedule the end of neural proliferation in Drosophila. *Cell*, *133*, 891–902.

Mellerick, D. M., Kassis, J. A., Zhang, S. D., & Odenwald, W. F. (1992). Castor encodes a novel zinc finger protein required for the development of a subset of CNS neurons in Drosophila. *Neuron*, *9*, 789–803.

Mettler, U., Vogler, G., & Urban, J. (2006). Timing of identity: Spatiotemporal regulation of hunchback in neuroblast lineages of Drosophila by Seven-up and Prospero. *Development*, *133*, 429–437.

Molyneaux, B. J., Arlotta, P., Menezes, J. R., & Macklis, J. D. (2007). Neuronal subtype specification in the cerebral cortex. *Nature Reviews Neuroscience*, *8*, 427–437.

Morante, J., & Desplan, C. (2008). The color-vision circuit in the medulla of Drosophila. *Current Biology*, *18*, 553–565.

Morante, J., Erclik, T., & Desplan, C. (2011). Cell migration in Drosophila optic lobe neurons is controlled by eyeless/Pax6. *Development*, *138*, 687–693.

Naka, H., Nakamura, S., Shimazaki, T., & Okano, H. (2008). Requirement for COUP-TFI and II in the temporal specification of neural stem cells in CNS development. *Nature Neuroscience*, *11*, 1014–1023.

Nakajima, A., Isshiki, T., Kaneko, K., & Ishihara, S. (2010). Robustness under functional constraint: The genetic network for temporal expression in Drosophila neurogenesis. *PLoS Computational Biology*, *6*, e1000760.

Novotny, T., Eiselt, R., & Urban, J. (2002). Hunchback is required for the specification of the early sublineage of neuroblast 7-3 in the Drosophila central nervous system. *Development*, *129*, 1027–1036.

Okano, H., & Temple, S. (2009). Cell types to order: Temporal specification of CNS stem cells. *Current Opinion in Neurobiology*, *19*, 112–119.

Park, D., Han, M., Kim, Y. C., Han, K. A., & Taghert, P. H. (2004). Ap-let neurons – A peptidergic circuit potentially controlling ecdysial behavior in Drosophila. *Developmental Biology*, *269*, 95–108.

Pearson, B. J., & Doe, C. Q. (2003). Regulation of neuroblast competence in Drosophila. *Nature*, *425*, 624–628.

Pearson, B. J., & Doe, C. Q. (2004). Specification of temporal identity in the developing nervous system. *Annual Review of Cell and Developmental Biology*, *20*, 619–647.

Peterson, C., Carney, G. E., Taylor, B. J., & White, K. (2002). Reaper is required for neuroblast apoptosis during Drosophila development. *Development*, *129*, 1467–1476.

Prokop, A., & Technau, G. M. (1991). The origin of postembryonic neuroblasts in the ventral nerve cord of Drosophila melanogaster. *Development*, *111*, 79–88.

Romani, S., Jimenez, F., Hoch, M., Patel, N. H., Taubert, H., & Jackle, H. (1996). Kruppel, a Drosophila segmentation gene, participates in the specification of neurons and glial cells. *Mechanisms of Development*, *60*, 95–107.

Schmid, A., Chiba, A., & Doe, C. Q. (1999). Clonal analysis of Drosophila embryonic neuroblasts: Neural cell types, axon projections and muscle targets. *Development*, *126*, 4653–4689.

Schmidt, H., Rickert, C., Bossing, T., Vef, O., Urban, J., & Technau, G. M. (1997). The embryonic central nervous system lineages of Drosophila melanogaster. II. Neuroblast lineages derived from the dorsal part of the neuroectoderm. *Developmental Biology*, *189*, 186–204.

Schuldt, A. J., & Brand, A. H. (1999). Mastermind acts downstream of notch to specify neuronal cell fates in the Drosophila central nervous system. *Developmental Biology*, *205*, 287–295.

Shen, Q., Wang, Y., Dimos, J. T., Fasano, C. A., Phoenix, T. N., Lemischka, I. R., et al. (2006). The timing of cortical neurogenesis is encoded within lineages of individual progenitor cells. *Nature Neuroscience*, *9*, 743–751.

Skeath, J. B., & Doe, C. Q. (1998). Sanpodo and Notch act in opposition to Numb to distinguish sibling neuron fates in the Drosophila CNS. *Development*, *125*, 1857–1865.

Skeath, J. B., & Thor, S. (2003). Genetic control of Drosophila nerve cord development. *Current Opinion in Neurobiology*, *13*, 8–15.

Spana, E. P., & Doe, C. Q. (1996). Numb antagonizes Notch signaling to specify sibling neuron cell fates. *Neuron*, *17*, 21–26.

Suzuki, T., Kaido, M., Takayama, R., & Sato, M. (2013). A temporal mechanism that produces neuronal diversity in the Drosophila visual center. *Developmental Biology*, *380*, 12–24.

Technau, G. M., Berger, C., & Urbach, R. (2006). Generation of cell diversity and segmental pattern in the embryonic central nervous system of Drosophila. *Developmental Dynamics*, *235*, 861–869.

Touma, J. J., Weckerle, F. F., & Cleary, M. D. (2012). Drosophila Polycomb complexes restrict neuroblast competence to generate motoneurons. *Development*, *139*, 657–666.

Tran, K. D., & Doe, C. Q. (2008). Pdm and Castor close successive temporal identity windows in the NB3-1 lineage. *Development, 135*, 3491–3499.
Tran, K. D., Miller, M. R., & Doe, C. Q. (2010). Recombineering Hunchback identifies two conserved domains required to maintain neuroblast competence and specify early-born neuronal identity. *Development, 137*, 1421–1430.
Truman, J. W., & Bate, M. (1988). Spatial and temporal patterns of neurogenesis in the central nervous system of Drosophila melanogaster. *Developmental Biology, 125*, 145–157.
Truman, J. W., Moats, W., Altman, J., Marin, E. C., & Williams, D. W. (2010). Role of Notch signaling in establishing the hemilineages of secondary neurons in Drosophila melanogaster. *Development, 137*, 53–61.
Tsuji, T., Hasegawa, E., & Isshiki, T. (2008). Neuroblast entry into quiescence is regulated intrinsically by the combined action of spatial Hox proteins and temporal identity factors. *Development, 135*, 3859–3869.
Udolph, G., Rath, P., & Chia, W. (2001). A requirement for Notch in the genesis of a subset of glial cells in the Drosophila embryonic central nervous system which arise through asymmetric divisions. *Development, 128*, 1457–1466.
Viktorin, G., Riebli, N., Popkova, A., Giangrande, A., & Reichert, H. (2011). Multipotent neural stem cells generate glial cells of the central complex through transit amplifying intermediate progenitors in Drosophila brain development. *Developmental Biology, 356*, 553–565.
Weng, M., Golden, K. L., & Lee, C. Y. (2010). dFezf/Earmuff maintains the restricted developmental potential of intermediate neural progenitors in Drosophila. *Developmental Cell, 18*, 126–135.
White, K., & Kankel, D. R. (1978). Patterns of cell division and cell movement in the formation of the imaginal nervous system in Drosophila melanogaster. *Developmental Biology, 65*, 296–321.
Wu, Y. C., Chen, C. H., Mercer, A., & Sokol, N. S. (2012). Let-7-complex microRNAs regulate the temporal identity of Drosophila mushroom body neurons via chinmo. *Developmental Cell, 23*, 202–209.
Yang, X., Yeo, S., Dick, T., & Chia, W. (1993). The role of a Drosophila POU homeo domain gene in the specification of neural precursor cell identity in the developing embryonic central nervous system. *Genes & Development*, 7, 504–516.
Yasugi, T., Umetsu, D., Murakami, S., Sato, M., & Tabata, T. (2008). Drosophila optic lobe neuroblasts triggered by a wave of proneural gene expression that is negatively regulated by JAK/STAT. *Development, 135*, 1471–1480.
Yu, H. H., Chen, C. H., Shi, L., Huang, Y., & Lee, T. (2009). Twin-spot MARCM to reveal the developmental origin and identity of neurons. *Nature Neuroscience, 12*, 947–953.
Yu, H. H., Kao, C. F., He, Y., Ding, P., Kao, J. C., & Lee, T. (2010). A complete developmental sequence of a Drosophila neuronal lineage as revealed by twin-spot MARCM. *PLoS Biology, 8*, e1000461.
Yu, F., Kuo, C. T., & Jan, Y. N. (2006). Drosophila neuroblast asymmetric cell division: Recent advances and implications for stem cell biology. *Neuron, 51*, 13–20.
Zhu, S., Barshow, S., Wildonger, J., Jan, L. Y., & Jan, Y. N. (2011). Ets transcription factor Pointed promotes the generation of intermediate neural progenitors in Drosophila larval brains. *Proceedings of the National Academy of Sciences of the United States of America, 108*, 20615–20620.
Zhu, S., Lin, S., Kao, C. F., Awasaki, T., Chiang, A. S., & Lee, T. (2006). Gradients of the Drosophila Chinmo BTB-zinc finger protein govern neuronal temporal identity. *Cell, 127*, 409–422.

CHAPTER FOUR

MicroRNAs as Components of Systemic Signaling Pathways in *Drosophila melanogaster*

Arthur Luhur[1], Geetanjali Chawla[1], Nicholas S. Sokol[2]
Department of Biology, Indiana University, Bloomington, Indiana, USA
[1]Co-first authors.
[2]Corresponding author: e-mail address: nsokol@indiana.edu

Contents

1. Introduction 98
2. Overview of miRNA Biogenesis and Function 99
3. Insulin and Nutritional Pathways 99
   3.1 Overview of ILPs and insulin signaling 100
   3.2 Identification of miRNAs involved in insulin signaling 102
   3.3 *miR-14* 102
   3.4 *miR-278* 102
   3.5 *miR-8* 102
   3.6 *miR-33* 103
   3.7 *dcr-1* 103
   3.8 *miR-1* 104
4. Lipophilic Hormone Signaling Pathways 104
   4.1 Overview of Ecdysone and JH signaling pathway 104
   4.2 Identification of Ecdysone- and JH-regulated miRNAs 106
   4.3 Molecular analysis of Ecdysone- and JH-control of *let-7-C*, *miR-34*, *miR-8*, and *miR-14* miRNAs 107
   4.4 Biological roles of Ecdysone regulated *let-7-C* miRNAs 108
   4.5 *miR-14* and *miR-8*: molecular links between steroid and insulin signaling pathways 111
   4.6 *miR-34* 112
5. Light Modulated Circadian Rhythm Pathways 113
   5.1 Molecular overview of circadian rhythm pathway 113
   5.2 Identification of miRNAs in clock pathways 114
   5.3 *bantam* 114
   5.4 *miR-279* 115
   5.5 *miR-959–964 cluster* 116
6. Stress Induced and Immune Response Pathways 116

Current Topics in Developmental Biology, Volume 105
ISSN 0070-2153
http://dx.doi.org/10.1016/B978-0-12-396968-2.00004-X

6.1 *miR-8* 117
6.2 *miR-959–964 cluster* 117
6.3 Bioinformatics predicts additional miRNAs involved in immune response 118
7. Summary 118
References 118

Abstract

MicroRNAs (miRNAs) ensure progression through development by synchronizing cell fate transitions in response to environmental cues. These cues are mediated at least in part by steroid hormones. Emerging evidence indicates that miRNAs are also components of additional systemic signaling pathways, including insulin, stress, immune, and circadian pathways. Thus, the roles that miRNAs play during development are reflected in their post-developmental functions, where they similarly function to coordinate cell behavior in response to environmental cues. In this review, we summarize current work highlighting the role of miRNAs in systemic signaling pathways in *Drosophila melanogaster* as a way of synthesizing the underlying roles of miRNAs in both animal developmental transitions and adult physiology.

1. INTRODUCTION

MicroRNAs (miRNAs) are short ~21-nucleotide (nt) noncoding RNAs that regulate the stability and translation of target messenger RNAs (mRNAs) and play essential roles in both animal development and physiology (Cohen, 2010; Mendell & Olson, 2012). Forward genetic analysis of *Caenorhabditis elegans* identified the first members of this class of post-transcriptional regulators, *lin-4* and *let-7*. These miRNAs are components of the heterochronic pathway, which is a developmental signaling pathway that ensures that cells adopt the appropriate stage-specific cell fates (for a review of this pathway, see chapter 6 by Rougvie and Moss in this volume). In response to environmental cues, like food availability and population density, steroid hormones activate miRNAs that repress cell fate determinants. The heterochronic pathway functions in cell lineages throughout the animal, synchronizing progression through developmental transitions with environmental conditions.

Emerging evidence indicates that this function of heterochronic miRNAs is evolutionarily conserved, and highlights a more general function of miRNAs to coordinate cell behavior in response to environmental stimuli. We explore this evidence here, reviewing recent work on miRNAs in *Drosophila melanogaster* that function within systemic signaling pathways.

These include the pathways regulated by small peptide hormones, like insulin, as well as pathways regulated by lipohilic hormones, like ecdysteroids and juvenile hormones. We also consider the role of miRNAs in other pathways that coordinate cell behaviors throughout the animal and in response to environmental cues, including circadian rhythm, stress, and immune response pathways.

2. OVERVIEW OF miRNA BIOGENESIS AND FUNCTION

Mature miRNAs are derived from longer transcripts via two cleavage steps (for a review of miRNA biology, see Krol, Loedige, & Filipowicz, 2010). The first cleavage involves the RNAse III-domain containing enzyme Drosha, which releases a ~70-nt hairpin structure from a longer primary miRNA (pri-miRNA) transcript. This hairpin is exported to the cytoplasm, where a second endoribonuclease, Dicer (Dcr), executes the next cleavage that releases the mature miRNA. Emerging evidence indicates that this post-transcriptional processing of miRNAs is highly regulated by a host of additional regulatory molecules (Siomi & Siomi, 2010). Once released, the mature miRNA enters the RNA-induced silencing complex (RISC) complex, which contains members of the Argonaute (Ago) family of proteins, and binds to target mRNAs, leading either to their translational repression or degradation. The precise mechanisms of miRNA-mediated gene silencing as well as the full repertoire of responsible cofactors remain to be identified.

3. INSULIN AND NUTRITIONAL PATHWAYS

An indispensable function of all living cells is to control the rates of metabolic processes in response to changes in the environment. The nutritional status of an organism is reflected by the levels of insulin, which in turn maintains a balance between circulating sugars and stored resources such as glycogen and fats. Another key role played by the insulin pathway is the control of cellular and organismal growth. Genetic manipulation of a number of components in this pathway in *Drosophila* has highlighted the role of insulin signaling in regulation of body size (Broughton et al., 2005; Rulifson, Kim, & Nusse, 2002). In this section, we first describe the molecular framework of the insulin-signaling pathway in *D. melanogaster* and then discuss the role of recently identified miRNAs in this pathway.

3.1. Overview of ILPs and insulin signaling

The insulin/insulin-like growth factor (IGF)-like signaling (IIS) pathway is an evolutionary conserved pathway with roles in tissue growth, development, metabolic homeostasis, stress resistance, and longevity (Broughton et al., 2005; Clancy et al., 2001; Holzenberger et al., 2003; Ikeya, Galic, Belawat, Nairz, & Hafen, 2002; Partridge, Alic, Bjedov, & Piper, 2011; Saltiel & Kahn, 2001). In *D. melanogaster*, insulin gene expression occurs in a group of 14 neurosecretory cells located in the brain (Ikeya et al., 2002). The *D. melanogaster* genome encodes seven different insulin-like peptides (ILPs) that are expressed and regulated differentially. Ligand binding to the *D. melanogaster* Insulin receptor (dInR) triggers autophosphorylation followed by either direct recruitment of phosphoinositol 3-kinase (PI3K) to the cell membrane or indirect recruitment through insulin receptor substrates (IRSs) such as Chico, Lnk, and Dreadlocks (Teleman, 2010; see Fig. 4.1). The PI3K is a heterodimer of a catalytic subunit Dp110 and a regulatory subunit Dp60. Together these catalyze conversion of phosphatidylinositol (4,5)-bisphosphate [PtdIns(4,5)P2] to phosphatidylinositol (1,4,5)-trisphosphate [PtdIns(1,4,5)P3]. A counteracting phosphatase Pten antagonizes PI3K activity and dephosphorylates PtdIns(1,4,5)P3 and PtdIns(4,5)P2. Accumulation of PtdIns(1,4,5)P3 activates a series of kinases such as phosphoinositide-dependent kinase 1 (PDK1) and Akt1/protein kinase B (PKB), which subsequently phosphorylate the transcription factor Forkhead Box, Sub-Group O (Foxo), leading to its cytoplasmic retention and thus inhibiting its transcriptional activity in the nucleus (Teleman, 2010). Foxo is a key transcriptional regulator that controls both downstream target genes responsible for growth as well as upstream feedback targets in the insulin-signaling pathway (Junger et al., 2003; Puig, Marr, Ruhf, & Tjian, 2003).

The nutrient responsive branch of the IIS pathway controls cellular protein synthesis via target of rapamycin (Tor). The Tor kinase phosphorylates S6 kinase and eukaryotic initiation factor 4E binding protein (4EBP) to stimulate protein translation. Akt1/PKB links the insulin and nutrient signaling branches by phosphorylation of tuberous sclerosis 2 (TSC2) in addition to Foxo (Hay & Sonenberg, 2004). In *D. melanogaster* larvae, the fat body is the tissue involved in sensing and relaying nutrient availability to coordinate growth of the whole organism. Low protein levels in the diet or amino acid deprivation results in inhibition of TOR kinase activity. This decreased TOR kinase signaling is relayed to other peripheral tissues like

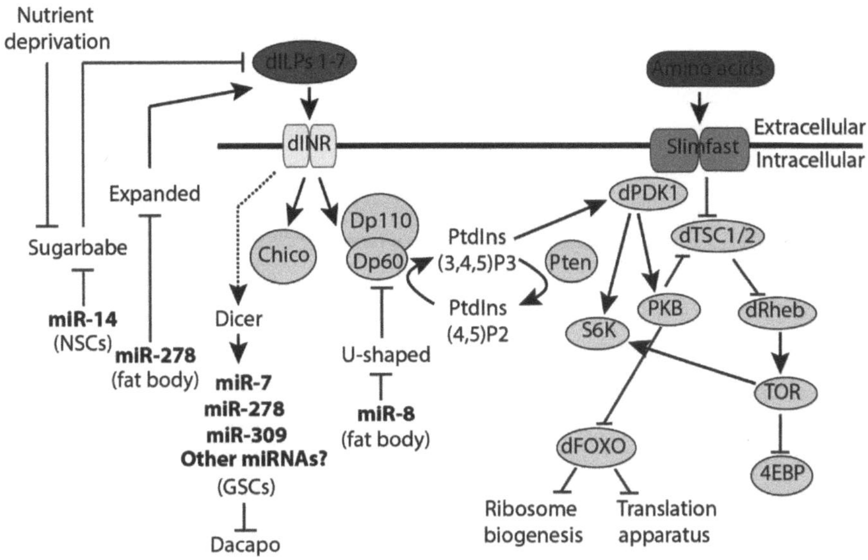

Figure 4.1 *Drosophila* miRNAs in the insulin signalling pathway. The figure summarizes the functional roles of *Drosophila* miRNAs involved in the IIS pathway. The functional interactions of the IIS components are described in the text. *miR-14* and nutrient levels function together in insulin producing neurosecretory cells (NSCs) in the adult *Drosophila* brain to regulate Sugarbabe, a zinc finger protein that regulates insulin gene expression (Varghese et al., 2010). *miR-278* controls energy homeostasis by regulating its target, *expanded*. De-regulation of Expanded in *miR-278* mutants results in elevated insulin production and causes leanness in flies (Teleman et al., 2006). *miR-8* activates PI3K by repressing U-shaped (Ush) in the fat body cells to regulate body size in *Drosophila*. U-shaped prevents formation of an active PI3K complex through its direct interaction with the adapter subunit Dp60 (Hyun et al., 2009). The miRNA pathway plays an important role in regulating cell division in the germline stem cells (GSCs) in response to environmental signals. One target that has been shown to mitigate this control is the cyclin-dependent kinase inhibitor, Dacapo (Dap) (Hatfield et al., 2005). Insulin signaling regulates expression of Dap in a miRNA dependent manner and multiple miRNAs (*miR-7*, *mir-278*, and *miR-309*) have been shown to regulate *dap* 3′UTR (Yu et al., 2009). Arrows (black) indicate activation and bar-ended lines (red) indicate inhibitory interactions. The IIS pathway components are shown in blue. Broken line indicates an interaction requiring further study. (For interpretation of the references to color in this figure legend, the reader is referred to the online version of this chapter.)

brain to suppress release of ILPs by the neurosecretory cells (Geminard, Rulifson, & Leopold, 2009). The fat body amino acid transporter, Slimfast (Slif), is an important component in nutritional sensing as knock down of *slif* in the fat body causes a non-autonomous reduction in insulin signaling in other tissues and a reduction in the total body size (Colombani et al., 2003).

3.2. Identification of miRNAs involved in insulin signaling

Over the past several years, miRNAs have been identified as key post-transcriptional players in the regulation of metabolic processes in response to environmental fluctuations (Krutzfeldt & Stoffel, 2006). Here, we discuss the roles of *D. melanogaster* miRNAs that impinge on insulin signaling and nutritional pathways. *D. melanogaster* miRNAs implicated in homeostasis include *miR-14*, *mir-278*, *miR-8*, *miR-33*, and *miR-1* (see Fig. 4.1).

3.3. *miR-14*

miR-14 was the first *D. melanogaster* miRNA implicated in metabolism. It was initially identified in a forward genetic screen for mutants that suppress apoptotic cell death (Xu, Vernooy, Guo, & Hay, 2003). *miR-14* mutants display a series of phenotypes, including reduced viability during development, reduced lifespan, elevated apoptosis, increased stress response, and increased diacylglycerol and triacylglycerol levels. This *miR-14* mutant fat metabolism defect leads to an obese phenotype, whereas extra copies of *miR-14* lead to lean flies (Xu et al., 2003). These metabolic phenotypes were attributed to a reduction in the expression of ILPs, since *miR-14* functions in the IPCs to regulate expression of a nutrient sensitive regulator of *dilp* expression, *sugarbabe* (Varghese, Lim, & Cohen, 2010). While *miR-14* performs its metabolic function by regulating *sugarbabe*, it also ensures viability during development and adulthood by regulating the expression of other mRNA targets including *Ecdysone Receptor* (*EcR*), as discussed below.

3.4. *miR-278*

miR-278 was originally identified in a genetic screen for genes whose forced expression causes robust overgrowth. *miR-278* mutants display a lean phenotype that is attributed to increased expression of ILPs (Teleman, Maitra, & Cohen, 2006). Moreover, elevation of insulin levels in *miR-278* mutants is unable to renormalize the increased circulating sugar levels in the organism. Overexpression of the *miR-278* target mRNA *expanded* in the fat body is sufficient to phenocopy this decreased insulin sensitivity (Teleman et al., 2006), suggesting that the main function of *miR-278* is to suppress expression of this single target mRNA in fat body cells.

3.5. *miR-8*

miR-8 is another miRNA that functions in the fat body to regulate organismal growth and metabolism (Hyun et al., 2009). Genetic elimination of *miR-8* results in small flies with defective insulin signaling. Fat body specific expression of *miR-8* restores body size and weight to near *wild type* levels. Both *miR-8* and its mammalian ortholog, *miR-200*, activate phosphatidylinositol 3-kinase (PI3K) by repressing the expression of *u-shaped (ush)* and *fog2*, respectively. Ush and Fog2 are orthologous proteins that interact directly with the regulatory subunits of PI3K and inhibit the formation of an active PI3K complex. Thus, *miR-8* can activate insulin signaling by suppressing *ush*, leading to the cell-autonomous growth of fat body as well as non-autonomous systemic growth of the organism (Hyun et al., 2009).

3.6. *miR-33*

Gain-of-function studies implicated *miR-33* in maintenance of lipid homeostasis in *D. melanogaster* (Davalos et al., 2011). Like its mammalian counterpart *miR-33a*, fly *miR-33* is located in the intron of *Sterol regulatory element-binding protein* (*Srebp*), a gene involved in fatty acid metabolism. Forced expression of *miR-33* in the fat body resulted in increased levels of triacylglycerols upon starvation as well as abnormal lipid accumulation. These phenotypes are consistent with inappropriate repression of the *miR-33* target *withered*, a mitochondrial enzyme that regulates fatty acid oxidation (Davalos et al., 2011).

3.7. *dcr-1*

The IIS signaling pathway has also been shown to regulate cell division in germline stem cells (GSCs) in a miRNA dependent manner (Yu et al., 2009). This regulation is dependent on the cyclin dependent kinase inhibitor and p21/p27 homolog, Dacapo (Dap) as the cell cycle defects of *InR*-deficient GSCs were partially rescued by a reduction in *dap*. Moreover, genetic as well as cell cycle analyses of Dicer-1 have highlighted the importance of the miRNA pathway in GSC maintenance and division. GSCs mutant for *dcr-1* displayed a Dap dependent reduction in germline cyst production and a defective cell cycle control (Hatfield et al., 2005). Subsequent work by Yu et al., has shown that the *dap* 3'UTR harbors binding sites for *miR-7*, *miR-278*, and *miR-309*. Although GSCs mutant for *miR-7* and *miR-278* displayed cell cycle defects, none of these defects were as dramatic as *dcr-1* mutant GSCs, thus implicating other unidentified miRNAs in this pathway (Yu et al., 2009). Thus, both Dicer 1 and Dap function in a nutrient dependent manner to regulate GSC homeostasis.

3.8. *miR-1*

Phenotypic evidence indicates that the muscle-specific *miR-1* miRNA also plays a role in tissue homeostasis in response to nutritional cues, although a direct molecular connection to the IIS pathway has not yet been identified. *miR-1* knockout larvae are much smaller than wildtype larvae, have severely deformed musculature, and become progressively compromised in movement (Sokol & Ambros, 2005). This paralysis and death of second instar *miR-1* mutant larvae is triggered by feeding and can be prevented by starving first instar mutants. Although relevant *miR-1* targets in this context have not been identified, these data reveal a critical role for *miR-1* in postmitotic growth of larval muscle (Sokol & Ambros, 2005). Owing to the high degree of evolutionary conservation of *miR-1*, it is likely that this miRNA plays an important role in maintaining muscle homeostasis in vertebrates.

4. LIPOPHILIC HORMONE SIGNALING PATHWAYS

Lipophilic hormones are endocrine signaling molecules that diffuse across cell membranes and bind to intracellular proteins, triggering coordinated and system-wide changes in cell behavior. *D. melanogaster* has two primary classes of lipophilic signaling molecules, ecdysteroids and juvenile hormones (JHs), and these hormones activate signaling pathways that transcriptionally control target genes (reviewed in Jindra, Palli, & Riddiford, 2013). Recent work has revealed roles for miRNAs in both ecdysteroid and JH pathways, indicating that these pathways also effect changes via post-transcriptional control of gene expression. In this section, we review the ecdysteroid and JH signaling pathways and discuss their relationships to miRNAs (see Fig. 4.2).

4.1. Overview of Ecdysone and JH signaling pathway

Twenty-hydroxy Ecdysone (20E) is the major *D. melanogaster* molting hormone that promotes progression through developmental transitions in response to environmental cues (Tennessen & Thummel, 2011). It is well known for its developmental role in initiating metamorphosis, but also functions post-developmentally to affect diverse adult process including reproduction, behavior, stress resistance, and lifespan (Schwedes & Carney, 2012). 20E is derived from a precursor ecdysteroid, termed Ecdysone, which is produced primarily in the prothoracic gland. Predictable release of Ecdysone from the prothoracic gland schedules the end of larval development. The exact timing

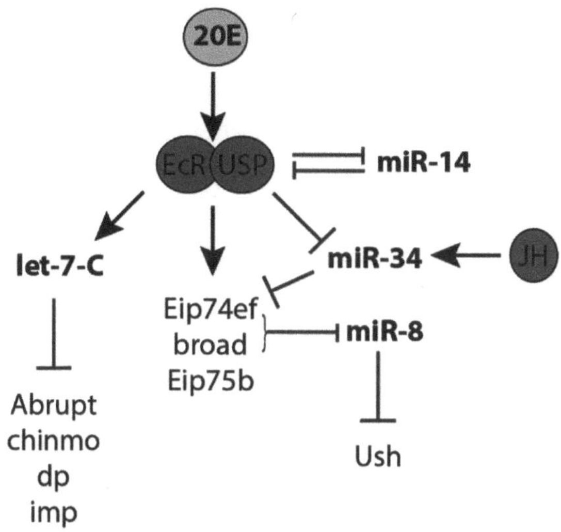

Figure 4.2 *Drosophila* miRNAs in the Ecdysone and JH pathways. The 20-hydroxyEcdysone (20E) steroid hormone regulates target genes via its receptor, a heterodimer composed of EcR and Usp. Direct targets of the 20E receptor include transcription factors *Eip74ef*, *Eip75b*, and *broad* as well as the *let-7-C* miRNA locus (Chawla & Sokol, 2012). These *let-7-C* miRNAs (*miR-100*, *let-7*, and *miR-125*) in turn repress the expression of *abrupt*, *chinmo*, *dp*, and *imp* to control neuronal fate, morphology, and maintenance as well as germline stem cell number (Caygill & Johnston, 2008; Gehrke et al., 2010; Toledano et al., 2012; Wu et al., 2012). While activating *let-7-C* miRNAs, 20E represses three additional miRNAs, *miR-8*, *miR-14*, and *miR-34* (Jin et al., 2012; Sempere et al., 2003; Varghese & Cohen, 2007). The regulation of *miR-8* is via the EcR/Usp targets 74ef and Broad (Jin et al., 2012). *miR-8* regulates Ush to control body size, while *miR-14* feeds back to regulate EcR to modulate 20E signaling (Jin et al., 2012; Varghese & Cohen, 2007). *miR-34* is also activated by JH and prevents neurodegeneration in the adult brain by repressing *Eip74ef* (Liu et al., 2012; Sempere et al., 2003). (For color version of this figure, the reader is referred to the online version of this chapter.)

of this release is determined by a complex interplay between neuropeptide and IIS signaling pathways and occurs in response to environmental cues like nutrition availability. Circulating Ecdysone is converted to 20E in peripheral tissues, and the active hormone then binds to a nuclear hormone receptor. This receptor in an obligate heterodimer composed of two subunits, EcR and ultraspiracle (Usp). A 20E-bound form of this receptor directly regulates the transcription of target genes, triggering changes in gene expression that lead to the whole-scale transformation of the larvae into the adult during metamorphosis.

JH is a lipid like hormone that functions together with 20E to regulate developmental timing and possibly adult processes as well. JH primarily functions like an anti-metamorphic hormone: presence of JH ensures that the periodic pulses of 20E produce another immature larvae. When the larvae have attained a critical size in the last larval stage, loss of JH coupled with a rise in JH metabolism, leads to 20E mediated metamorphosis (for review Jindra et al., 2013). Other processes affected by JH include behavior, reproduction, diapause, stress resistance, and aging (Flatt, Tu, & Tatar, 2005). Unlike 20E, much less is known regarding the mechanism of JH signaling as the receptors involved in JH mediated signaling have until recently eluded identification (Jindra et al., 2013).

4.2. Identification of Ecdysone- and JH-regulated miRNAs

20E and JH promote developmental transitions by activating pathways that lead to global changes in gene expression, including the expression of particular miRNAs as indicated by developmental profiling of miRNAs. These include *miR-100*, *let-7*, and *miR-125*, whose expression is upregulated at the larval-to-pupal transition, as well as *miR-8* and *miR-34*, whose expression is downregulated during the larval-to-pupal transition (Aravin et al., 2003; Jin, Kim, & Hyun, 2012; Leaman et al., 2005; Sempere, Sokol, Dubrovsky, Berger, & Ambros, 2003). Treatment of cultured *Drosophila* cells with purified 20E triggers the expression of *miR-100*, *let-7*, and *miR-125* and represses the expression of *miR-8* and *miR-34* along with an additional miRNA, *miR-14* (Bashirullah et al., 2003; Garbuzov & Tatar, 2010; Jin et al., 2012; Sempere, Dubrovsky, Dubrovskaya, Berger, & Ambros, 2002; Sempere et al., 2003; Varghese & Cohen, 2007). While *miR-34* is repressed by 20E treatment, it is activated by the JH analog, methoprene, though this increase in *miR-34* expression is completely suppressed by simultaneous treatment of S2 cells with 20E. Moreover, *miR-34* expression increases in mutant larvae containing mutations in either *Ecdysoneless* (*ecd*), a gene required for Ecdysone biosynthesis, or the 20E primary response gene *broad* (Sempere et al., 2003). Thus, 20E and JH pathways appear to have opposite effects on expression of *miR-34*, and interplay of these two signaling pathways may be required for the temporal downregulation of *miR-34* during metamorphosis and its subsequent upregulation in adult flies. Taken together, this data indicates that endocrine signaling pathways activate one class of miRNAs and repress another, and that these miRNAs may play important roles in promoting progression through developmental transitions.

4.3. Molecular analysis of Ecdysone- and JH-control of *let-7-C*, *miR-34*, *miR-8*, and *miR-14* miRNAs

A key question has been the identity of the transcriptional regulators of these miRNAs. Ecdysone induction of *miR-100*, *let-7*, and *miR-125* suggested that these miRNAs might be regulated either by the 20E receptor itself or by one of its three well-established direct targets, *Ecdysone-induced protein 74EF* (*Eip74ef*), *Ecdysone-induced protein 75B* (*Eip75b*), and *broad* (*br*). The co-regulation of these miRNAs is not surprising, since these miRNAs are co-transcribed from a single locus termed the *let-7-Complex* (*let-7-C*) (Sokol, Xu, Jan, & Ambros, 2008). Two pieces of additional data, however, suggested a potentially more complex relationship between Ecdysone and *let-7-C* miRNAs (Bashirullah et al., 2003). First, expression of mature processed *let-7-C* miRNAs is delayed relative to known direct targets, suggesting that the *let-7-C* locus is not a direct target of the 20E receptor. Second, silencing of *EcR* using a heat-shock inducible RNAi transgene during late larval development did not affect *let-7-C* miRNA expression *in vivo*, even though the expression of known direct targets was severely reduced. These data raised the possibility that the Ecdysone responsiveness of the *let-7-C* locus was independent of 20E and the 20E receptor.

Chawla and Sokol, however, provide clear supporting evidence that the *let-7-C* locus is a direct target of the 20E receptor (Chawla & Sokol, 2012). Consistent with this notion, the primary *let-7-C* transcript is detected within 30 min of 20E treatment of cultured cells. The delayed expression of mature *miR-100*, *let-7*, and *miR-125* observed in previous experiments is likely due to their post-transcriptional processing. The 20E responsiveness of the *let-7-C* locus is mediated by three conserved sequence motifs that matched the known 20E responsive elements site and that are bound by recombinant EcR and Usp heterodimers *in vitro*. In addition, these same sequences are responsible for *let-7-C* expression *in vivo*, since rescuing transgenes in which they were deleted displayed reduced nervous system expression and rescuing activity. Taken together, this data indicates the 20E-receptor directly activate the *let-7-C* locus during development. The observation by Bashirullah et al. that *let-7-C* miRNAs are activated even when *EcR* is knocked down by RNAi may indicate that these miRNAs are responsive to low levels of EcR, since *EcR* RNAi is unlikely to completely eliminate EcR (Bashirullah et al., 2003). *let-7-C* expression and activity are not completely eliminated in transgenes in which the 20E-receptor binding sites are deleted, however, indicating that the *let-7-C* locus is independently regulated by additional unknown factors that await identification.

While activating *let-7-C* miRNAs on the one hand, Ecdysone signaling also represses *miR-8*, *miR-14*, and *miR-34* on the other. Repression of *miR-8* and *miR-14* occurs at the transcriptional level, since Ecdysone treatment of cultured cells and tissues leads to a decrease in the levels of *pri-miR-8* as well as *pri-miR-14* (Jin et al., 2012; Varghese & Cohen, 2007). Endogenous EcR activity is required for this repression and, at least in the case of *miR-8*, acts through its direct transcriptional targets, *Eip74ef* and *br*. A *miR-8* enhancer fragment containing multiple predicted *Eip74ef* and *br* binding sites mediates repression of a reporter gene, and mutational analysis indicates that two of these *Eip74ef* binding sites are at least partially responsible (Jin et al., 2012). *miR-8* and *miR-14* regulate components of the IIS pathway, offering a direct molecular connection between the Ecdysone and IIS pathways as discussed further below. In addition, *miR-14* feeds back to directly repress the expression of EcR. The EcR/*miR-14* circuit provides a mechanism for amplifying EcR expression in response to Ecdysone. The molecular basis for Ecdysone regulation of *miR-34* is less clear. The locus encoding *miR-34* is likely to be complex, since profiling arrays indicate that *miR-34* is co-transcribed with two other miRNAs, *miR-277* and *miR-317* (Kadener, Rodriguez, et al., 2009). Determining whether Ecdysone affects *miR-277* and *miR-317* levels should shed light on whether the hormonal regulation of *miR-34* occurs at the transcriptional or post-transcriptional levels.

4.4. Biological roles of Ecdysone regulated *let-7-C* miRNAs

Steroid hormone induction of *let-7* family miRNAs appears to be conserved feature of animal development, since it is shared by both flies and nematodes (for additional review, see Sokol, 2012). *C. elegans* stage-specific miRNAs are activated by the dafachronic acid steroid hormone to ensure that cells within lineages throughout the animal adopt stage specific cell fates (Bethke, Fielenbach, Wang, Mangelsdorf, & Antebi, 2009). Disruption of this pathway leads to clear temporal cell fate transformations that are apparent through the transparent cuticle of the adult. In miRNA mutants, for example, cells inappropriately reiterate mitotic programs of previous developmental stages and this frequently leads to inappropriate proliferation of cells that were scheduled to terminally differentiate. Genetic elimination of *Drosophila let-7-C* miRNAs has little effect on the morphology and timing of development at the organismal level: *let-7-C* mutant animals proceed through development on schedule and eclose with wildtype appearance (Caygill & Johnston, 2008; Sokol et al., 2008). But newly eclosed *let-7-C* mutant flies are unable to perform many basic adult behaviors, including walking, climbing, and flying, indicating that these miRNAs ensure the proper formation of internal adult structures during metamorphosis. 20E-receptor regulated expression of *let-7-C* is critical because a *let-7-C* rescuing transgene in which the 20E-receptor binding sites are missing is unable to rescue the adult behavioral defects (Chawla & Sokol, 2012). As summarized below, recent evidence indicates that *Drosophila let-7-C* miRNAs control temporal cell fate decisions during developmental transitions and suggests that defects in steroid hormone regulated temporal cell fate transitions lead to the behavioral phenotypes displayed by *let-7-C* mutant flies.

The temporal wave of *let-7-C* expression during the larval-to-pupal transition plays important roles in ensuring the cells adopt appropriate stage specific behaviors and morphologies. For example, genetic elimination of *let-7* and *miR-125* leads to a delay in cell-cycle exit in the pupal wing disc, where cells usually stop dividing early in metamorphosis (Caygill & Johnston, 2008). Thus, mutant adult wings contain more cells than control flies, although the relevant target in this case remains unknown.

In addition, *let-7-C* miRNAs ensure that both centrally located mushroom body (MB) neurons as well as peripherally located motoneurons adopt appropriate stage-specific morphologies by regulating a pair of *bric-a-brac/tramtrack/broad-complex* (BTB)-domain containing zinc finger (BTB-ZF) transcription factors. In the MBs, *let-7-C* miRNAs synchronize cell fate transitions amongst a set of eight neuronal linages by downregulating the BTB-ZF *chronologically inappropriate morphogenesis* (*chinmo*) (Wu, Chen, Mercer, & Sokol, 2012). The adult fly brain contains two MBs, which are composed of four distinct neuronal subtypes (γ, α'/β', pioneer α/β, and α/β neurons) that are generated by eight neural progenitor cells (Ito & Awasaki, 2008; Lee, Lee, & Luo, 1999; Zhu et al., 2006). These progenitors generate the γ, α'/β', pioneer α/β, and α/β subtypes sequentially during development, so that γ neurons are born during early larval stages while the α'/β', pioneer α/β, and α/β neurons are born at progressively later timepoints during the larval-to-pupal transition, respectively (reviewed in Lin & Lee, 2012). Although many genes are expressed in the MB and have effects on MB morphology, Chinmo is unique because its expression is gradually downregulated as transitions between γ, α'/β', pioneer α/β, and α/β production occur suggesting a mechanism for the ordered production of neuronal subtypes (Zhu et al., 2006). Indeed, genetic analysis indicates that *chinmo* has a dosage sensitive effect on MB cell fate, and that its post-transcriptional downregulation during the larval-to-pupal transition is responsible for the $\gamma \rightarrow \alpha'/\beta' \rightarrow$ pioneer $\alpha/\beta \rightarrow \alpha/\beta$ cell fate transitions

(Zhu et al., 2006). A mechanism for this post-transcriptional regulation was suggested by a previously unannotated *chinmo* 3'UTR that contains multiple binding sites for both *let-7* and *miR-125,* since *let-7-C* onset coincides with Chinmo offset and the $\alpha'/\beta' \rightarrow$ pioneer $\alpha/\beta \rightarrow \alpha/\beta$ cell fate transitions (Wu et al., 2012). These transitions are delayed in *let-7-C* mutants, leading to an increase in the number of pioneer α/β neurons. Elevated Chinmo is indeed responsible for the *let-7-C* mutant phenotype, since this increase in pioneer α/β neurons is suppressed by genetic reduction of *chinmo* levels. *let-7-C* miRNAs likely control Chinmo levels directly; Chinmo expression is eliminated in MBs of early larvae when *let-7-C* miRNAs are prematurely expressed, Chinmo expression inappropriately persists in *let-7-C* mutant pupae and adults, and the predicted *let-7* and *miR-125* sites in the *chinmo* 3'UTR mediate reporter repression in cultured cells. Other as yet unknown factors must also be responsible for the regulation of Chinmo expression since cell fate transitions and Chinmo downregulation are delayed but not eliminated in *let-7-C* mutants. Nevertheless, activation of *let-7-C* miRNAs during the larval-to-pupal transition shapes a temporal gradient of Chinmo in order to specify cell fates of a mulitpotent neuronal lineage over time.

Ecdysone is very likely involved in the activation of *let-7-C* miRNAs in the MB, where *let-7-C* miRNAs apparently also regulate the expression of a second BTB-ZF Abrupt to control MB morphology. *let-7* expression is generally dependent on Ecdysone in the adult brain, since the miRNA is virtually eliminated in *ecd* mutants (Kucherenko, Barth, Fiala, & Shcherbata, 2012). However, *let-7-C* control of temporal identity in the MB appears to be partially independent of the 20E-receptor, since the normal number of pioneer α/β neurons is found in *let-7-C* mutant adults that harbor a rescuing transgene devoid of known 20E-receptor binding sites (Yen-chi Wu and Nick Sokol, unpublished observations). Genetic depletion experiments indicate that the 20E-receptor plays important roles in controlling the remodeling and final morphology of γ and α/β lobes (Kucherenko et al., 2012; Lin, Huang, & Lee, 2009). Part of this function may be mediated by *let-7* regulation of the previously known *let-7*-specific target Abrupt, since genetic depletion of *abrupt* rescues a *let-7* dependent defect in α/β lobe morphology (Burgler & Macdonald, 2005; Kucherenko et al., 2012).

let-7-C regulation of *abrupt* also ensures that post-mitotic motoneurons adopt their appropriate developmental stage-specific morphologies. *let-7-C* mutants exhibit persistence into adulthood of pupal muscles and their corresponding innervations, and a failure to fully develop certain adult muscles and their respective innervations (Caygill & Johnston, 2008; Sokol et al., 2008). Genetic depletion of *abrupt* suppresses the *let-7* mutant phenotype, indicating that ectopic Abrupt contributes to the *let-7* phenotype (Caygill & Johnston, 2008). The finding that *let-7-C* adults display both persistent juvenile and immature adult characteristics indicates that the essential function of *let-7-C* as a regulator of developmental timing appears to be evolutionarily conserved from worms to flies, and provides a cellular basis for the behavioral defects displayed by *let-7-C* adults.

let-7-C miRNAs have also been found to regulate dopaminergic neuron and GSC behavior in adults, although it remains unclear whether hormonal control of *let-7-C* is important for these post-developmental functions. *let-7* mutants adults, for example, have a reduced number of dopaminergic neurons (Gehrke, Imai, Sokol, & Lu, 2010). This phenotype may indicate a role for *let-7* in maintaining this post-mitotic neuron subtype, since injection of *let-7* antisense RNA into the adult leads to a similar phenotype. *let-7* appears to mediate this role by repressing translation of the *dp transcription factor* (*dp*), since Dp expression is elevated in *let-7* mutants and expression of a *let-7* insensitive transgene also leads to dopaminergic neuron loss. *let-7* has been implicated in the age-dependent loss of GSCs in males (Toledano, D'Alterio, Czech, Levine, & Jones, 2012). The number of GSCs is controlled in part by their rate of self-renewal, which is modulated by the janus kinase/signal transducer and activator of transcription (JAK-STAT) pathway. *let-7* can indirectly affect the expression of the secreted activating ligand of this pathway by directly controlling the levels of the RNA-binding protein Igf-II mRNA-binding protein (Imp). Overexpression of *let-7* in the hub cells that surround the GSCs can repress Imp via a binding site in the *imp* 3'UTR, leading to a reduction in GSC number. Thus, an observed increase of *let-7* expression in the hub cells during adulthood could be responsible for the age-dependent loss of GSCs by regulating the JAK-STAT pathway.

4.5. *miR-14* and *miR-8*: molecular links between steroid and insulin signaling pathways

Coordination between the 20E and insulin signaling pathways regulates the rate of growth during development, and may also function post-developmentally to link nutritional status with adult processes like reproduction, behavior, and stress resistance (Schwedes & Carney, 2012; Tennessen & Thummel, 2011). Both *miR-14* and *miR-8* represent molecular connections between these pathways, since they are repressed by 20E and, in turn, repress genes involved in the insulin-signaling pathway (Hyun et al., 2009; Jin et al., 2012; Varghese & Cohen, 2007; Varghese et al., 2010). As mentioned above, genetic analysis has shown that *miR-14* promotes fat metabolism and developmental and adult viability by regulating *sugarbabe* and *EcR* expression, respectively (Varghese & Cohen, 2007; Varghese et al., 2010). Additional *miR-14* mutant phenotypes, including elevated apoptosis, could be due to the misregulation of both *sugarbabe* and *EcR* or may involve additional unidentified *miR-14* targets. While current evidence suggests that the 20E/*miR-14*/*sugarbabe* circuit may regulate ILPs to control adult processes, *miR-8* integrates 20E and insulin signaling to control body size during development (Jin et al., 2012). 20E signaling regulates pupal body size, since feeding larvae a 20E-supplemented diet decreases animal size while inactivating the 20E-receptor increases animal size (Colombani et al., 2003; Jin et al., 2012). Consistent with the negative regulation of *miR-8* by 20E, ectopic expression of *miR-8* also leads to increased body size. Genetic evidence indicates that 20E controls body size via *miR-8*, since genetic depletion of *miR-8* suppresses the effects of 20E-receptor knockdown. Likewise, the effects of ectopic 20E are genetically suppressed by reduction in the *miR-8* target and negative regulator of insulin signaling, *ush*. Thus, 20E represses *miR-8* levels to promote insulin signaling, thereby modulating organismal growth rate.

4.6. *miR-34*

Despite its intriguing developmental profile and hormonally regulated expression, a recent study showed that *D. melanogaster miR-34* plays an important neuroprotective role in adults by regulating the Ecdysone inducible developmental regulator, *Eip74ef*. This finding is consistent with previous genetics linking the miRNA pathway to long term brain integrity (for review Abe & Bonini, 2013). *miR-34* is a brain-enriched miRNA whose expression increases with age, possibly due to hormonal control (Liu et al., 2012). Loss of function mutants of *miR-34* display premature brain deterioration, early onset impaired motor behavior, altered chaperone accumulation, reduced lifespan, and susceptibility to stress. While loss of *miR-34* triggered a transcriptional program seen during accelerated brain aging, its upregulation extended median lifespan and mitigated neurodegeneration induced by human pathogenic polyglutamine disease protein (Liu et al., 2012). *miR-34* and *Eip74ef* exhibit mutually exclusive expression patterns in the aging adult brain, and age associated *mir-34* phenotypes are suppressed by an *Eip74ef* hypomorphic allele. Although *miR-34* expression has been linked to aging and disease in higher eukaryotes, the *miR-34–Eip74ef* relationship does not appear to be conserved (Aw & Cohen, 2012). Thus, other mechanisms and targets may be involved in mediating the age related *miR-34* function in these organisms. Determination of whether *miR-34* regulation of *Eip74ef* plays an important role during developmental transitions awaits further investigation.

5. LIGHT MODULATED CIRCADIAN RHYTHM PATHWAYS

The physiology and behavior of many organisms are under the control of circadian rhythm (Hall, 2003). In *D. melanogaster,* the internal clock receives external input and can be entrained by environmental stimuli (Dubruille & Emery, 2008). The intrinsic clock synchronizes a multitude of events, including adult eclosion time, rest/activity cycle, courtship, feeding, metabolism, and immune response (Hall, 2003). Studies using *D. melanogaster* have provided deep molecular insights on how the light/dark cycle feeds into the internal clock, how the intrinsic clock runs on an approximately 24h rhythm, and also how the circadian clock regulates the many physiological and behavioral outputs.

5.1. Molecular overview of circadian rhythm pathway

Transcriptional feedback loops are characteristic features of the circadian clock, which include the central loop (CL) and interlocking loop (IL). In the CL, expression of the heterodimeric activator complex composed of clock (Clk) and cycle (Cyc) is transcriptionally repressed by its targets, *period* (*per*) and *timeless* (*tim*) (Allada, White, So, Hall, & Rosbash, 1998; Hardin, Hall, & Rosbash, 1990; Rutila et al., 1998; Sehgal, Price, Man, & Young, 1994). The IL stabilizes the CL, and therefore the circadian clock. The components of the IL include the Par domain protein 1 (Pdp1), which activates *clk* transcription, Vrille (Vri), which represses *clk* transcription, and Clockwork Orange (Cwo), which represses Clk-mediated transcription (Hardin, 2005; Kadener, Stoleru, McDonald, Nawathean, & Rosbash, 2007). In addition to transcriptional control, post-translational modifications also regulate clock progression (Chiu, Ko, & Edery, 2011; Yu, Zheng, Houl, Dauwalder, & Hardin, 2006). While the core transcriptional feedback loops lead to rhythmic mRNA abundance, post-translational modification such as ubiquitylation and phosphorylation controls circadian protein abundance and sub-cellular location.

5.2. Identification of miRNAs in clock pathways

Post-transcriptional control also likely regulates the circadian clock, and recent analysis identified a set of miRNAs that may be involved (see Fig. 4.3). A microarray study, for example, detected two mature miRNAs *miR-263a* and *miR-263b* that display a circadian expression profile in adult heads (Yang, Lee, Padgett, & Edery, 2008). This study also identified miRNAs whose expression profile is altered in *cyc* mutants, including *miR-31a, miR-124, miR-133, miR-184, miR-210*, and *miR-276b*. In a separate study, tiling array analysis of adult heads similarly identified a set of pri-miRNA transcripts whose expression increases when Dicer processing is disrupted specifically in circadian neurons. These include primary transcripts encoding *bantam*, *let-7-Complex* miRNAs, *miR-8, miR-9a, miR-14, miR-12, miR-34, miR-184, miR-252, miR-263a, miR-274, miR-275, miR-276a, miR-277, miR-278, miR-279, miR-282, miR-283, miR-284, miR-285, miR-286a, miR-304, miR-305, miR-317, miR-986*, and *miR-996* (Kadener, Menet, et al., 2009). Disruption of additional miRNA processing components, like Drosha, leads to elevation of a subset of these, including *pri-bantam*. MiRNAs are present in circadian neurons and may regulate the stability or translation of core clock components *clk*, *vri*, and *cwo,* since these mRNAs are enriched in immunoprecipitations of the RISC component Ago-1 from fly heads harvested at specific times (Kadener, Menet, et al., 2009). Such identification of circadian miRNAs is consistent with computational simulations suggesting that miRNAs can alter both the amplitude and frequency/periodicity of circadian oscillations (Nandi, Vaz, Bhattacharya, & Ramaswamy, 2009). Genetic analyses have already illuminated a role for *bantam*, *miR-279*, and the *miR-959–964* cluster miRNAs, and we review this evidence below.

5.3. *bantam*

While *bantam* does not exhibit a circadian pattern of expression at the organismal level (Vodala et al., 2012), its expression may cycle in adult circadian neurons. A role for *bantam* within the central clock is suggested by the presence of a binding site for this miRNA in the 3'UTR of *clk* (Kadener, Menet, et al., 2009). Rescue experiments indicate the importance of this site, since it is required in a *clk* rescuing transgene to restore rhythmicity to an arrhythmic *clk* mutant. Consistent with a role in regulating Clk, overexpression of *bantam* leads to significantly longer period rhythms. Detailed analysis of bantam's clock function is challenging due to its essential function during development, and awaits its specific removal in adult circadian neurons.

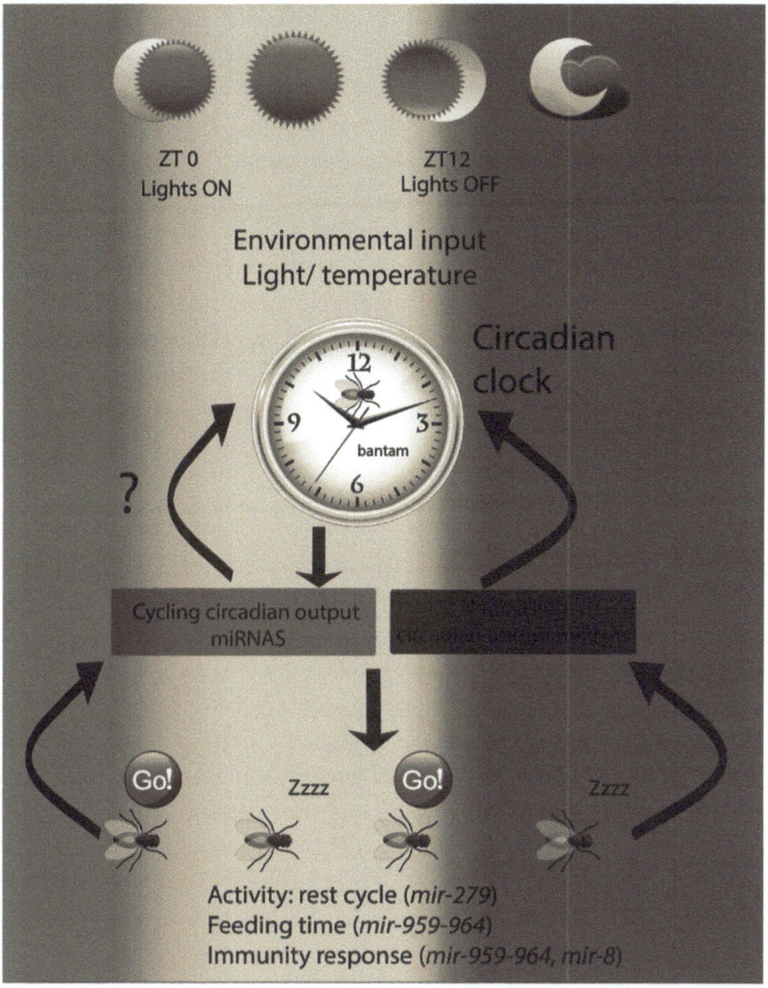

Figure 4.3 Environmental stimuli in the wild or in laboratory conditions entrain the *Drosophila* circadian clock. miRNAs function in both the core circadian clock and also in clock output pathway to regulate organismal behavior, including sleep/wake cycle, feeding, and foraging time and immune response to pathogenic infections. The miRNAs can be further divided into two sub-categories: (1) miRNAs with fluctuating expression levels and (2) miRNAs with constant daily levels. Recent studies provide evidence for animal behavior providing feedback to the expression of the miRNAs. (For color version of this figure, the reader is referred to the online version of this chapter.)

5.4. *miR-279*

miR-279 coordinates multiple functions in the output pathway of the core clock, synchronizing animal behavior during the daily temporal transitions (Luo & Sehgal, 2012). Although *miR-279* does not exhibit a circadian expression pattern, genetic elimination of *miR-279* leads to an arrhythmic activity cycle. RNAi knockdown studies of predicted *miR-279* targets identified a bona fide target, Unpaired (Upd), the JAK-STAT pathway secreted ligand. The *upd* 3'UTR contains *mir-279* responsive binding sites. Flies in which *upd* is knocked down also exhibit impaired locomotor rhythm, suggesting the requirement of the JAK-STAT pathway for normal rhythm. It turns out that altering the levels of other JAK-STAT pathway components (Hop, Dome) also disrupted locomotor rhythm. *MiR-279* is detected in neurons that can receive direct input from circadian neurons in their immediate proximity. These cells also express Upd, whose expression *miR-279* regulates. This modulation of Upd regulates the oscillating rest: activity cycle via circadian STAT92E output (Luo & Sehgal, 2012).

5.5. *miR-959–964 cluster*

The *miR-959–964 cluster* represents another example of miRNAs involved in the circadian output. MiRNAs encoded by this cluster have short half-lives and display a circadian expression that peaks at dusk (ZT12) and has its lowest level at dawn (ZT0) (Vodala et al., 2012; Yang et al., 2008). The *pri-miR-959–964* transcript also cycles, suggesting the involvement of the circadian transcription control on the locus. In fact, *clk* mutants abolished the circadian expression of the cluster miRNAs. Transcription of the cluster is also regulated by nutritional status, since feeding activates transcription of the cluster while starvation represses cluster transcription. Analysis of a mutant that eliminates four of the six miRNAs revealed many roles in circadian governed behavior, feeding and metabolism. Elimination of the cluster reduces the period by half an hour, while cluster over-expression mutants extend the period by an hour. Low levels of cluster miRNAs correlate with feeding time. Gradual increase in miRNA cluster expression prepares the animal for normal physiology post-feeding, until the cluster miRNAs begin to decay and reach their trough levels in the morning. The targets of these miRNAs remain to be determined.

6. STRESS INDUCED AND IMMUNE RESPONSE PATHWAYS

Innate immunity in flies arises from a number of immune-responsive tissues which are capable of sensing pathogen and mount a defensive reaction (Lemaitre & Hoffmann, 2007). The epithelial layer underneath the body cuticle, in the intestinal tract acts as a physical barrier against infection. This layer is also capable of generating different classes of antimicrobial peptides (AMPs) and reactive oxygen species (ROS) locally (Lemaitre & Hoffmann, 2007). The hemolymph contains immune cells that can phagocytose and isolate pathogens. Finally, the fat body is the center for humoral immunity, where a number of conserved immune signaling pathways, including the Toll pathway, Imd pathway, the JAK-STAT pathway, JNK, and p38 pathways are known to act (Lemaitre & Hoffmann, 2007). Here, we focus on the role of miRNAs as local or systemic factors in the *D. melanogaster* immune system.

6.1. *miR-8*

miR-8 is expressed in the fat body and has been shown to regulate the production of AMPs (Choi & Hyun, 2012). In *miR-8* null flies, the basal level of antifungal peptide Drosomycin and antimicrobial peptide Diptericin, which is effective against Gram-negative bacteria, are increased in non-pathogen challenged flies. The increased Drosomycin levels are observed specifically in the fat body. Furthermore, *miR-8* flies exhibit a higher occurrence of melanization, an indicator of auto-immune reaction. Genetic rescue experiments introducing *miR-8* back in the fat body of null flies reduced the elevated AMP levels, and also the incidence of melanization. Thus, *miR-8* is necessary to function to keep the basal immune activity low in non-pathogen challenged flies, although the role of *miR-8* during pathogenic infection remains unclear.

6.2. *miR-959–964 cluster*

In addition to their roles in synchronizing circadian behaviors including feeding and metabolism, members of the *miR-959–964 cluster* also ensure circadian control of immunity (Vodala et al., 2012). The *miR-959–964* cluster is expressed in the pericerebral fat body, a site implicated in local immune response (Lee & Edery, 2008), and the levels of many mRNAs involved in immune functions are altered in the mutant. Cluster mutants display a temporally shifted survival curve when infected with *Pseudomonas aeruginosa* (Vodala et al., 2012). Physiologically relevant targets of this cluster of miRNAs remain unidentified, but may include the metallopeptidase *ance-4* and the antifungal AMP *drosomycin* since these were identified as responsive targets of the *miR-959–964 Cluster* in cell culture reporter experiments (Vodala et al., 2012).

6.3. Bioinformatics predicts additional miRNAs involved in immune response

A bioinformatic based screen predicted seven additional miRNAs to be involved in the immune system (Fullaondo & Lee, 2012), although verification of their involvement is lacking due to the absence of genetic mutants. The implicated miRNAs include *miR-12*, *miR-31b*, *miR-33*, *miR-283*, *miR-304*, *miR-1003*, and *miR-1016*. These miRNAs were identified because they are predicted to target immunity-related mRNAs, and are expressed in immune related cells or organs. Although this screen initially identified *mir-8* due to its potential targeting of GNBP3 (Gram Negative Binding Protein 3), Pvf1 and Pvf2, it was ultimately excluded from the "immune miRNA" list is due to the absence of information regarding its expression pattern.

7. SUMMARY

Much progress has been made in the recent past to illuminate the functions of miRNAs within systemic signaling pathways, such as insulin, steroid hormone, stress, immune, and circadian pathways. During both development and adulthood, these miRNA components function to coordinate cell growth, metabolism, fate, and morphology with changing environmental conditions. In so doing, these miRNAs ensure that the animal undergoes appropriate developmental and post-developmental transitions in response to environmental cues, and is consistent with the possibility that adult physiology and metabolism involve the redeployment of pathways that controlled cell fate progression during development. Future work in flies and other animals will determine how faithful this redeployment is, and whether the same miRNA/mRNA regulatory relationships persist from development through adulthood.

REFERENCES

Abe, M., & Bonini, N. M. (2013). MicroRNAs and neurodegeneration: Role and impact. *Trends in Cell Biology*, *23*(1), 30–36. http://dx.doi.org/10.1016/j.tcb.2012.08.013.

Allada, R., White, N. E., So, W. V., Hall, J. C., & Rosbash, M. (1998). A mutant Drosophila homolog of mammalian Clock disrupts circadian rhythms and transcription of period and timeless. *Cell*, *93*(5), 791–804.

Aravin, A. A., Lagos-Quintana, M., Yalcin, A., Zavolan, M., Marks, D., Snyder, B., et al. (2003). The small RNA profile during *Drosophila melanogaster* development. *Developmental Cell*, *5*(2), 337–350.

Aw, S., & Cohen, S. M. (2012). Time is of the essence: MicroRNAs and age-associated neurodegeneration. *Cell Research*, *22*(8), 1218–1220. http://dx.doi.org/10.1038/cr.2012.59.

Bashirullah, A., Pasquinelli, A. E., Kiger, A. A., Perrimon, N., Ruvkun, G., & Thummel, C. S. (2003). Coordinate regulation of small temporal RNAs at the onset of *Drosophila metamorphosis*. *Developmental Biology*, *259*(1), 1–8.

Bethke, A., Fielenbach, N., Wang, Z., Mangelsdorf, D. J., & Antebi, A. (2009). Nuclear hormone receptor regulation of microRNAs controls developmental progression. *Science*, *324*(5923), 95–98. http://dx.doi.org/10.1126/science.1164899.

Broughton, S. J., Piper, M. D., Ikeya, T., Bass, T. M., Jacobson, J., Driege, Y., et al. (2005). Longer lifespan, altered metabolism, and stress resistance in Drosophila from ablation of cells making insulin-like ligands. *Proceedings of the National Academy of Sciences of the United States of America*, *102*(8), 3105–3110. http://dx.doi.org/10.1073/pnas.0405775102.

Burgler, C., & Macdonald, P. M. (2005). Prediction and verification of microRNA targets by MovingTargets, a highly adaptable prediction method. *BMC Genomics*, *6*, 88. http://dx.doi.org/10.1186/1471-2164-6-88.

Caygill, E. E., & Johnston, L. A. (2008). Temporal regulation of metamorphic processes in Drosophila by the let-7 and miR-125 heterochronic microRNAs. *Current Biology*, *18*(13), 943–950. http://dx.doi.org/10.1016/j.cub.2008.06.020.

Chawla, G., & Sokol, N. S. (2012). Hormonal activation of let-7-C microRNAs via EcR is required for adult *Drosophila melanogaster* morphology and function. *Development*, *139*(10), 1788–1797.

Chiu, J. C., Ko, H. W., & Edery, I. (2011). NEMO/NLK phosphorylates PERIOD to initiate a time-delay phosphorylation circuit that sets circadian clock speed. *Cell*, *145*(3), 357–370. http://dx.doi.org/10.1016/j.cell.2011.04.002.

Choi, I. K., & Hyun, S. (2012). Conserved microRNA miR-8 in fat body regulates innate immune homeostasis in Drosophila. *Developmental and Comparative Immunology*, *37*(1), 50–54. http://dx.doi.org/10.1016/j.dci.2011.12.008.

Clancy, D. J., Gems, D., Harshman, L. G., Oldham, S., Stocker, H., Hafen, E., et al. (2001). Extension of life-span by loss of CHICO, a Drosophila insulin receptor substrate protein. *Science*, *292*(5514), 104–106. http://dx.doi.org/10.1126/science.1057991.

Cohen, S. (2010). MicroRNAs in animal development. Editorial. *Seminars in Cell &Developmental Biology*, *21*(7), 727. http://dx.doi.org/10.1016/j.semcdb.2010.07.004.

Colombani, J., Raisin, S., Pantalacci, S., Radimerski, T., Montagne, J., & Leopold, P. (2003). A nutrient sensor mechanism controls Drosophila growth. *Cell*, *114*(6), 739–749.

Davalos, A., Goedeke, L., Smibert, P., Ramirez, C. M., Warrier, N. P., Andreo, U., et al. (2011). miR-33a/b contribute to the regulation of fatty acid metabolism and insulin signaling. *Proceedings of the National Academy of Sciences of the United States of America*, *108*(22), 9232–9237. http://dx.doi.org/10.1073/pnas.1102281108.

Dubruille, R., & Emery, P. (2008). A plastic clock: How circadian rhythms respond to environmental cues in Drosophila. *Molecular Neurobiology*, *38*(2), 129–145. http://dx.doi.org/10.1007/s12035-008-8035-y.

Flatt, T., Tu, M. P., & Tatar, M. (2005). Hormonal pleiotropy and the juvenile hormone regulation of Drosophila development and life history. *Bioessays*, *27*(10), 999–1010. http://dx.doi.org/10.1002/bies.20290.

Fullaondo, A., & Lee, S. Y. (2012). Identification of putative miRNA involved in *Drosophila melanogaster* immune response. *Developmental and Comparative Immunology*, *36*(2), 267–273. http://dx.doi.org/10.1016/j.dci.2011.03.034.

Garbuzov, A., & Tatar, M. (2010). Hormonal regulation of Drosophila microRNA let-7 and miR-125 that target innate immunity. *Fly*, *4*(4), 306–311.

Gehrke, S., Imai, Y., Sokol, N., & Lu, B. (2010). Pathogenic LRRK2 negatively regulates microRNA-mediated translational repression. *Nature*, *466*(7306), 637–641. http://dx.doi.org/10.1038/nature09191.

Geminard, C., Rulifson, E. J., & Leopold, P. (2009). Remote control of insulin secretion by fat cells in Drosophila. *Cell Metabolism*, *10*(3), 199–207. http://dx.doi.org/10.1016/j.cmet.2009.08.002.

Hall, J. C. (2003). Genetics and molecular biology of rhythms in Drosophila and other insects. *Advances in Genetics*, *48*, 1–280.
Hardin, P. E. (2005). The circadian timekeeping system of Drosophila. *Current Biology*, *15*(17), R714–R722. http://dx.doi.org/10.1016/j.cub.2005.08.019.
Hardin, P. E., Hall, J. C., & Rosbash, M. (1990). Feedback of the Drosophila period gene product on circadian cycling of its messenger RNA levels. *Nature*, *343*(6258), 536–540. http://dx.doi.org/10.1038/343536a0.
Hatfield, S. D., Shcherbata, H. R., Fischer, K. A., Nakahara, K., Carthew, R. W., & Ruohola-Baker, H. (2005). Stem cell division is regulated by the microRNA pathway. *Nature*, *435*(7044), 974–978. http://dx.doi.org/10.1038/nature03816.
Hay, N., & Sonenberg, N. (2004). Upstream and downstream of mTOR. *Genes & Development*, *18*(16), 1926–1945. http://dx.doi.org/10.1101/gad.1212704.
Holzenberger, M., Dupont, J., Ducos, B., Leneuve, P., Geloen, A., Even, P. C., et al. (2003). IGF-1 receptor regulates lifespan and resistance to oxidative stress in mice. *Nature*, *421*(6919), 182–187. http://dx.doi.org/10.1038/nature01298.
Hyun, S., Lee, J. H., Jin, H., Nam, J., Namkoong, B., Lee, G., et al. (2009). Conserved MicroRNA miR-8/miR-200 and its target USH/FOG2 control growth by regulating PI3K. *Cell*, *139*(6), 1096–1108. http://dx.doi.org/10.1016/j.cell.2009.11.020.
Ikeya, T., Galic, M., Belawat, P., Nairz, K., & Hafen, E. (2002). Nutrient-dependent expression of insulin-like peptides from neuroendocrine cells in the CNS contributes to growth regulation in Drosophila. *Current Biology*, *12*(15), 1293–1300.
Ito, K., & Awasaki, T. (2008). Clonal unit architecture of the adult fly brain. *Advances in Experimental Medicine and Biology*, *628*, 137–158. http://dx.doi.org/10.1007/978-0-387-78261-4_9.
Jin, H., Kim, V. N., & Hyun, S. (2012). Conserved microRNA miR-8 controls body size in response to steroid signaling in Drosophila. *Genes & Development*, *26*(13), 1427–1432. http://dx.doi.org/10.1101/gad.192872.112.
Jindra, M., Palli, S. R., & Riddiford, L. M. (2013). The juvenile hormone signaling pathway in insect development. *Annual Review of Entomology*, *58*, 181–204. http://dx.doi.org/10.1146/annurev-ento-120811-153700.
Junger, M. A., Rintelen, F., Stocker, H., Wasserman, J. D., Vegh, M., Radimerski, T., et al. (2003). The Drosophila forkhead transcription factor FOXO mediates the reduction in cell number associated with reduced insulin signaling. *Journal of Biology*, *2*(3), 20. http://dx.doi.org/10.1186/1475-4924-2-20.
Kadener, S., Menet, J. S., Sugino, K., Horwich, M. D., Weissbein, U., Nawathean, P., et al. (2009). A role for microRNAs in the Drosophila circadian clock. *Genes & Development*, *23*(18), 2179–2191. http://dx.doi.org/10.1101/gad.1819509.
Kadener, S., Rodriguez, J., Abruzzi, K. C., Khodor, Y. L., Sugino, K., Marr, M. T., 2nd., et al. (2009). Genome-wide identification of targets of the drosha-pasha/DGCR8 complex. *RNA Journal*, *15*(4), 537–545. http://dx.doi.org/10.1261/rna.1319309.
Kadener, S., Stoleru, D., McDonald, M., Nawathean, P., & Rosbash, M. (2007). Clockwork Orange is a transcriptional repressor and a new Drosophila circadian pacemaker component. *Genes & Development*, *21*(13), 1675–1686. http://dx.doi.org/10.1101/gad.1552607.
Krol, J., Loedige, I., & Filipowicz, W. (2010). The widespread regulation of microRNA biogenesis, function and decay. *Nature Reviews Genetics*, *11*(9), 597–610. http://dx.doi.org/10.1038/nrg2843.
Krutzfeldt, J., & Stoffel, M. (2006). MicroRNAs: A new class of regulatory genes affecting metabolism. *Cell Metabolism*, *4*(1), 9–12. http://dx.doi.org/10.1016/j.cmet.2006.05.009.
Kucherenko, M. M., Barth, J., Fiala, A., & Shcherbata, H. R. (2012). Steroid-induced microRNA let-7 acts as a spatio-temporal code for neuronal cell fate in the developing

Drosophila brain. *EMBO Journal*, *31*(24), 4511–4523. http://dx.doi.org/10.1038/emboj.2012.298.

Leaman, D., Chen, P. Y., Fak, J., Yalcin, A., Pearce, M., Unnerstall, U., et al. (2005). Antisense-mediated depletion reveals essential and specific functions of microRNAs in Drosophila development. *Cell*, *121*(7), 1097–1108. http://dx.doi.org/10.1016/j.cell.2005.04.016.

Lee, J. E., & Edery, I. (2008). Circadian regulation in the ability of Drosophila to combat pathogenic infections. *Current Biology*, *18*(3), 195–199. http://dx.doi.org/10.1016/j.cub.2007.12.054.

Lee, T., Lee, A., & Luo, L. (1999). Development of the Drosophila mushroom bodies: Sequential generation of three distinct types of neurons from a neuroblast. *Development*, *126*(18), 4065–4076.

Lemaitre, B., & Hoffmann, J. (2007). The host defense of *Drosophila melanogaster*. *Annual Review of Immunology*, *25*, 697–743. http://dx.doi.org/10.1146/annurev.immunol.25.022106.141615.

Lin, S., Huang, Y., & Lee, T. (2009). Nuclear receptor unfulfilled regulates axonal guidance and cell identity of Drosophila mushroom body neurons. *PLoS One*, *4*(12), e8392. http://dx.doi.org/10.1371/journal.pone.0008392.

Lin, S., & Lee, T. (2012). Generating neuronal diversity in the Drosophila central nervous system. *Developmental Dynamics*, *241*(1), 57–68. http://dx.doi.org/10.1002/dvdy.22739.

Liu, N., Landreh, M., Cao, K., Abe, M., Hendriks, G. J., Kennerdell, J. R., et al. (2012). The microRNA miR-34 modulates ageing and neurodegeneration in Drosophila. *Nature*, *482*(7386), 519–523. http://dx.doi.org/10.1038/nature10810.

Luo, W., & Sehgal, A. (2012). Regulation of circadian behavioral output via a MicroRNA-JAK/STAT circuit. *Cell*, *148*(4), 765–779. http://dx.doi.org/10.1016/j.cell.2011.12.024.

Mendell, J. T., & Olson, E. N. (2012). MicroRNAs in stress signaling and human disease. *Cell*, *148*(6), 1172–1187. http://dx.doi.org/10.1016/j.cell.2012.02.005.

Nandi, A., Vaz, C., Bhattacharya, A., & Ramaswamy, R. (2009). miRNA-regulated dynamics in circadian oscillator models. *BMC Systems Biology*, *3*, 45–47. http://dx.doi.org/10.1186/1752-0509-3-45.

Partridge, L., Alic, N., Bjedov, I., & Piper, M. D. (2011). Ageing in Drosophila: The role of the insulin/Igf and TOR signalling network. *Experimental Gerontology*, *46*(5), 376–381. http://dx.doi.org/10.1016/j.exger.2010.09.003.

Puig, O., Marr, M. T., Ruhf, M. L., & Tjian, R. (2003). Control of cell number by Drosophila FOXO: Downstream and feedback regulation of the insulin receptor pathway. *Genes & Development*, *17*(16), 2006–2020. http://dx.doi.org/10.1101/gad.1098703.

Rulifson, E. J., Kim, S. K., & Nusse, R. (2002). Ablation of insulin-producing neurons in flies: Growth and diabetic phenotypes. *Science*, *296*(5570), 1118–1120. http://dx.doi.org/10.1126/science.1070058.

Rutila, J. E., Suri, V., Le, M., So, W. V., Rosbash, M., & Hall, J. C. (1998). CYCLE is a second bHLH-PAS clock protein essential for circadian rhythmicity and transcription of Drosophila period and timeless. *Cell*, *93*(5), 805–814.

Saltiel, A. R., & Kahn, C. R. (2001). Insulin signalling and the regulation of glucose and lipid metabolism. *Nature*, *414*(6865), 799–806. http://dx.doi.org/10.1038/414799a.

Schwedes, C. C., & Carney, G. E. (2012). Ecdysone signaling in adult *Drosophila melanogaster*. *Journal of Insect Physiology*, *58*(3), 293–302. http://dx.doi.org/10.1016/j.jinsphys.2012.01.013.

Sehgal, A., Price, J. L., Man, B., & Young, M. W. (1994). Loss of circadian behavioral rhythms and per RNA oscillations in the Drosophila mutant timeless. *Science*, *263*(5153), 1603–1606.

Sempere, L. F., Dubrovsky, E. B., Dubrovskaya, V. A., Berger, E. M., & Ambros, V. (2002). The expression of the let-7 small regulatory RNA is controlled by ecdysone during metamorphosis in *Drosophila melanogaster*. *Developmental Biology*, *244*(1), 170–179. http://dx.doi.org/10.1006/dbio.2002.0594.

Sempere, L. F., Sokol, N. S., Dubrovsky, E. B., Berger, E. M., & Ambros, V. (2003). Temporal regulation of microRNA expression in *Drosophila melanogaster* mediated by hormonal signals and broad-Complex gene activity. *Developmental Biology*, *259*(1), 9–18.

Siomi, H., & Siomi, M. C. (2010). Posttranscriptional regulation of microRNA biogenesis in animals. *Molecular Cell*, *38*(3), 323–332. http://dx.doi.org/10.1016/j.molcel.2010.03.013.

Sokol, N. S. (2012). Small temporal RNAs in animal development. *Current Opinion in Genetics and Development*, *22*(4), 368–373. http://dx.doi.org/10.1016/j.gde.2012.04.001.

Sokol, N. S., & Ambros, V. (2005). Mesodermally expressed Drosophila microRNA-1 is regulated by Twist and is required in muscles during larval growth. *Genes & Development*, *19*(19), 2343–2354. http://dx.doi.org/10.1101/gad.1356105.

Sokol, N. S., Xu, P., Jan, Y. N., & Ambros, V. (2008). Drosophila let-7 microRNA is required for remodeling of the neuromusculature during metamorphosis. *Genes & Development*, *22*(12), 1591–1596. http://dx.doi.org/10.1101/gad.1671708.

Teleman, A. A. (2010). Molecular mechanisms of metabolic regulation by insulin in Drosophila. *Biochemical Journal*, *425*(1), 13–26. http://dx.doi.org/10.1042/BJ20091181.

Teleman, A. A., Maitra, S., & Cohen, S. M. (2006). Drosophila lacking microRNA miR-278 are defective in energy homeostasis. *Genes & Development*, *20*(4), 417–422. http://dx.doi.org/10.1101/gad.374406.

Tennessen, J. M., & Thummel, C. S. (2011). Coordinating growth and maturation—Insights from Drosophila. *Current Biology*, *21*(18), R750–R757. http://dx.doi.org/10.1016/j.cub.2011.06.033.

Toledano, H., D'Alterio, C., Czech, B., Levine, E., & Jones, D. L. (2012). The let-7-Imp axis regulates ageing of the Drosophila testis stem-cell niche. *Nature*, *485*(7400), 605–610. http://dx.doi.org/10.1038/nature11061.

Varghese, J., & Cohen, S. M. (2007). microRNA miR-14 acts to modulate a positive autoregulatory loop controlling steroid hormone signaling in Drosophila. *Genes & Development*, *21*(18), 2277–2282. http://dx.doi.org/10.1101/gad.439807.

Varghese, J., Lim, S. F., & Cohen, S. M. (2010). Drosophila miR-14 regulates insulin production and metabolism through its target, sugarbabe. *Genes & Development*, *24*(24), 2748–2753. http://dx.doi.org/10.1101/gad.1995910.

Vodala, S., Pescatore, S., Rodriguez, J., Buescher, M., Chen, Y. W., Weng, R., et al. (2012). The oscillating miRNA 959–964 cluster impacts drosophila feeding time and other circadian outputs. *Cell Metabolism*, *16*(5), 601–612. http://dx.doi.org/10.1016/j.cmet.2012.10.002.

Wu, Y. C., Chen, C. H., Mercer, A., & Sokol, N. S. (2012). Let-7-complex microRNAs regulate the temporal identity of Drosophila mushroom body neurons via chinmo. *Developmental Cell*, *23*(1), 202–209. http://dx.doi.org/10.1016/j.devcel.2012.05.013.

Xu, P., Vernooy, S. Y., Guo, M., & Hay, B. A. (2003). The Drosophila microRNA Mir-14 suppresses cell death and is required for normal fat metabolism. *Current Biology*, *13*(9), 790–795.

Yang, M., Lee, J. E., Padgett, R. W., & Edery, I. (2008). Circadian regulation of a limited set of conserved microRNAs in Drosophila. *BMC Genomics*, *9*, 83. http://dx.doi.org/10.1186/1471-2164-9-83.

Yu, J. Y., Reynolds, S. H., Hatfield, S. D., Shcherbata, H. R., Fischer, K. A., Ward, E. J., et al. (2009). Dicer-1-dependent Dacapo suppression acts downstream of Insulin receptor in regulating cell division of Drosophila germline stem cells. *Development*, *136*(9), 1497–1507. http://dx.doi.org/10.1242/dev.025999.

Yu, W., Zheng, H., Houl, J. H., Dauwalder, B., & Hardin, P. E. (2006). PER-dependent rhythms in CLK phosphorylation and E-box binding regulate circadian transcription. *Genes & Development*, *20*(6), 723–733. http://dx.doi.org/10.1101/gad.1404406.

Zhu, S., Lin, S., Kao, C. F., Awasaki, T., Chiang, A. S., & Lee, T. (2006). Gradients of the Drosophila Chinmo BTB-zinc finger protein govern neuronal temporal identity. *Cell*, *127*(2), 409–422. http://dx.doi.org/10.1016/j.cell.2006.08.045.

CHAPTER FIVE

Vegetative Phase Change and Shoot Maturation in Plants

R. Scott Poethig[1]
Department of Biology, University of Pennsylvania, Philadelphia, Pennsylvania, USA
[1]Corresponding author: e-mail address: spoethig@sas.upenn.edu

Contents

1. Introduction 126
2. Terminology 126
3. Heteroblasty and Vegetative Phase Change 127
4. miR156 and miR157: Master Regulators of Vegetative Phase Change 129
5. miR156 Targets 130
6. Molecular Insights into the Phenomenology of Vegetative Phase Change 134
   6.1 How are vegetative phases specified? 134
   6.2 The relationship between vegetative and reproductive maturation 135
   6.3 How is timing of vegetative phase change regulated? 139
7. Conclusion 144
Acknowledgments 144
References 144

Abstract

As a plant shoot develops, it produces different types of leaves, buds, and internodes, and eventually acquires the capacity to produce structures involved in sexual reproduction. Morphological and anatomical traits that change in coordinated fashion at a predictable time in vegetative development allow this process to be divided into several more-or-less discrete phases; the transition between these phases is termed "vegetative phase change." Vegetative phase change is regulated by a decrease in the expression of the related microRNAs, miR156, and miR157, which act by repressing the expression of squamosa promoter binding protein/SBP-like (SBP/SPL) transcription factors. SBP/SPL proteins regulate a wide variety of processes in shoot development, including flowering time and inflorescence development. Answers to long-standing questions about the relationship between vegetative and reproductive maturation have come from genetic analyses of the transcriptional and posttranscriptional regulatory networks in which these proteins are involved. Studies conducted over several decades indicate that carbohydrates have a significant effect on phase-specific leaf traits, and recent research suggests that sugar may be the leaf signal that promotes vegetative phase change.

Current Topics in Developmental Biology, Volume 105
ISSN 0070-2153
http://dx.doi.org/10.1016/B978-0-12-396968-2.00005-1

1. INTRODUCTION

During its postembryonic development, a plant becomes established as a young seedling, increases in size and complexity, undergoes sexual reproduction, and eventually senesces and dies. This process involves gradual quantitative changes, as well as more rapid qualitative changes that occur at particular times in shoot development. The most obvious and best understood of these transitions is the switch from vegetative to reproductive development, which is accompanied by the production of novel reproductive structures, such as flowers or cones (Amasino, 2010; Andres & Coupland, 2012; Huijser & Schmid, 2011; Wilkie, Sedgley, & Olesen, 2008). This transition is preceded by a change in the competence of the shoot to respond to stimuli that induce reproductive development, and by changes in a variety of other traits, including leaf and stem morphology, growth rate and orientation, branching patterns, and disease or herbivore resistance. Variation in these latter traits has been described in many different species, starting with the observations of Goethe and Knight in the eighteenth century (Goethe, 1790; Knight, 1795). But—in contrast to the vegetative-to-reproductive transition—the molecular mechanism of these vegetative changes is still largely unknown. Here, the author presents some of the major questions about this process that remain to be answered, focusing on the phenomenon of vegetative phase change in herbaceous plants. The extensive literature on this topic in vines and woody plants has been discussed in several excellent reviews (Day, Greenwood, & Diaz-Sala, 2002; Doorenbos, 1965; Greenwood, 1995; Hackett, 1985; Lee & Richards, 1991; Schaffalitzky de Muckadell, 1954; Zotz, Wilhelm, & Becker, 2011), and readers should consult these for other perspectives on this problem.

2. TERMINOLOGY

Hildebrand (1875) and Goebel (1889) were the first to recognize that shoot development can be divided into juvenile and adult stages on the basis of species-specific vegetative traits such as leaf shape, the orientation of branch growth, and the capacity for sexual reproduction. Goebel noted that the degree of variation in these traits varied considerably between species, and coined the term "heteroblasty" to describe species that undergo major morphological changes, and "homoblasty" to refer to plants that display more modest changes (Goebel, 1900). Heteroblasty has since acquired a broader meaning, and is now often used to describe any morphological variation along the length of a shoot, independent of the nature or degree of this change (Zotz et al., 2011). An additional terminological complication is that the terms "juvenile" and "adult" were initially employed to describe different stages of vegetative development (Goebel, 1889; Hildebrand, 1875), but are now more often used to refer to vegetative versus reproductive (flowering) shoots. Indeed, some authors have criticized the use of these terms to describe changes in vegetative morphology (Jones, 1999, 2001; Zotz et al., 2011), arguing that they should be used exclusively to describe a change in reproductive competence. As described below, reproductive competence is controlled by multiple pathways, which interact with the pathway controlling vegetative maturation at various points. Because vegetative and reproductive maturation are equally important aspects of shoot development and are regulated in concert, it is not obvious why the developmental state of the shoot should only be defined by one of these processes. To avoid confusion it has been suggested that vegetative and reproductive maturation be described using terms specific to each process, for example, juvenile and adult phases of vegetative development and juvenile and adult phases of reproductive development (Poethig, 1990). In this chapter the terms "juvenile" and "adult" will refer specifically to phases of vegetative development, that is, developmental changes that occur prior to floral induction.

3. HETEROBLASTY AND VEGETATIVE PHASE CHANGE

Variation in the morphology or physiology of a shoot can occur for many reasons. Traits that change in a coordinated fashion at a predictable time prior to flowering and which are not readily modified by environmental conditions, are the basis for the division of shoot development into juvenile and adult vegetative phases. The process responsible for this type of variation has been termed "ontogenetic maturation" (Wareing, 1959) or "phase change" (Brink, 1962; Poethig, 1990). A second type of variation is evident in traits that change gradually over the life of the shoot, and which can be modified by rerooting or by grafting shoots onto a more vigorous root stock (Day et al., 2002; Greenwood, Day, & Schatz, 2010; Mencuccini, Martinez-Vilalta, Hamid, Korakaki, & Vanderklein, 2007). This type of heteroblasty has been termed "physiological aging" (Wareing, 1959), and is thought to be a consequence of an increase in the size of the shoot (Day et al., 2002; Mencuccini et al., 2007). A third type of variation, termed "seasonal heterophylly" (Godley, 1985), is represented by the reproducible changes in shoot morphology that occur during a growing season along the primary axis of the shoot and in newly formed branches of some herbaceous and woody perennials (Kozlowski, 1971; Moriuchi & Winn, 2005; Winn, 1999). These changes sometimes resemble those that occur early in the growth of the shoot (Critchfield, 1960), but differ in that they occur regularly, rather than only once. Changes in vegetative morphology may also arise from environmental heterogeneity, such as variation in light quality, temperature, growth substrate, and humidity (Bruni, Young, & Dengler, 1995; Cutri et al., 2013; Deschamp & Cooke, 1985; Fisher, Posluszny, & Lee, 2002; Goliber & Feldman, 1990; Jones, 1995; Lee & Richards, 1991; Ray, 1987), and may also be induced by damage from herbivory or disease (Boege & Marquis, 2005). Changes produced by these environmental conditions can resemble the changes that occur during phase change, but it is unclear if they are mediated by the same regulatory mechanism.

In higher plants, the transition from vegetative to reproductive development is marked by the production of a completely new structure specialized for gamete production (e.g., a flower or cone). In contrast, vegetative phase change involves heterogeneous and sometimes quite subtle changes in the character of leaves, stems, and buds. These changes are species-specific and include leaf shape and size, branching patterns, epicuticular wax, patterns of trichome production, cell shape, vascular patterns, histological staining, phyllotaxy, the capacity for adventitious root production, and the presence or absence of anthocyanin or other phytochemicals, as well as disease or insect resistance (Allsopp, 1967; Boege & Marquis, 2005; Greenwood, 1995; Hackett, 1985; Kerstetter & Poethig, 1998; Poethig, 1990; Whalen, 2005). Depending on the species and the trait under investigation, vegetative transitions may occur quickly and encompass only a few nodes, or occur gradually and encompass many nodes. This greatly complicates the analysis of vegetative phase change because different temporal patterns and traits may be controlled by different mechanisms in different species (Borchert, 1976; Jones, 1999). Fortunately, this problem has been resolved by the discovery of major regulators of vegetative phase change—the microRNAs, miR156, and miR157, and their direct targets, the SBP/SPL family of transcription factors. Recent studies demonstrating that these evolutionarily conserved genes regulate vegetative phase change in a range of plants, and are differentially expressed in juvenile and adult tissues of herbaceous as well as woody plants, makes it possible to study the underlying molecular mechanism of vegetative phase change across the plant kingdom.

4. miR156 AND miR157: MASTER REGULATORS OF VEGETATIVE PHASE CHANGE

miR156 was initially discovered as a regulator of vegetative phase change in *Arabidopsis* (Wu & Poethig, 2006) and maize (Chuck, Cigan, Saeteurn, & Hake, 2007), and has since been shown to regulate this process in a number of other herbaceous (Fu et al., 2012; Salinas, Xing, Hohmann, Berndtgen, & Huijser, 2012; Shikata, Yamaguchi, Sasaki, & Ohtsubo, 2012; Xie et al., 2012; Xie, Wu, & Xiong, 2006; Zhang et al., 2011) and woody (Wang et al., 2011) species. miR156 is present in all land plants including moss (Axtell & Bowman, 2008), where it has been implicated in the transition from the protonemal to the leafy gametophore stage of development (Cho, Coruh, & Axtell, 2012). miR157 differs by three nucleotides from miR156, but is frequently misannotated as miR156; it is therefore difficult to determine its taxonomic distribution from the literature. Sequence searches of various databases demonstrate that it is not present in the moss *Physcomitrella patens*, but is present as a single nucleotide variant in the lycopod *Selaginella moellendorffii*. It is also present in the gymnosperm *Taxus chinensis* (Qiu et al., 2009), in some monocots, and in most if not all eudicots (www.mirbase.org). The overexpression phenotype of miR157 is quite similar to that of miR156 (Shikata, Koyama, Mitsuda, & Ohme-Takagi, 2009; Shikata et al., 2012), but its expression pattern and function have not been as well characterized.

Evidence for the involvement of miR156 in vegetative phase change has come primarily from the phenotype of plants overexpressing this miRNA. Plants expressing miR156 under the regulation of a strong constitutive promoter have a common phenotype consisting of a prolonged juvenile phase, increased branching, accelerated leaf production, and delayed flowering. This phenotype has been observed in *Arabidopsis* (Schwab et al., 2005; Shikata et al., 2009; Wang, Czech, & Weigel, 2009; Wang, Schwab, Czech, Mica, & Weigel, 2008; Wu et al., 2009; Wu & Poethig, 2006), maize (Chuck et al., 2007), rice (Xie et al., 2006), switchgrass (Chuck et al., 2011; Fu et al., 2012), poplar (Wang et al., 2011), tomato (Zhang et al., 2011), and *Torenia fournieri* (Shikata et al., 2012). In *Arabidopsis*, inactivation of miR156 with a target site mimic (*35S::MIM156*) (Franco-Zorrilla et al., 2007; Todesco, Rubio-Somoza, Paz-Ares, & Weigel, 2010; Wu et al., 2009), as well as loss-of-function mutations of *MIR156A* and *MIR156C* (Yang, Xu, Koo, He, & Poethig, 2013; Yu et al., 2013), produce the opposite phenotype: these plants express adult leaf traits precociously, are slow growing, and flower with many fewer leaves than normal. These results demonstrate that miR156 is both necessary and sufficient for the expression of the juvenile vegetative phase, but has other functions as well.

The expression pattern of miR156 is consistent with its role in promoting juvenile development. In *Arabidopsis* (Jung, Ju, Seo, Lee, & Park, 2012; Jung, Seo, Kang, & Park, 2011; Wahl et al., 2013; Wang et al., 2009; Wu et al., 2009; Wu & Poethig, 2006), maize (Chuck et al., 2007), and rice (Xie et al., 2006), miR156 is expressed at high levels in young seedlings and at lower levels in older plants. An analysis of miR156 levels in fully expanded leaves of species that undergo significant changes in leaf morphology during vegetative phase change (specifically, *Acacia confusa*, *Acacia colei*, *Eucalyptus globulus*, *Hedera helix*, and *Quercus acutissima*) demonstrated that juvenile leaves have significantly higher levels of miR156 and miR157 than adult leaves (Wang et al., 2011), and showed that variation in leaf morphology in the transition zone between these two phases was correlated with intermediate levels of these miRNAs (Fig. 5.1). Along with the results of the genetic analyses mentioned above, these results strongly suggest that vegetative phase change is mediated by a decline in the level of miR156/miR157. From a practical standpoint, these observations also indicate that the relative levels of miR156/miR157 can be used as molecular markers for shoot identity in situations in which it is difficult to identify these phases on the basis of morphological criteria.

5. miR156 TARGETS

miR156 acts by repressing the expression of *SQUAMOSA PROMOTER BINDING PROTEIN* (*SBP/SPL*) genes, which encode plant-specific transcription factors (Cardon et al., 1999; Klein, Saedler, & Huijser, 1996). Three of the 13 *SBP* genes in moss (Axtell, Snyder, & Bartel, 2007; Riese, Hohmann, Saedler, Munster, & Huijser, 2007), 11 of the 19 *SBP* genes in rice (Xie et al., 2006), and 11 of the 17 *SPL* genes in the Columbia accession of *Arabidopsis thaliana* (Gandikota et al., 2007; Rhoades et al., 2002) have sequences complementary to this miRNA. miR156 represses the expression of *SPL* genes by directing the cleavage of their transcripts and by promoting translational repression. SPL transcripts with sequences complementary to miR156 are cleaved at the expected

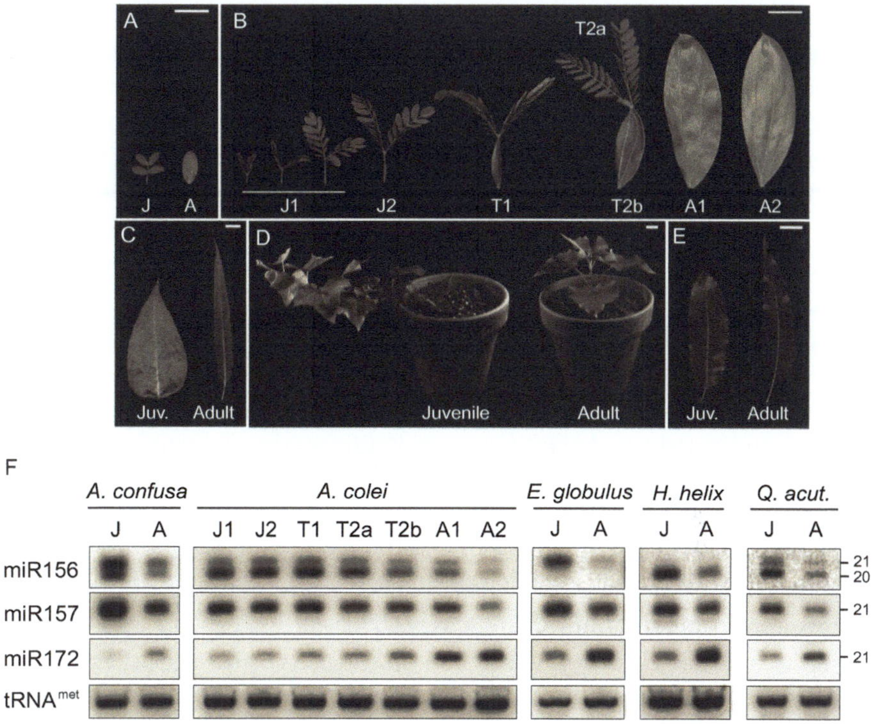

Figure 5.1 The abundance of miR156, miR157, and miR172 is correlated with the vegetative phase of woody species. (A–E) Juvenile and adult leaves of (A) *Acacia confusa*, (B) *Acacia colei*, (C) *Eucalyptus globulus*, (D) *Hedera helix* (English ivy), and (E) *Quercus acutissima* (sawtooth oak). The leaves shown in (A) and (B) are successive leaves on the same shoot. (F) Northern blots of small RNA from the leaves shown in (A–E), hybridized with probes to the indicated transcripts. Scale bars correspond to 2 cm. *Modified from Wang et al. (2011).* (For color version of this figure, the reader is referred to the online version of this chapter.)

position within this sequence (Addo-Quaye, Eshoo, Bartel, & Axtell, 2008; German et al., 2008), and increased levels of miR156 typically lead to a decrease in the abundance of these transcripts (Chuck et al., 2007; Kim et al., 2012; Sanchez, Cagnola, Crepy, Yanovsky, & Casal, 2011; Schwab et al., 2005; Wu & Poethig, 2006; Xie et al., 2006). Conversely, mutations that block miRNA activity (Ronemus, Vaughn, & Martienssen, 2006; Smith et al., 2009), as well overexpression of a miR156 target site mimic (Franco-Zorrilla et al., 2007; Wang et al., 2008; Wu et al., 2009), produce an increase in SPL transcript levels. Evidence that miR156 is also capable of inhibiting translation initially came from the observation that the abundance of the SPL3 protein was not correlated with the abundance of the SPL3 mRNA in lines transformed with *35S::SPL3*; in particular, it was noted that transgenic seedlings with elevated levels of the SPL3 mRNA displayed no increase in the SPL3 protein (Gandikota et al., 2007). Subsequent studies demonstrating that mutations in the microtubule-binding protein, katanin (Brodersen et al., 2008), and the GW-repeat protein, SUO (Yang, Wu, & Poethig, 2012), increase the abundance of the SPL3 and SPL9 proteins without affecting the level of their mRNAs, provided additional evidence that plant miRNAs are capable of mediating translational repression. Although the mechanism by which katanin promotes miRNA function is still unknown, GW-repeat proteins in animals have well-characterized roles in miRNA-mediated translational repression and mRNA turnover (Braun, Huntzinger, & Izaurralde, 2013; Ding & Han, 2007) and it may be that SUO acts in a similar fashion. In any case, it is significant that the phenotype of *suo* mutations is largely attributable to a reduction in the activity of miR156 (Yang et al., 2012). This result demonstrates that translational repression is important for the normal function of miR156, and implies that the activity of *SPL* genes cannot be accurately predicted from the levels of their transcripts in cells in which miR156 is expressed.

Evidence that the effect of miR156 on vegetative phase change is attributable to its effect on *SPL* expression comes from the phenotype of *SPL* genes with mutations in the miR156 target site. Transgenic plants constitutively expressing wild-type SPL3, SPL4, and SPL5 transcripts have a nearly normal phenotype, whereas plants transformed with miR156-resistant versions of these transcripts flower extremely early, and express some adult vegetative traits precociously (Cardon, Hohmann, Nettesheim, Saedler, & Huijser, 1997; Gandikota et al., 2007; Jung et al., 2011; Wang et al., 2009; Wu & Poethig, 2006; Yamaguchi et al., 2009). Similarly, transgenic plants expressing miR156-resistant SPL9, SPL10, or SPL13 transcripts have a much stronger phenotype than plants expressing the corresponding wild-type versions of these transcripts (Martin et al., 2010; Shikata et al., 2009; Wang et al., 2008; Wu et al., 2009). However, the best evidence that miR156 regulates phase change through its effect on *SPL* gene expression is provided by an Ethyl methane sulfonate-induced mutation of *SPL15*, whose precocious phenotype is attributable to a single nucleotide change in the miR156 target site that leads to an increase in the expression of the SPL15 transcript (Usami, Horiguchi, Yano, & Tsukaya, 2009). These results demonstrate that miR156 has a major effect on the expression of *SPL* genes and suggest that it acts primarily, if not exclusively, by repressing the expression of this gene family.

Phylogenetic analyses indicate that *SBP/SPL* genes with miR156 target sites can be grouped into five clades (Guo et al., 2008; Salinas et al., 2012). The phenotypes of plants transformed with miR156-resistant forms of these genes, as well as their loss-of-function phenotypes, suggest that they regulate many different aspects of plant development. These include glume architecture (Wang et al., 2005), inflorescence branching/bract development (Chuck, Whipple, Jackson, & Hake, 2010), and ligule differentiation (Moreno, Harper, Krueger, Dellaporta, & Freeling, 1997) in maize; panicle (Jiao et al., 2010; Miura et al., 2010) and grain morphology (Wang et al., 2012) in rice; fruit ripening in tomato (Manning et al., 2006); and leaf initiation (Martin et al., 2010; Wang et al., 2008), embryo development (Nodine & Bartel, 2010), pollen development (Xing, Salinas, Hohmann, Berndtgen, & Huijser, 2010), and trichome (Gan, Xia, Chen, & Wang, 2011) and anthocyanin patterning (Gou, Felippes, Liu, Weigel, & Wang, 2011) in the inflorescence in *Arabidopsis*. Thus, SBP/SPL genes regulate a variety of processes in addition to those associated with vegetative phase change.

The divergent functions of SPL genes are apparent from the phenotypes of plants overexpressing these genes (Cardon et al., 1997; Martin et al., 2010; Usami et al., 2009; Wang et al., 2008; Wu et al., 2009; Wu & Poethig, 2006). However, their normal functions have been difficult to establish because of the high degree of functional redundancy within this family. *SPL8* (Unte et al., 2003; Zhang, Schwarz, Saedler, & Huijser, 2007), *SPL14* (Stone, Liang, Nekl, & Stiers, 2005), and *SPL9* are the only members of this family in *Arabidopsis* that have a loss-of-function phenotype on their own, and of these only *SPL9* is regulated by miR156. *SPL9* and *SPL15* are members of the same clade. *spl9* mutations produce a slight increase in the rate of leaf initiation and a very weak juvenilized phenotype, and *spl15* mutations are nearly wild type in appearance. However, plants mutant for both genes have a prolonged juvenile phase, are late flowering, have an increased number of branches, and an increased rate of leaf initiation (Schwarz, Grande, Bujdoso, Saedler, & Huijser, 2008; Wang et al., 2008). *spl9* and *spl15* also enhance the male sterility phenotype of *spl8*, demonstrating that they act redundantly with this gene to promote pollen development (Xing et al., 2010). *SPL2*, *SPL10*, and *SPL11* are also members of a single clade in *Arabidopsis*. *spl2* mutations have no apparent loss-of-function phenotype, but slightly enhance the delayed phase change phenotype of the *spl9 spl15* double mutant (Schwarz et al., 2008) and the male sterile phenotype of the *spl9 spl15 spl8* triple mutant (Xing et al., 2010). *SPL10* and

SPL11 are closely related, tandemly duplicated genes, and have little or no loss-of-function phenotype alone, or in combination with *spl2* (Shikata et al., 2009). To get around this problem, Shikata and colleagues (Shikata et al., 2009) fused SPL10 to the SRDX transcription repression domain, and expressed this protein in transgenic plants using both the constitutive *35S* promoter and the endogenous *SPL10* promoter. This approach was expected to produce a phenotype similar to the loss-of-function phenotype of *SPL10* on the assumption that SPL10 normally acts to promote gene expression. Transgenic plants had narrow rosette leaves, stunted inflorescences, reduced apical dominance, defective flowers and siliques, and abnormally round cauline leaves with excessive numbers of trichomes, but had a normal pattern of abaxial trichome production and flowered normally. Unfortunately, this phenotype is largely inconsistent with the phenotype of plants expressing miR156-resistant SPL10 transcripts, whose major feature is the accelerated expression of adult traits (Wu et al., 2009). Given that both these phenotypes are produced by mutant transgenes, it is difficult to know which phenotype more accurately reflects the normal function of *SPL10*.

6. MOLECULAR INSIGHTS INTO THE PHENOMENOLOGY OF VEGETATIVE PHASE CHANGE

6.1. How are vegetative phases specified?

Vegetative development is commonly divided into two phases—a juvenile and an adult phase. However, patterns of heteroblastic variation are often more complex than this. For example, in *Eucalyptus* shoot development is typically divided into seedling, juvenile, transitional, and adult phases (Blake, 1953; Boland et al., 2006), and similar categories have been described in several species native to New Zealand (Day, Gould, & Jameson, 1997; Gould, 1993). Even in weakly heteroblastic species, such as *Arabidopsis* (Telfer, Bollman, & Poethig, 1997) and maize (Bongard-Pierce, Evans, & Poethig, 1996), the first two leaves differ from other juvenile leaves in a variety of ways, and could easily be considered a distinct leaf type. Can these patterns be reconciled with a model of shoot maturation based on changes in the abundance of miR156?

Northern analysis shows that miR156 is present throughout shoot development, but is present at different levels at different times in development. In *Arabidopsis*, miR156 declines rapidly within 2 weeks after germination (Jung et al., 2012; Wu et al., 2009; Wu & Poethig, 2006), and then declines slowly over several weeks in plants grown in short days to delay flowering (Wahl et al., 2013; Wang et al., 2009). If quantitative variation in the abundance of miR156 produces corresponding variation in *SPL* gene expression, it is not difficult to imagine how continuous variation in the expression of this miRNA could lead to multiple, apparently discrete leaf types. Preliminary evidence for this scenario is provided by *Acacia colei*, where the abundance miR156 and miR157 have been measured in individual juvenile, transition, and adult leaves (Fig. 5.1). In this species, miR156 and miR157 are present at high levels in juvenile leaves, at intermediate levels in transition leaves, and at low levels in adult leaves (Wang et al., 2011). In other species, miR156 expression has only been measured in entire plants, in shoot apices containing multiple leaves, or in completely juvenile or adult leaves, so it is unknown how tightly its expression is correlated with the expression of the morphological traits that are used to define vegetative phases. The precise relationship between miR156 levels, target gene expression, and leaf morphology will need to be assessed in order to determine if this regulatory system is capable of generating the diversity of heteroblastic patterns seen in nature. It is of particular interest to determine how this system operates to specify intermediate developmental states. At present, the juvenile phase is probably best defined as the period during which miR156 expression is sufficiently high to completely, or nearly completely suppress the production of SPL proteins.

6.2. The relationship between vegetative and reproductive maturation

One of the most confusing issues in shoot maturation is the relationship between vegetative and reproductive aspects of this process (Jones, 1999). Most plants only flower when they are in an adult vegetative phase, but the timing of flower production varies widely relative to the timing of vegetative phase change: some plants flower immediately after this transition, others remain in an adult vegetative phase for a long time before flowering, and some species routinely flower in a juvenile vegetative phase (Brown, Ariati, Murphy, Miller, & Ladiges, 2006; Hopper & Maslin, 1978; Potts & Wiltshire, 1997; Wiltshire, Potts, & Reid, 1991; Zimmerman, Hackett, & Pharis, 1985). Furthermore, genetic analyses of vegetative phase change and floral induction indicate that these developmental transitions are inherited independently in *Eucalyptus* (Jordan, Potts, & Wiltshire, 1999; Wiltshire, Potts, & Reid, 1998), *Pisum* (Wiltshire, Murfet, & Reid, 1994), maize (Abedon, Revilla, & Tracy, 1996), and *Arabidopsis* (Telfer et al., 1997).

Although these observations suggest that vegetative phase change and floral induction may have little to do with each other, they are readily explained by the architecture of the regulatory pathways that control these transitions. Genetic analyses of flowering time in *Arabidopsis* have shown that floral induction is regulated by multiple inputs ("pathways") that regulate the expression of genes involved in the transformation of the vegetative meristem into an inflorescence meristem (reviewed in (Amasino, 2010; Wellmer & Riechmann, 2010)). In *Arabidopsis*, these inputs include photoperiod, prolonged cold temperature (vernalization), ambient temperature, gibberellic acid (GA), sugar, the autonomous pathway, and the vegetative phase change pathway. The ability of a plant to respond to any one of these pathways depends on the state of the other pathways. A good example is the interaction between the vernalization and photoperiod pathways. Under laboratory conditions—but not necessarily in the field (Wilczek et al., 2009)—plants that contain functional alleles of *FRI* and *FLC* require several weeks of exposure to cold in order to respond to floral inductive long-day (LD) conditions; that is, in the absence of a cold treatment, the vernalization pathway overrules the photoperiod pathway. On the other hand, vernalized plants only flower early if they are grown under LD, demonstrating that vernalization by itself does not induce flowering. Rather, it creates a permissive state in which other factors can operate.

The vegetative phase change pathway resembles the vernalization pathway in the sense that it regulates the competence of the shoot to respond to various conditions that promote flowering rather than by directly mediating flower production. High levels of miR156 presumably set a threshold that buffers fluctuation in the abundance of SPL transcripts, preventing premature floral induction. Evidence that this pathway impacts reproductive competence was initially obtained in maize, where it was found that *Teopod2*—a hypermorphic mutation of miR156—delays the photosensitive period for floral induction (Bassiri, Irish, & Poethig, 1992). In *Arabidopsis*, overexpressing miR156 delays flowering, whereas reduced levels of miR156 accelerate flowering under LD (Schwab et al., 2005; Schwarz et al., 2008; Wang et al., 2009). However, plants with reduced levels of miR156 do not flower earlier than normal in a noninductive short-day (SD) photoperiod (Wang et al., 2009). These results suggest that miR156 represses flowering early in development when it is highly expressed, and that the decline in its expression creates a permissive state for floral induction. During the adult phase, floral induction is dependent on other factors—such as photoperiod, vernalization or GA signaling—which regulate the transcription of

SPL genes and other genes involved in flowering, but have no effect on the expression of miR156 (Jung et al., 2012, 2011; Wang et al., 2009; Yu et al., 2012) (Fig. 5.2).

One of the ways in which miR156 represses flowering is through its effect on miR172, a miRNA that represses several AP2-like transcription factors that inhibit flowering. High levels of miR172 promote flowering by repressing the expression of these floral repressors (reviewed in (Zhu & Helliwell, 2011). miR172 is strongly upregulated when adult plants are transferred from SD to LD (Jung et al., 2007; Schmid et al., 2003), but its expression also increases early in development in the absence of a LD stimulus (Wu et al., 2009). This early increase in miR172 is complementary to the decrease in miR156 expression, and is thought to be a consequence of this decrease because plants overexpressing miR156 have reduced levels of miR172, whereas plants with reduced levels of miR156 have elevated levels of mIR172 (Jung et al., 2011; Wu et al., 2009). miR156 represses miR172 through its effect on the expression of *SPL9*, *SPL10*, and related *SPL* genes. Transgenic plants expressing miR156-resistant SPL9 transcripts under the regulation of the endogenous *SPL9* promoter have elevated levels of miR172 and flower early (Wang et al., 2009; Wu et al., 2009). This

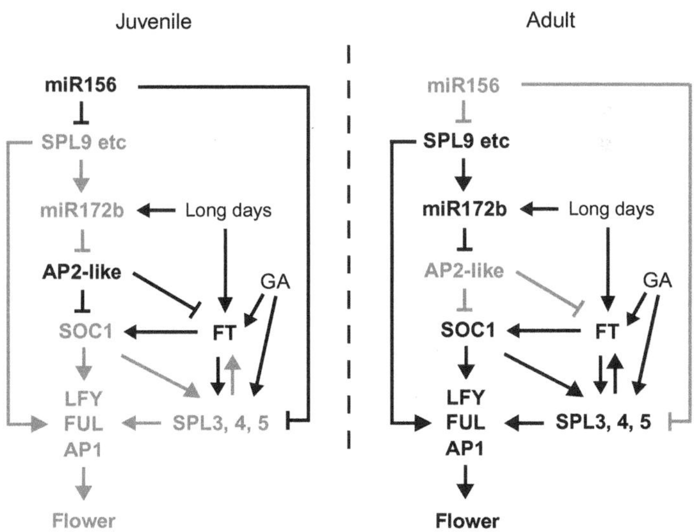

Figure 5.2 miR156 controls reproductive competence by repressing the expression of genes required for flower production. The pathways illustrated here are based on validated miRNA-target interactions and transcription factor binding studies described in the text. Black indicates genes that are expressed; gray represents repressed genes.

phenotype is attributable to the increased transcription of *MIR172B*, which is a direct target of SPL9 (Wu et al., 2009). A reasonable hypothesis is that high levels of miR156 during the juvenile phase reduce the sensitivity of plants to a photoperiodic stimulus (i.e., reduce reproductive competence) by repressing the expression of *SPL9* and other *SPL* genes that promote the transcription of *MIR172B*. This hypothesis is supported by the observation that *spl9 spl15* double mutants are late flowering and display a delayed sensitivity to photoperiod (Schwarz et al., 2008).

Three other targets of miR156 that have been implicated in flowering are *SPL3*, *SPL4*, and *SPL5*. Indeed, the first evidence that the miR156–SPL interaction plays a role in shoot maturation was the observation that plants overexpressing miR156-resistant forms of these genes flower early (Cardon et al., 1997; Gandikota et al., 2007; Wu & Poethig, 2006). In addition to being directly repressed by miR156, *SPL3/4/5* are transcriptionally repressed by the AP2-like proteins targeted by miR172, and are thus indirectly repressed by miR156 through its effect on *miR172B* expression (Jung et al., 2011; Wu et al., 2009; Fig. 5.2). *SPL3/4/5* are expressed at relatively low levels in SD, and at much higher levels in LD (Cardon et al., 1999; Schmid et al., 2003). This response to photoperiod is independent of miR156 because miR156 levels are identical in plants grown in LD and SD (Jung et al., 2012); furthermore, mutations in genes required for the photoperiodic response do not affect miR156 levels (Wang et al., 2009). Photoperiodic regulation of *SPL3/4/5* is mediated by FT (Schmid et al., 2003) through its effect on the transcription of *SOC1* and *FD*, both of which directly promote the transcription of *SPL3* (Jung et al., 2012). SPL3, in turn, directly regulates the transcription of several genes involved in flowering, including *FUL*, *LFY*, *AP1*, and *FT* (Wang et al., 2009; Yamaguchi et al., 2009).

Given that photoperiod is capable of inducing *SPL3/4/5* expression independently of miR156, what role does miR156 play in the regulation of these genes during shoot development? One possibility is that miR156 overrides the inductive effects of LD during the juvenile phase, and this hypothesis is supported by the observation that plants overexpressing miR156-sensitive SPL3/4/5 transcripts flower only slightly earlier than normal in LD (Wu & Poethig, 2006). Apparently, miR156 levels are sufficiently high at this stage that even elevated levels of SPL3/4/5 transcripts cannot overcome repression by this miRNA. miR156 has also been postulated to play a role in floral induction in SD (Wahl et al., 2013; Wang et al., 2009). Under these conditions, miR156 levels decline gradually at the shoot apex, and SPL3/4/5 transcripts increase in a reciprocal fashion. Unfortunately, there is still no evidence that the decline in miR156 is responsible for the increase in SPL3/4/6 transcripts under SD, nor is there conclusive evidence that these genes play a role in floral induction under either LD or SD conditions. For example, the phenotype of plants expressing miR156-resistant versions of SPL3/4/5 transcripts under the regulation of their endogenous promoters has yet to be described. If miR156 indeed plays an important role in the regulation of *SPL3/4/5*, these plants should have an early flowering phenotype. More importantly, plants with reduced levels of miR156 do not flower early under SD (Wang et al., 2009). They produce fewer leaves than wild-type plants because they initiate leaves more slowly (Wang et al., 2008), but they produce flowers at the same time as wild-type plants. Another major problem is that the loss-of-function phenotype of *SPL3*, *SPL4*, and *SPL5* is still unknown. The only existing loss-of-function mutation of *SPL3* has no obvious phenotype (Wu & Poethig, 2006). Whether this is because the mutation is hypomorphic (Kim et al., 2012), because of functional redundancy with *SPL4* and *SPL5*, or because *SPL3* has only a minor role in shoot development, remains to be determined. Viral-induced gene silencing of a homolog of *SPL3/4/5* in *Antirrhinum majus* delays flowering (Preston & Hileman, 2010), which is consistent with the early flowering phenotype of plants overexpressing *SPL3/4/5*. But it will take additional work to establish the function of these genes in *Arabidopsis*.

6.3. How is timing of vegetative phase change regulated?

A remarkably comprehensive but little known study of vegetative phase change in *Eucalyptus tenuiramus* (Wiltshire & Reid, 1992) illustrates some of the common features of this process. One long-standing question is whether the timing of this process is dependent on the age or the size of the shoot (Day et al., 2002; Greenwood et al., 2010; Mencuccini et al., 2007; Robinson & Wareing, 1969; Vanderklein, Martinez-Vilalta, Lee, & Mencuccini, 2007). Wiltshire and Reid (1992) showed that the node at which vegetative phase change occurs is more closely correlated with the total number of nodes produced up until the transition point than with the height of the shoot or the duration of shoot growth, and is more strongly correlated with the cumulative amount of light received by the shoot than with day length or temperature. These results suggest that vegetative phase change is controlled by a factor that changes with leaf number, and suggest that the production of this factor is light dependent. Wiltshire and Reid addressed the question of whether phase change is regulated by factors from the root system, by reciprocally grafting *E. tenuiramis* and *E. risdonii*. These cross-compatible sister species differ primarily in the timing of vegetative phase change; *E. tenuiramis* undergoes vegetative phase change at the 36th node, on average, whereas *E. risdonii* remains permanently juvenile. Grafting adult shoots from *E. tenuiramis* onto an *E. risdonii* root stock or juvenile shoots of *E. risdonii* onto an *E. tenuiramis* root stock had no effect on the timing of vegetative phase change in the scions, suggesting that the timing mechanism is localized to the shoot.

In *E. tenuiramis* (Wiltshire & Reid, 1992), as in other plants (Schaffalitzky de Muckadell, 1954; Telfer et al., 1997), the phase identity of a branch matches the phase identity of the primary node from which it arises. Branches at juvenile nodes produce approximately the same number of juvenile leaves as the primary shoot subsequently produces, whereas branches at adult nodes immediately produce adult leaves. Because the lateral buds that produce branches arise *de novo* from stem/leaf tissue, this phenomenon suggests that the developmental identity of a lateral bud is determined by the identity of the cells from which it originates; that is, the identity of leaf or stem cells determines the fate of a lateral bud. An alternative possibility is that the factors responsible for vegetative phase change act globally throughout the shoot, simultaneously transforming both the primary shoot apex and lateral buds and immature branches. For this to be true, branches at different positions on the shoot would have to be at different growth stages at the time of vegetative phase change in order to account for the gradient in the number of juvenile leaves produced by these branches.

The most striking result to emerge from this study (Wiltshire & Reid, 1992) concerns the behavior of epicormic buds. Epicormic buds are developmentally arrested accessory meristems that form at the base of primary lateral buds, and only emerge after the primary shoot is damaged or decapitated. In contrast to primary lateral buds—which produce branches with slightly more juvenile leaves than the primary shoot—epicormic buds from the base of mature trees produce fewer juvenile nodes than the primary shoot. Wiltshire and Reid noted that these buds were present in an arrested state while the shoot was undergoing phase change and concluded "although apical meristems have an innate timing . . . other meristematic tissue (such as accessory buds) can be influenced by the status of the whole plant." This is the first indication that vegetative phase change is associated with changes in the character of organs and tissues throughout the entire shoot, not just within the shoot apex. It raises the question of whether the signals that regulate this process act specifically on or within the shoot apical meristem (SAM), or operate more generally on organs and tissues throughout the shoot.

This question has been addressed in maize, where adult shoots can be readily rejuvenated. Adult shoot apices with one or two leaf primordia undergo complete rejuvenation *in vitro* (Irish & Karlen, 1998). The plants that develop from these cultured shoot apices produce the same number of juvenile leaves and flower with the same number of leaves as seed-derived plants. In contrast, explanted shoot apices with five or more leaf primordia undergo partial rejuvenation (Orkwiszewski & Poethig, 2000). The leaf primordia present on the apex at the time it was explanted into culture, develop juvenile tissue at their base, and these partially rejuvenated leaves are followed by one or two completely juvenile leaves before the shoot reverts to the adult phase and produces flowers. Importantly, the number of leaves produced by explanted shoots is identical to the number of leaves they would have produced *in situ*. That is, the fate of the SAM is unaffected by this treatment. These results indicate that the phase identity of a leaf is specified independently of the phase identity of the SAM, and implies that the factors that regulate juvenile versus adult leaf identity act directly on leaf primordia, and can modify their identity even after they have been initiated.

To identify the source of these factors, Yang and colleagues deleted the root system, cotyledons, and leaf primordia of juvenile *Arabidopsis* seedlings, and monitored the effect of these treatments on leaf identity and the expression of miR156 (Yang, Conway, & Poethig, 2011). They found that phase change occurred normally in the absence of the root system and in plants lacking cotyledons, but was delayed by leaf ablation. This effect was associated with an increase in miR156 expression, and was dependent on miR156. miR156 levels also increased significantly in adult shoot apices of maize cultured *in vitro*, which is consistent with the effect of this treatment on shoot identity (Irish & Karlen, 1998; Orkwiszewski & Poethig, 2000), and with the observation that *SPL* gene expression decreases in rejuvenating maize shoots (Strable, Borsuk, Nettleton, Schnable, & Irish, 2008). The conclusion of this study is that vegetative phase change is promoted by a factor or factors produced by leaves. This conclusion is consistent with earlier studies demonstrating that defoliation and severe pruning prolongs the production of juvenile leaves in both herbaceous (Njoku, 1956) and woody plants (Libby & Hood, 1976; Schaffalitzky de Muckadell, 1954).

Efforts to identify the endogenous factors that regulate vegetative phase change have focused on carbohydrates and on the hormone, GA. Depending on the species, GA promotes or suppresses vegetative phase change and flowering (Zimmerman et al., 1985). For example, exogenous GA promotes rejuvenation in English ivy (Rogler & Hackett, 1975) and *Acacia melanoxylon* (Borchert, 1965), but accelerates vegetative phase change and flowering in maize (Evans & Poethig, 1995) and *Arabidopsis* (Telfer et al., 1997; Wilson et al., 1992). GA has no effect on the expression miR156 in *Arabidopsis*, but promotes the expression of some *SPL* genes (Galvao, Horrer, Kuttner, & Schmid, 2012; Jung et al., 2012; Wang et al., 2009; Yu et al., 2012). This response contributes to GA-mediated flowering under SD conditions because plants overexpressing miR156 (i.e., plants in which *SPL* expression is suppressed) are less sensitive to GA than wild-type plants (Yu et al., 2012). The extent to which GA promotes vegetative phase change via its effect on *SPL* expression is less clear because this hormone has nearly the same effect on vegetative phase change in wild-type plants, plants over-expressing miR156, and plants doubly mutant for *spl9* and *spl15* (Schwarz et al., 2008). In this respect, it is interesting that GA levels do not increase in the SAM within the first 2 weeks after germination (Eriksson, Bohlenius, Moritz, & Nilsson, 2006); GA would be expected to increase during this period if it played a major role in vegetative phase change.

Evidence for the involvement of carbohydrates in vegetative phase change has been accumulating for over 100 years. Goebel (Goebel, 1900) was the first to propose that nutrients play a crucial role in this transition, and he performed many experiments to test this hypothesis (Goebel, 1908). A particularly comprehensive analysis of the effect of specific nutrients on leaf morphology was conducted by Allsopp using the water fern *Marsilea drummondii* (Allsopp, 1954, 1963). Allsopp showed that plants grown in the absence of exogenous sugar produced only simple, juvenile leaves (Allsopp, 1952), and found that supplementing the growth medium with any of several metabolizable sugars accelerated the production of adult leaves (Allsopp, 1953b). Adult shoots transferred to a medium lacking exogenous carbohydrates reverted to the juvenile form (Allsopp, 1953a). Similar results have been obtained with isolated shoot apices (Feldman & Cutter, 1970; Njoku, 1971), and isolated leaf primordia (Sussex & Clutter, 1960) cultured on media containing varying amounts of sugar.

Recent studies in *Arabidopsis* suggest that the striking effect of sugar on heteroblastic development is attributable to its effect on the expression of miR156 (Yang et al., 2013; Yu et al., 2013). Mutations in the chlorophyll biosynthetic gene, *ch1*, as well as several other yellow-green mutations, prolong the juvenile phase (Röbbelen, 1957; Yang et al., 2013; Yu et al., 2013).

ch1 has elevated levels of miR156, and its effect on vegetative phase change is dependent on miR156, as demonstrated by the observation that a reduction in the level of miR156 blocks the vegetative phase change phenotype of *ch1*. The possibility that the phenotype of *ch1* is attributable to reduced carbohydrates was tested by examining the effect of various sugars on the level of miR156 and vegetative phase change in mutant and wild-type plants. Glucose or sucrose accelerated vegetative phase change, and reduced the abundance of miR156 in *ch1* and wild-type plants when applied to whole seedlings, isolated leaf primordia, or to petiole stubs on defoliated seedlings (Yang et al., 2013; Yu et al., 2013). These studies also showed that *MIR156A* and *MIR156C* are the major sources of miR156 in Arabidopsis seedlings, and are specifically downregulated by sugar. These results therefore support the long-standing hypothesis that vegetative phase change is mediated by an increase in the nutritional status of the shoot.

The identity of the endogenous sugars and signaling pathways involved in vegetative phase change is still largely unknown, although there has been some progress on this front. HEXOKINASE1 (HXK1)—a regulator of glucose signaling in Arabidopsis—promotes the accumulation of miR156 in seedlings grown in the absence of exogenous sugar and is required for the effect of glucose on miR156 expression, but does not block the decline in the abundance of this miRNA during shoot development (Yang et al., 2013). This result implies that glucose promotes vegetative phase change via the HXK1 signaling pathway, but is not completely responsible for this developmental transition. Another sugar that may be involved in vegetative phase change is trehalose-6-phosphate (T6P). In Arabidopsis, T6P is synthesized from glucose-6-phosphate by TREHALOSE PHOSPHATE SYNTHETASE 1 (TPS1) (Blazquez et al., 1998). Null alleles of *TPS1* are embryo lethal, but this phenotype can be corrected by transiently expressing TPS1 during embryo development (van Dijken, Schluepmann, & Smeekens, 2004), or avoided by using hypomorphic alleles (Gomez, Gilday, Feil, Lunn, & Graham, 2010). *tps1* plants are extremely late flowering due to the reduced expression of the floral inducer FT (Wahl et al., 2013). In addition to having low levels of FT, *tps1* mutants have elevated levels of miR156 and reduced levels of at least three miR156-regulated transcripts—SPL3, SPL4, and SPL5 (Wahl et al., 2013). Although the functional significance of the effect of *tps1* on miR156 and SPL3/4/5 expression has not been determined, this phenotype suggests that T6P may play a role in vegetative phase as well as in floral induction.

7. CONCLUSION

Genetic analyses of vegetative phase change and flowering in model organisms such as *Arabidopsis*, maize, and rice have produced significant advances in our understanding of the molecular mechanisms of shoot maturation in plants. These studies have shown that miR156/miR157 coordinate events in vegetative and reproductive development by posttranscriptionally repressing the expression of SBP/SPL transcription factors. This regulatory mechanism not only explains how temporal changes in phase-specific traits are coordinated, but also provides an explanation for the apparent independence of many of these traits. In particular, the identification of *SBP/SPL* genes that regulate both vegetative traits and floral induction reveals the basis for the association between vegetative phase change and reproductive competence. At the same time, the fact that *SBP/SPL* genes are regulated independently of each other at transcriptional level explains how traits that are normally coordinated can become dissociated in time or space. For example, increased transcription of *SBP/SPL* genes that promote floral induction without a corresponding increase in the transcription of *SBP/SPL* genes that regulate vegetative phase change may explain the phenomenon of flowering during the juvenile phase. One major unanswered question is how the timing of vegetative phase change is regulated. Leaves produce a signal that promotes vegetative phase change, and there is good evidence that sugar is a component of this signal. But it remains to be determined if this leaf signal is an active or permissive regulator of vegetative phase change, and whether other factors also play important roles in the timing of this developmental transition.

ACKNOWLEDGMENTS

The author is grateful to Li Yang, who provided helpful comments about this manuscript. Research in the Poethig lab is supported by grants from the NIH (GM051893) and the NSF 2010 program.

REFERENCES

Abedon, B. G., Revilla, P., & Tracy, W. F. (1996). Vegetative phase change in sweet corn populations: Genetics and relationship with agronomic traits. *Maydica*, *41*, 77–82.

Addo-Quaye, C., Eshoo, T. W., Bartel, D. P., & Axtell, M. J. (2008). Endogenous siRNA and miRNA targets identified by sequencing of the *Arabidopsis* degradome. *Current Biology*, *18*, 758–762.

Allsopp, A. (1952). Experimental and analytical studies of Pteridophytes XVII. The effect of various physiologically active substances on the development of *Marsilea* in sterile culture. *Annals of Botany*, *16*, 165–183.
Allsopp, A. (1953a). Experimental and analytical studies of Pteridophytes XIX. Investigations on *Marsilea* 2. Induced reversion to juvenile stages. *Annals of Botany*, *17*, 37–55.
Allsopp, A. (1953b). Experimental and analytical studies of Pteridophytes XXI. Investigations on *Marsilea* 3. The effect of various sugars on development and morphology. *Annals of Botany*, *17*, 447–463.
Allsopp, A. (1954). Juvenile stages of plants and the nutritional status of the shoot apex. *Nature*, *173*, 1032–1035.
Allsopp, A. (1963). Morphogenesis in *Marsilea*. *Journal of the Linnean Society of London, Botany*, *58*, 417–427.
Allsopp, A. (1967). Heteroblastic development in vascular plants. *Advances in Morphogenesis*, *6*, 127–171.
Amasino, R. (2010). Seasonal and developmental timing of flowering. *The Plant Journal*, *61*, 1001–1013.
Andres, F., & Coupland, G. (2012). The genetic basis of flowering responses to seasonal cues. *Nature Reviews. Genetics*, *13*, 627–639.
Axtell, M. J., & Bowman, J. L. (2008). Evolution of plant microRNAs and their targets. *Trends in Plant Science*, *13*, 343–349.
Axtell, M. J., Snyder, J. A., & Bartel, D. P. (2007). Common functions for diverse small RNAs of land plants. *The Plant Cell*, *19*, 1750–1769.
Bassiri, A., Irish, E. E., & Poethig, R. S. (1992). Heterochronic effects of *Teopod2* on the growth and photosensitivity of the maize shoot. *The Plant Cell*, *4*, 497–504.
Blake, S. T. (1953). Botanical contributions of the Northern Australian Regional Survey. I. Studies on northern Australian species of *Eucalyptus*. *Australian Journal of Botany*, *1*, 185–352. with 136 plates.
Blazquez, M. A., Santos, E., Flores, C. L., Martinez-Zapater, J. M., Salinas, J., & Gancedo, C. (1998). Isolation and molecular characterization of the *Arabidopsis TPS1* gene, encoding trehalose-6-phosphate synthase. *The Plant Journal*, *13*, 685–689.
Boege, K., & Marquis, R. J. (2005). Facing herbivory as you grow up: The ontogeny of resistance in plants. *Trends in Ecology & Evolution*, *20*, 441–448.
Boland, D. J., Brooker, M. I. H., Chippendale, G. M., Hall, N., Hyland, B. P. M., Johnson, R. D., et al. (2006). *Forest trees of Australia* (5th ed.). Collingwood, Australia: CSIRO.
Bongard-Pierce, D. K., Evans, M. M. S., & Poethig, R. S. (1996). Heteroblastic features of leaf anatomy in maize and their genetic regulation. *International Journal of Plant Sciences*, *157*, 331–340.
Borchert, R. (1965). Gibberellic acid and rejuvenation of apical meristems in *Acacia melanoxylon*. *Naturwissenschaften*, *52*, 65–66.
Borchert, R. (1976). The concept of juvenility in woody plants. *Acta Horticulturae*, *56*, 21–36.
Braun, J. E., Huntzinger, E., & Izaurralde, E. (2013). The role of GW182 proteins in miRNA-mediated gene silencing. *Advances in Experimental Medicine and Biology*, *768*, 147–163.
Brink, R. A. (1962). Phase change in higher plants and somatic cell heredity. *The Quarterly Review of Biology*, *37*, 1–22.
Brodersen, P., Sakvarelidze-Achard, L., Bruun-Rasmussen, M., Dunoyer, P., Yamamoto, Y. Y., Sieburth, L., et al. (2008). Widespread translational inhibition by plant miRNAs and siRNAs. *Science*, *320*, 1185–1190.
Brown, G. K., Ariati, S. R., Murphy, D. J., Miller, J. T. H., & Ladiges, P. Y. (2006). Bipinnate acacias (*Acacia* subg. Phyllodineae sect. Botrycephalae) of eastern Australia are polyphyletic based on DNA sequence data. *Australian Systematic Botany*, *19*, 315–326.

Bruni, N., Young, J. P., & Dengler, N. G. (1995). Leaf developmental plasticity of *Ranunculus flabellaris* in response to terrestrial and submerged environments. *Canadian Journal of Botany, 74,* 823–837.
Cardon, G., Hohmann, S., Klein, J., Nettesheim, K., Saedler, H., & Huijser, P. (1999). Molecular characterisation of the *Arabidopsis* SBP-box genes. *Gene, 237,* 91–104.
Cardon, G. H., Hohmann, S., Nettesheim, K., Saedler, H., & Huijser, P. (1997). Functional analysis of the *Arabidopsis thaliana* SBP-box gene *SPL3*: A novel gene involved in the floral transition. *The Plant Journal, 12,* 367–377.
Cho, S. H., Coruh, C., & Axtell, M. J. (2012). miR156 and miR390 regulate tasiRNA accumulation and developmental timing in *Physcomitrella patens. The Plant Cell, 24,* 4837–4849.
Chuck, G., Cigan, A. M., Saeteurn, K., & Hake, S. (2007). The heterochronic maize mutant *Corngrass1* results from overexpression of a tandem microRNA. *Nature Genetics, 39,* 544–549.
Chuck, G. S., Tobias, C., Sun, L., Kraemer, F., Li, C., Dibble, D., et al. (2011). Overexpression of the maize *Corngrass1* microRNA prevents flowering, improves digestibility, and increases starch content of switchgrass. *Proceedings of the National Academy of Sciences of the United States of America, 108,* 17550–17555.
Chuck, G., Whipple, C., Jackson, D., & Hake, S. (2010). The maize SBP-box transcription factor encoded by *tasselsheath4* regulates bract development and the establishment of meristem boundaries. *Development, 137,* 1243–1250.
Critchfield, W. B. (1960). Leaf dimorphism in *Populus trichocarpa. American Journal of Botany, 47,* 699–711.
Cutri, L., Nave, N., Ben Ami, M., Chayut, N., Samach, A., & Dornelas, M. C. (2013). Evolutionary, genetic, environmental and hormonal-induced plasticity in the fate of organs arising from axillary meristems in *Passiflora* spp. *Mechanisms of Development, 130,* 61–69.
Day, J. S., Gould, K. S., & Jameson, P. E. (1997). Vegetative architecture of *Elaeocarpus hookerianus.* Transition from juvenile to adult. *Annals of Botany, 79,* 617–624.
Day, M. E., Greenwood, M. S., & Diaz-Sala, C. (2002). Age- and size-related trends in woody plant shoot development: Regulatory pathways and evidence for genetic control. *Tree Physiology, 22,* 507–513.
Deschamp, P. A., & Cooke, T. J. (1985). Leaf dimorphism in the aquatic angiosperm *Callitriche heterophylla. American Journal of Botany, 72,* 1377–1387.
Ding, L., & Han, M. (2007). GW182 family proteins are crucial for microRNA-mediated gene silencing. *Trends in Cell Biology, 17,* 411–416.
Doorenbos, J. (1965). Juvenile and adult phases in woody plants. *Encyclopedia of Plant Physiology, 15,* 1222–1235.
Eriksson, S., Bohlenius, H., Moritz, T., & Nilsson, O. (2006). GA4 is the active gibberellin in the regulation of LEAFY transcription and *Arabidopsis* floral initiation. *The Plant Cell, 18,* 2172–2181.
Evans, M. M., & Poethig, R. S. (1995). Gibberellins promote vegetative phase change and reproductive maturity in maize. *Plant Physiology, 108,* 475–487.
Feldman, L. J., & Cutter, E. G. (1970). Regulation of leaf form in *Centaurea solstitialis* L. I. Leaf development on whole plants in sterile culture. *Botanical Gazette, 131,* 31–39.
Fisher, J. B., Posluszny, U., & Lee, D. W. (2002). Shade promotes thorn development in a tropical liana, *Artabotrys hexapetalus* (Annonaceae). *International Journal of Plant Sciences, 163,* 295–300.
Franco-Zorrilla, J. M., Valli, A., Todesco, M., Mateos, I., Puga, M. I., Rubio-Somoza, I., et al. (2007). Target mimicry provides a new mechanism for regulation of microRNA activity. *Nature Genetics, 39,* 1033–1037.
Fu, C., Sunkar, R., Zhou, C., Shen, H., Zhang, J. Y., Matts, J., et al. (2012). Overexpression of miR156 in switchgrass (*Panicum virgatum* L.) results in various morphological

alterations and leads to improved biomass production. *Plant Biotechnology Journal, 10,* 443–452.
Galvao, V. C., Horrer, D., Kuttner, F., & Schmid, M. (2012). Spatial control of flowering by DELLA proteins in *Arabidopsis thaliana. Development, 139,* 4072–4082.
Gan, L., Xia, K., Chen, J. G., & Wang, S. (2011). Functional characterization of TRICHOMELESS2, a new single-repeat R3 MYB transcription factor in the regulation of trichome patterning in *Arabidopsis. BMC Plant Biology, 11,* 176.
Gandikota, M., Birkenbihl, R. P., Hohmann, S., Cardon, G. H., Saedler, H., & Huijser, P. (2007). The miRNA156/157 recognition element in the 3′ UTR of the *Arabidopsis* SBP box gene *SPL3* prevents early flowering by translational inhibition in seedlings. *The Plant Journal, 49,* 683–693.
German, M. A., Pillay, M., Jeong, D.-H. , Hetawal, A., Luo, S., Janardhanan, P., et al. (2008). Global identification of microRNA–target RNA pairs by parallel analysis of RNA ends. *Nature Biotechnology, 26,* 941–946.
Godley, E. J. (1985). Paths to maturity. *New Zealand Journal of Botany, 23,* 687–706.
Goebel, K. (1889). Ueber die Jungendzustånde der Pflanzen. *Flora, 72,* 1–45.
Goebel, K. (1900). *Organography of plants part I. General organography. (English translation by I. B. Balfour).* Oxford: Clarendon Press.
Goebel, K. (1908). *Einleitung in die Experimentelle Morphologie der Pflanzen.* Leipzig: B. G. Teubner.
Goethe, J. W. V. (1790). *Versuch die Metamorphose der Pflanzen zu erklären.* Germany: Gotha.
Goliber, T. E., & Feldman, L. J. (1990). Developmental analysis of leaf plasticity in the heterophyllous aquatic plant *Hippuris vulgaris. American Journal of Botany, 77,* 399–412.
Gomez, L. D., Gilday, A., Feil, R., Lunn, J. E., & Graham, I. A. (2010). AtTPS1-mediated trehalose 6-phosphate synthesis is essential for embryogenic and vegetative growth and responsiveness to ABA in germinating seeds and stomatal guard cells. *The Plant Journal, 64,* 1–13.
Gou, J. Y., Felippes, F. F., Liu, C. J., Weigel, D., & Wang, J. W. (2011). Negative regulation of anthocyanin biosynthesis in *Arabidopsis* by a miR156-targeted SPL transcription factor. *The Plant Cell, 23,* 1512–1522.
Gould, K. S. (1993). Leaf heteroblasty in *Pseudopanax crassifolius*: Functional significance of leaf morphology and anatomy. *Annals of Botany, 71,* 61–70.
Greenwood, M. S. (1995). Juvenility and maturation in conifers: Current concepts. *Tree Physiology, 15,* 433–438.
Greenwood, M. S., Day, M. E., & Schatz, J. (2010). Separating the effects of tree size and meristem maturation on shoot development of grafted scions of red spruce (*Picea rubens* Sarg.). *Tree Physiology, 30,* 459–468.
Guo, A. Y., Zhu, Q. H., Gu, X., Ge, S., Yang, J., & Luo, J. (2008). Genome-wide identification and evolutionary analysis of the plant specific SBP-box transcription factor family. *Gene, 418,* 1–8.
Hackett, W. P. (1985). Juvenility, maturation and rejuvenation in woody plants. *Horticultural Reviews, 7,* 109–155.
Hildebrand, F. (1875). Ueber die Jungendzustände solcher Pflanzen, welche im Alter vom vegetativen Charakter ihrer Verwandten abweichen. *Flora, 21,* 321–330.
Hopper, S. D., & Maslin, B. R. (1978). Phytogeography of *Acacia* in Western Australia. *Australian Journal of Botany, 26,* 63–78.
Huijser, P., & Schmid, M. (2011). The control of developmental phase transitions in plants. *Development, 138,* 4117–4129.
Irish, E. E., & Karlen, S. (1998). Restoration of juvenility in maize shoots by meristem culture. *International Journal of Plant Sciences, 159,* 695–701.
Jiao, Y., Wang, Y., Xue, D., Wang, J., Yan, M., Liu, G., et al. (2010). Regulation of OsSPL14 by OsmiR156 defines ideal plant architecture in rice. *Nature Genetics, 42,* 541–544.

Jones, C. S. (1995). Does shade prolong juvenile development? A morphological analysis of leaf shape changes in *Cucurbita argyrosperma* Subsp. *sororia* (Cucurbitaceae). *American Journal of Botany, 82*, 346–359.

Jones, C. S. (1999). An essay on juvenility, phase change, and heteroblasty in seed plants. *International Journal of Plant Sciences, 160*, S105–S111.

Jones, C. S. (2001). The functional correlates of heteroblastic variation in leaves: Changes in form and ecophysiology with whole plant ontogeny. *Boletín de la Sociedad Argentina de Botánica, 36*, 171–184.

Jordan, G. J., Potts, B. M., & Wiltshire, R. J. E. (1999). Strong, independent, quantitative genetic control of the timing of vegetative phase change and first flowering in *Eucalyptus globulus* ssp *globulus* (Tasmanian Blue Gum). *Heredity, 83*, 179–187.

Jung, J. H., Ju, Y., Seo, P. J., Lee, J. H., & Park, C. M. (2012). The SOC1-SPL module integrates photoperiod and gibberellic acid signals to control flowering time in *Arabidopsis*. *The Plant Journal, 69*, 577–588.

Jung, J. H., Seo, P. J., Kang, S. K., & Park, C. M. (2011). miR172 signals are incorporated into the miR156 signaling pathway at the SPL3/4/5 genes in *Arabidopsis* developmental transitions. *Plant Molecular Biology, 76*, 35–45.

Jung, J. H., Seo, Y. H., Seo, P. J., Reyes, J. L., Yun, J., Chua, N. H., et al. (2007). The GIGANTEA-regulated microRNA172 mediates photoperiodic flowering independent of CONSTANS in Arabidopsis. *The Plant Cell, 19*, 2736–2748.

Kerstetter, R. A., & Poethig, R. S. (1998). The specification of leaf identity during shoot development. *Annual Review of Cell and Developmental Biology, 14*, 373–398.

Kim, J. J., Lee, J. H., Kim, W., Jung, H. S., Huijser, P., & Ahn, J. H. (2012). The microRNA156-SQUAMOSA PROMOTER BINDING PROTEIN-LIKE3 module regulates ambient temperature-responsive flowering via FLOWERING LOCUS T in Arabidopsis. *Plant Physiology, 159*, 461–478.

Klein, J., Saedler, H., & Huijser, P. (1996). A new family of DNA binding proteins includes putative transcriptional regulators of the *Antirrhinum majus* floral meristem identity gene *SQUAMOSA*. *Molecular & General Genetics, 250*, 7–16.

Knight, T. A. (1795). Observations on the grafting of trees. In a letter from Thomas Andrew Knight, Esq. to Sir Joseph Banks, Bart. P. R. S. *Philosophical Transactions of the Royal Society of London, 85*, 290–295.

Kozlowski, T. T. (Ed.) (1971). *Growth and development in trees* (Vol. 1). New York, NY: Academic Press.

Lee, D. W., & Richards, J. H. (1991). Heteroblastic development in vines. In F. E. Putz & H. A. Mooney (Eds.), *The biology of vines* (pp. 205–243). New York, NY: Cambridge University Press.

Libby, W. J., & Hood, J. V. (1976). Juvenility in hedged radiata pine. *Acta Horticulturae, 56*, 91–98.

Manning, K., Tor, M., Poole, M., Hong, Y., Thompson, A. J., King, G. J., et al. (2006). A naturally occurring epigenetic mutation in a gene encoding an SBP-box transcription factor inhibits tomato fruit ripening. *Nature Genetics, 38*, 948–952.

Martin, R. C., Asahina, M., Liu, P. P., Kristof, J. R., Coppersmith, J. L., Pluskota, W. E., et al. (2010). The regulation of post-germinative transition from the cotyledon- to vegetative-leaf stages by microRNA-targeted SQUAMOSA PROMOTER-BINDING PROTEIN LIKE13 in Arabidopsis. *Seed Science Research, 20*, 89–96.

Mencuccini, M., Martinez-Vilalta, J., Hamid, H. A., Korakaki, E., & Vanderklein, D. (2007). Evidence for age- and size-mediated controls of tree growth from grafting studies. *Tree Physiology, 27*, 463–473.

Miura, K., Ikeda, M., Matsubara, A., Song, X.-J. , Ito, M., Asano, K., et al. (2010). OsSPL14 promotes panicle branching and higher grain productivity in rice. *Nature Genetics, 42*, 545–549.

Moreno, M. A., Harper, L. C., Krueger, R. W., Dellaporta, S. L., & Freeling, M. (1997). *liguleless1* encodes a nuclear-localized protein required for induction of ligules and auricles during maize leaf organogenesis. *Genes & Development, 11*, 616–628.

Moriuchi, K. S., & Winn, A. A. (2005). Relationships among growth, development and plastic response to environment quality in a perennial plant. *The New Phytologist, 166*, 149–158.

Njoku, E. (1956). The effect of defoliation on leaf shape in *Ipomoea caerulea*. *The New Phytologist, 55*, 213–228.

Njoku, E. (1971). Effect of sugars and applied chemicals on heteroblastic development in *Ipomoea purpurea* grown in aseptic culture. *American Journal of Botany, 58*, 61–64.

Nodine, M. D., & Bartel, D. P. (2010). MicroRNAs prevent precocious gene expression and enable pattern formation during plant embryogenesis. *Genes & Development, 24*, 2678–2692.

Orkwiszewski, J. A., & Poethig, R. S. (2000). Phase identity of the maize leaf is determined after leaf initiation. *Proceedings of the National Academy of Sciences of the United States of America, 97*, 10631–10636.

Poethig, R. S. (1990). Phase change and the regulation of shoot morphogenesis in plants. *Science, 250*, 923–930.

Potts, B. M., & Wiltshire, R. J. E. (1997). *Eucalyptus* genetics and genecology. In J. Williams & J. Woinarski (Eds.), *Eucalypt ecology: Individuals to ecosystems* (pp. 56–91). Cambridge: Cambridge University Press.

Preston, J. C., & Hileman, L. C. (2010). SQUAMOSA-PROMOTER BINDING PROTEIN 1 initiates flowering in *Antirrhinum majus* through the activation of meristem identity genes. *The Plant Journal, 62*, 704–712.

Qiu, D., Pan, X., Wilson, I. W., Li, F., Liu, M., Teng, W., et al. (2009). High throughput sequencing technology reveals that the taxoid elicitor methyl jasmonate regulates microRNA expression in Chinese yew (*Taxus chinensis*). *Gene, 436*, 37–44.

Ray, T. S. (1987). Cyclic heterophylly in *Syngonium* (Araceae). *American Journal of Botany, 74*, 16–26.

Rhoades, M. W., Reinhart, B. J., Lim, L. P., Burge, C. B., Bartel, B., & Bartel, D. P. (2002). Prediction of plant microRNA targets. *Cell, 110*, 513–520.

Riese, M., Hohmann, S., Saedler, H., Munster, T., & Huijser, P. (2007). Comparative analysis of the SBP-box gene families in P. patens and seed plants. *Gene, 401*, 28–37.

Röbbelen, G. (1957). Uber heterophyllie bei *Arabidopsis thaliana* (L.). *Heynh Berichte der Deutschen Botanischen Gesellschaft, 70*, 39–44.

Robinson, L. W., & Wareing, P. F. (1969). Experiments on the juvenile-adult phase change in some woody species. *The New Phytologist, 69*, 67–78.

Rogler, C. E., & Hackett, W. P. (1975). Phase change in *Hedera helix*: Induction of the mature to juvenile phase change by gibberellin A3. *Physiologia Plantarum, 34*, 141–147.

Ronemus, M., Vaughn, M. W., & Martienssen, R. A. (2006). MicroRNA-targeted and small interfering RNA-mediated mRNA degradation is regulated by argonaute, dicer, and RNA-dependent RNA polymerase in *Arabidopsis*. *The Plant Cell, 18*, 1559–1574.

Salinas, M., Xing, S., Hohmann, S., Berndtgen, R., & Huijser, P. (2012). Genomic organization, phylogenetic comparison and differential expression of the SBP-box family of transcription factors in tomato. *Planta, 235*, 1171–1184.

Sanchez, S. E., Cagnola, J. I., Crepy, M., Yanovsky, M. J., & Casal, J. J. (2011). Balancing forces in the photoperiodic control of flowering. *Photochemical & Photobiological Sciences, 10*, 451–460.

Schaffalitzky de Muckadell, M. (1954). Juvenile stages in woody plants. *Physiologia Plantarum, 7*, 782–796.

Schmid, M., Uhlenhaut, N. H., Godard, F., Demar, M., Bressan, R., Weigel, D., et al. (2003). Dissection of floral induction pathways using global expression analysis. *Development, 130*, 6001–6012.

Schwab, R., Palatnik, J. F., Riester, M., Schommer, C., Schmid, M., & Weigel, D. (2005). Specific effects of microRNAs on the plant transcriptome. *Developmental Cell*, *8*, 517–527.

Schwarz, S., Grande, A. V., Bujdoso, N., Saedler, H., & Huijser, P. (2008). The microRNA regulated SBP-box genes *SPL9* and *SPL15* control shoot maturation in *Arabidopsis*. *Plant Molecular Biology*, *67*, 183–195.

Shikata, M., Koyama, T., Mitsuda, N., & Ohme-Takagi, M. (2009). *Arabidopsis* SBP-box genes *SPL10*, *SPL11* and *SPL2* control morphological change in association with shoot maturation in the reproductive phase. *Plant & Cell Physiology*, *50*, 2133–2145.

Shikata, M., Yamaguchi, H., Sasaki, K., & Ohtsubo, N. (2012). Overexpression of Arabidopsis miR157b induces bushy architecture and delayed phase transition in *Torenia fournieri*. *Planta*, *236*, 1027–1035.

Smith, M. R., Willmann, M. R., Wu, G., Berardini, T. Z., Moller, B., Weijers, D., et al. (2009). Cyclophilin 40 is required for microRNA activity in Arabidopsis. *Proceedings of the National Academy of Sciences of the United States of America*, *106*, 5424–5429.

Stone, J. M., Liang, X., Nekl, E. R., & Stiers, J. J. (2005). Arabidopsis AtSPL14, a plant-specific SBP-domain transcription factor, participates in plant development and sensitivity to fumonisin B1. *The Plant Journal*, *41*, 744–754.

Strable, J., Borsuk, L., Nettleton, D., Schnable, P. S., & Irish, E. E. (2008). Microarray analysis of vegetative phase change in maize. *The Plant Journal*, *56*, 1045–1057.

Sussex, I. M., & Clutter, M. E. (1960). A study of the effect of externally supplied sucrose on the morphology of excised fern leaves *in vitro*. *Phytomorphology*, *10*, 87–99.

Telfer, A., Bollman, K. M., & Poethig, R. S. (1997). Phase change and the regulation of trichome distribution in *Arabidopsis thaliana*. *Development*, *124*, 645–654.

Todesco, M., Rubio-Somoza, I., Paz-Ares, J., & Weigel, D. (2010). A collection of target mimics for comprehensive analysis of microRNA function in *Arabidopsis thaliana*. *PLoS Genetics*, *6*, e1001031.

Unte, U. S., Sorensen, A. M., Pesaresi, P., Gandikota, M., Leister, D., Saedler, H., et al. (2003). *SPL8*, an SBP-box gene that affects pollen sac development in Arabidopsis. *The Plant Cell*, *15*, 1009–1019.

Usami, T., Horiguchi, G., Yano, S., & Tsukaya, H. (2009). The more and smaller cells mutants of Arabidopsis thaliana identify novel roles for *SQUAMOSA PROMOTER BINDING PROTEIN-LIKE* genes in the control of heteroblasty. *Development*, *136*, 955–964.

van Dijken, A. J., Schluepmann, H., & Smeekens, S. C. (2004). Arabidopsis trehalose-6-phosphate synthase 1 is essential for normal vegetative growth and transition to flowering. *Plant Physiology*, *135*, 969–977.

Vanderklein, D., Martinez-Vilalta, J., Lee, S., & Mencuccini, M. (2007). Plant size, not age, regulates growth and gas exchange in grafted Scots pine trees. *Tree Physiology*, *27*, 71–79.

Wahl, V., Ponnu, J., Schlereth, A., Arrivault, S., Langenecker, T., Franke, A., et al. (2013). Regulation of flowering by trehalose-6-phosphate signaling in *Arabidopsis thaliana*. *Science*, *339*, 704–707.

Wang, J.-W. , Czech, B., & Weigel, D. (2009). miR156-regulated SPL transcription factors define an endogenous flowering pathway in *Arabidopsis thaliana*. *Cell*, *138*, 738–749.

Wang, H., Nussbaum-Wagler, T., Li, B., Zhao, Q., Vigouroux, Y., Faller, M., et al. (2005). The origin of the naked grains of maize. *Nature*, *436*, 714–719.

Wang, J.-W. , Park, M. Y., Wang, L. J., Koo, Y., Chen, X. Y., Weigel, D., et al. (2011). MiRNA control of vegetative phase change in trees. *PLoS Genetics*, 7, e1002012.

Wang, J.-W. , Schwab, R., Czech, B., Mica, E., & Weigel, D. (2008). Dual effects of miR156-targeted SPL genes and CYP78A5/KLUH on plastochron length and organ size in *Arabidopsis thaliana*. *The Plant Cell*, *20*, 1231–1243.

Wang, S., Wu, K., Yuan, Q., Liu, X., Liu, Z., Lin, X., et al. (2012). Control of grain size, shape and quality by OsSPL16 in rice. *Nature Genetics*, *44*, 950–954.

Wareing, P. (1959). Problems of juvenility and flowering in trees. *Journal of the Linnean Society of London, Botany*, *56*, 282–289.

Wellmer, F., & Riechmann, J. L. (2010). Gene networks controlling the initiation of flower development. *Trends in Genetics*, *26*, 519–527.

Whalen, M. C. (2005). Host defence in a developmental context. *Molecular Plant Pathology*, *6*, 347–360.

Wilczek, A. M., Roe, J. L., Knapp, M. C., Cooper, M. D., Lopez-Gallego, C., Martin, L. J., et al. (2009). Effects of genetic perturbation on seasonal life history plasticity. *Science*, *323*, 930–934.

Wilkie, J. D., Sedgley, M., & Olesen, T. (2008). Regulation of floral initiation in horticultural trees. *Journal of Experimental Botany*, *59*, 3215–3228.

Wilson, R. N., Heckman, J. W., & Somerville, C. R. (1992). Gibberellin is required for flowering in *Arabidopsis* under short days. *Plant Physiology*, *100*, 403–408.

Wiltshire, R. J. E., Murfet, I. C., & Reid, J. B. (1994). The genetic control of heterochrony: Evidence from developmental mutants of *Pisum sativum* L. *Journal of Evolutionary Biology*, *7*, 447–465.

Wiltshire, R. J. E., Potts, B. M., & Reid, J. B. (1991). A paedomorphocline in *Eucalyptus*: Natural variation in the *E. risdonii—E. tenuiramis* complex. *Australian Journal of Botany*, *39*, 545–566.

Wiltshire, R. J. E., Potts, B. M., & Reid, J. B. (1998). Genetic control of reproductive and vegetative phase change in the *Eucalyptus risdonii–E. tenuiramis* complex. *Australian Journal of Botany*, *46*, 45–63.

Wiltshire, R. J. E., & Reid, J. B. (1992). The pattern of juvenility within *Eucalyptus tenuiramis* Miq. saplings. In *Mass production technology for genetically improved fast growing forest tree species. AFOCEL-IUFRO symposium, Bordeaux* (pp. 37–49). Nangis-Paris: Association Foret Cellulose.

Winn, A. A. (1999). The functional significance and fitness consequences of heterophylly. *International Journal of Plant Sciences*, *160*, S113–S121.

Wu, G., Park, M. Y., Conway, S. R., Wang, J. W., Weigel, D., & Poethig, R. S. (2009). The sequential action of miR156 and miR172 regulates developmental timing in *Arabidopsis*. *Cell*, *138*, 750–759.

Wu, G., & Poethig, R. S. (2006). Temporal regulation of shoot development in *Arabidopsis thaliana* by miR156 and its target *SPL3*. *Development*, *133*, 3539–3547.

Xie, K., Shen, J., Hou, X., Yao, J., Li, X., Xiao, J., et al. (2012). Gradual increase of miR156 regulates temporal expression changes of numerous genes during leaf development in rice. *Plant Physiology*, *158*, 1382–1394.

Xie, K., Wu, C., & Xiong, L. (2006). Genomic organization, differential expression, and interaction of SQUAMOSA promoter-binding-like transcription factors and microRNA156 in rice. *Plant Physiology*, *142*, 280–293.

Xing, S., Salinas, M., Hohmann, S., Berndtgen, R., & Huijser, P. (2010). miR156-targeted and nontargeted SBP-box transcription factors act in concert to secure male fertility in *Arabidopsis*. *The Plant Cell*, *22*, 3935–3950.

Yamaguchi, A., Wu, M. F., Yang, L., Wu, G., Poethig, R. S., & Wagner, D. (2009). The microRNA-regulated SBP-Box transcription factor SPL3 is a direct upstream activator of *LEAFY*, *FRUITFULL*, and *APETALA1*. *Developmental Cell*, *17*, 268–278.

Yang, L., Conway, S. R., & Poethig, R. S. (2011). Vegetative phase change is mediated by a leaf-derived signal that represses the transcription of miR156. *Development*, *138*, 245–249.

Yang, L., Wu, G., & Poethig, R. S. (2012). Mutations in the GW-repeat protein SUO reveal a developmental function for microRNA-mediated translational repression in

Arabidopsis. *Proceedings of the National Academy of Sciences of the United States of America*, *109*, 315–320.
Yang, L., Xu, M., Koo, Y., He, J., & Poethig, R. S. (2013). Sugar promotes vegetative phase change in *Arabidopsis thaliana* by repressing the expression of MIR156A and MIR156C. *eLife*, *2*, e00260.
Yu, S., Cao, L., Zhou, C. M., Zhang, T. Q., Lian, H., Sun, Y., et al. (2013). Sugar is an endogenous cue for juvenile-to-adult phase transition in plants. *eLife*, *2*, e00269.
Yu, S., Galvao, V. C., Zhang, Y. C., Horrer, D., Zhang, T. Q., Hao, Y. H., et al. (2012). Gibberellin regulates the Arabidopsis floral transition through miR156-targeted SQUAMOSA promoter binding-like transcription factors. *The Plant Cell*, *24*, 3320–3332.
Zhang, Y., Schwarz, S., Saedler, H., & Huijser, P. (2007). *SPL8*, a local regulator in a subset of gibberellin-mediated developmental processes in Arabidopsis. *Plant Molecular Biology*, *63*, 429–439.
Zhang, X., Zou, Z., Zhang, J., Zhang, Y., Han, Q., Hu, T., et al. (2011). Over-expression of sly-miR156a in tomato results in multiple vegetative and reproductive trait alterations and partial phenocopy of the *sft* mutant. *FEBS Letters*, *585*, 435–439.
Zhu, Q. H., & Helliwell, C. A. (2011). Regulation of flowering time and floral patterning by miR172. *Journal of Experimental Botany*, *62*, 487–495.
Zimmerman, R. H., Hackett, W. P., & Pharis, R. P. (1985). Hormonal aspects of phase change and precocious flowering. In R. P. Pharis & D. M. Reid (Eds.), *Encyclopedia of plant physiology, new series* (pp. 79–115). Berlin, New York: Springer-Verlag.
Zotz, G., Wilhelm, K., & Becker, A. (2011). Heteroblasty-a review. *The Botanical Review*, 77, 109–151.

CHAPTER SIX

Developmental Transitions in *C. elegans* Larval Stages

Ann E. Rougvie*[,1], Eric G. Moss[†]

*Department of Genetics, Cell Biology and Development, University of Minnesota, Minneapolis, Minnesota, USA

[†]Department of Molecular Biology, Rowan University, Stratford, New Jersey, USA

[1]Corresponding author: e-mail address: rougvie@umn.edu

Contents

1. Introduction 154
2. A Worm Well-Suited to Timing Studies 154
3. Heterochronic Mutants 156
4. microRNA Switches 159
5. LIN-28: A microRNA Regulator at the Pathway's Center 160
6. The Heterochronic Pathway 161
7. Developmental Transitions: The Molts 162
8. LIN-42: A Link Between Molting and Developmental Decisions 164
9. Reversible Interruption of Developmental Progression: DAF-12 167
10. Interrupted Development: Reprogramming and Resynchronization 169
11. Launching Larval Development 171
12. A Heterochronic Pathway for Larger Animals? 173
13. Conclusions and Future Directions 173

Acknowledgments 174

References 175

Abstract

Molecular mechanisms control the timing, sequence, and synchrony of developmental events in multicellular organisms. In *Caenorhabditis elegans*, these mechanisms are revealed through the analysis of mutants with "heterochronic" defects: cell division or differentiation patterns that occur in the correct lineage, but simply at the wrong time. Subsets of cells in these mutants thus express temporal identities normally restricted to a different life stage. A seminal finding arising from studies of the heterochronic genes was the discovery of miRNAs; these tiny miRNAs are now a defining feature of the pathway. A series of sequentially expressed miRNAs guide larval transitions through stage-specific repression of key effector molecules. The wild-type lineage patterns are executed as discrete modules programmed between temporal borders imposed by the molting cycles. How these successive events are synchronized with the oscillatory molting cycle is just beginning to come to light. Progression through larval stages

Current Topics in Developmental Biology, Volume 105
ISSN 0070-2153
http://dx.doi.org/10.1016/B978-0-12-396968-2.00006-3

can be specifically, yet reversibly, halted in response to environmental cues, including nutrient availability. Here too, heterochronic genes and miRNAs play key roles. Remarkably, developmental arrest can, in some cases, either mask or reveal timing defects associated with mutations. In this chapter, we provide an overview of how the *C. elegans* heterochronic gene pathway guides developmental transitions during continuous and interrupted larval development.

1. INTRODUCTION

Developmental transitions can be seen at the level of morphology and widespread cellular changes, as in insect metamorphosis, or by more subtle shifts in cell growth and behavior, as when neural cells produce neurons and then glia. These transitions are fascinating because they imply the existence of molecular mechanisms that coordinate the behaviors of many cells in time. We can expect that such mechanisms are not simple, and consist of many parts with complex interactions and dynamics. To work out how such transitions are regulated and executed, we need experimental systems that permit multiple approaches and provide unusual access to the underlying molecular phenomena.

Two kinds of developmental transitions exemplified by the *Caenorhabditis elegans* larva are the four stage-to-stage molts on the way to adulthood and an optional diapause state that can be initiated midway in that course in response to harsh environmental conditions. The regulation of these two types of transitions intersects when the normal progress of development is interrupted and then resumes where it left off. The molecular mechanisms and regulatory networks underlying these transitions are not simple, but sufficient molecular details have been elucidated that a few principles have emerged. Persistent use of classical genetics, combined with one of the first sequenced genomes, RNA interference, and molecular approaches has revealed a dynamic network of molecular regulators, switches, and oscillators, that ensure that cells do the right thing at the right time.

2. A WORM WELL-SUITED TO TIMING STUDIES

C. elegans is a soil-dwelling nematode, a microscopic worm with an extremely streamlined anatomy. It emerges from eggshell with exactly 558 cells, and having the necessary nervous system and musculature, it is competent to forage and feed (see http://www.wormbook.org/). What it lacks at the beginning of postembryonic development is the ability to reproduce. Over four larval stages, 53 cells divide and differentiate to ultimately form the gonad and reproductive system, as well as additions and rearrangements to the epidermis, muscles, and nervous system. Each larval stage ends with a molt when the epidermis secretes a collagenous cuticle, and the old cuticle is shed. At the last transition, somatic differentiation is complete and the animal becomes sexually mature: either a self-fertilizing hermaphrodite or a male.

The stage-to-stage transitions during larval development appear relatively subtle. The larva grows roughly continuously in size with little change in overall morphology. Each molt is preceded by a period of lethargus, when the worm seems to sleep while its cuticle changeover takes place. Emerging from the old cuticle, the worm resumes foraging and pumping food into its pharynx, and growth continues. By the fourth stage, about 400 cells have been added to the epidermis, nervous system, gonad, and muscle. The adult hermaphrodite that emerges from the last molt consists of 959 somatic cells and scores of germ cells, and is about 1 mm long. The adult will increase in size, but will only add germ cells, not somatic cells, throughout the rest of its roughly 2–3-week life.

If environmental conditions deteriorate, larval development can undergo a controlled and reversible suspension that is coordinated throughout the animal. The trigger is an increased population density, exacerbated by diminishing food supply and slightly warmer temperatures. Under such conditions, early stage larvae have the opportunity to transition into a morphologically distinct form, called the dauer larva, which is specialized for long-term survival under less than optimal growth conditions. The decision to make this choice occurs during the first larval stage (L1) but the morphological shift is not complete until midway through larval development, at the end of the second stage (L2). Survival as a dauer extends far beyond the normal lifespan of the worm due to numerous physiological adjustments. As the dauer disperses, the crowding dissipates (sensed by the drop in a worm-produced pheromone), and if nutritional conditions improve, the larva morphs back into the feeding/foraging condition and picks up development where it left off. In this instance, the transition into and out of dauer must be coordinated with the transition that normally occurs in continuous development from the L2 to the third stage (L3).

Two features of *C. elegans* make it a particularly powerful experimental system for investigating developmental transitions. First, the position and division of every cell is observable in living specimens due to the worm's transparency. Because cell behavior is remarkably the same from animal to animal, the lineage of every cell in the animal is traceable through larval development. But what makes *C. elegans* exceptionally informative is the genetics: forward genetic screens have identified a diversity of mutants and these mutants can be combined to tell the tale of how the genes function together.

3. HETEROCHRONIC MUTANTS

Among the many *C. elegans* mutants isolated in genetic screens were a few that eventually were grouped together because they altered the succession of larval developmental events. The gross phenotypes of these mutants are not very different from those of many developmental mutants in the lin or "cell lineage abnormal" class: They are unable to lay eggs, having either a missing or defective egg-laying system (Ferguson & Horvitz, 1985). But cell lineage analysis revealed that this subset displayed features not found in other mutants: the normal order of stage-specific lineage patterns is altered throughout the animal (Ambros & Horvitz, 1984). In one class of mutants, development was precocious, where stage-specific events were skipped, and in another development was retarded, where events were reiterated (Fig. 6.1). The whole group was termed "heterochronic mutants," for their abnormal developmental timing. Initially, the mutations in this collection defined just four genes: *lin-4*, *lin-14*, *lin-28*, and *lin-29*. Many others were to follow.

The mere existence of these mutants allowed two insights: First, the temporal progress of many cell lineages is under direct genetic control, rather than being an emergent property of spatial patterning and tissue-specific regulation. Secondly, the mutants revealed that the stage-specific sublineages were modular, that is, what happens at one stage does not necessarily depend on what happened at a previous stage. One commentator compared the modularity of the *C. elegans* larval stage patterns to parasegments in *Drosophila* embryonic development, with the parasegment boundaries being represented by the molts (Thummel, 2001). Indeed, there is likely a mechanistic connection between temporal patterning and the molting cycle.

The epidermis replaces its cuticle four times, once per larval stage. During the intermolt period, certain epidermal blast cells divide in patterns that are characteristic of the lineage at each stage. For instance, lateral epidermal blast cells V1 to V4 each divide once in the first larval stage producing a daughter that joins the epidermal syncytium and one that remains a blast cell

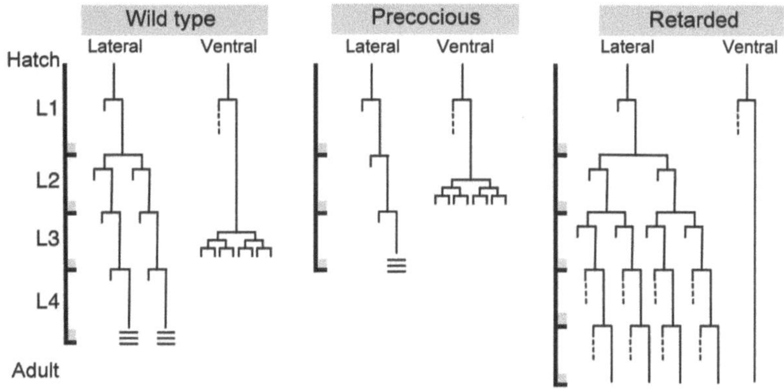

Figure 6.1 Cell lineages of precocious and retarded heterochronic mutants. Postembryonic development is indicated on the left from hatching to adulthood, encompassing four larval stages. Thick lines represent the molting cycle with gray boxes indicating the lethargus period and the short horizontal lines indicating the molts. Thin lines are lineage diagrams of representative lateral and ventral epidermal cells. Lateral epidermal blast cells terminally differentiate into seam cells of the adult, indicated by three horizontal lines. Ventral cells give rise to the vulva. Dashed lines indicate additional cells not shown. Lineages of representative precocious and retarded mutants are shown. Precocious mutants skip stage-specific lineage events (in this case, those of the L2) and may cease the molting cycle prematurely. Retarded mutants reiterate stage-specific events (e.g., L2), and, as in the example shown, cease molting indeterminantly after five or more cycles.

(Fig. 6.1). This blast cell divides twice in the second larval stage, and once again in the third and fourth stages. At adulthood, these cells fuse to form another syncytium, the lateral seam, which secretes a cuticle structure called adult lateral alae. Tracking of these and other lineages in heterochronic mutants reveals their temporal patterning defects in detail (Ambros & Horvitz, 1984).

lin-4 null mutants reiterate the L1 pattern in subsequent stages and terminal differentiation of the lateral epidermal blast cells never occurs. In the ventral epidermis, the three "vulva precursor cells" (VPCs) normally divide in the L3 to produce the 22 cells of the vulva, but do not do so in *lin-4* mutant animals. A recessive mutation in *lin-14*, on the other hand, causes L1 patterns to be skipped, and all subsequent nongonadal lineages, including the VPCs, advance precociously. In these animals, the lateral epidermal blasts divide twice in the first stage, which they would normally do in the second stage, and differentiation occurs at the end of the L3, one stage early. Thus, *lin-4* and *lin-14* mutants show opposite effects on the L1, resulting in retarded and precocious development, respectively.

lin-28 mutants, like *lin-14* mutants, are precocious, but instead of skipping first-stage events, they skip second-stage patterns (some cells also skip third-stage events). *lin-29* mutants are retarded like *lin-4* animals, but their problems do not begin until the L4 stage. The animal becomes adult, but its epidermis remains larval. Eventually, other heterochronic mutants were identified that showed precocious or retarded development starting at one stage or another. Some heterochronic mutants have stronger null phenotypes than others, and a surprising number affect the L2, and most affect the transition to adulthood, directly or indirectly. The overall theme, however, is that their loss-of-function phenotypes are either precocious or retarded, and the defect begins at a particular stage.

The organization of the heterochronic pathway emerges when individual mutants are combined into multiply-mutant strains (Ambros, 1989). For example, when a null allele of *lin-4*, which reiterates the L1 fates, is put together with a null allele of *lin-14*, which skips the L1 fates, the resulting double-mutant animal resembles the *lin-14* mutant in all respects. By the logic of epistasis analysis, this result suggests that *lin-4* normally negatively regulates *lin-14* in some way. When a *lin-29* mutant is combined with any known precocious mutant, the resulting animal is always retarded for late-stage developmental events, indicating that *lin-29* is the most downstream gene in the pathway.

In this way, the heterochronic pathway has been built up as new genes are added. In most cases, the mutations are strong and the double phenotypes clear. These results give us the core pathway (Fig. 6.2, top). In other cases, however, the gene mutations are weaker (either incompletely penetrant, incompletely expressive, or both), or, the mutations affect a subset of cells or tissues affected by the stronger genes. In these cases, how the genes fit into or onto the core pathway is much less clear. Nevertheless, certain themes of organization have emerged from the genetics. Combined with molecular knowledge gained from cloning and expressing the genes, certain mechanisms that seem particular to developmental timing have emerged.

Among the first mutant animals with heterochronic phenotypes were some that resembled the *lin-4* loss-of-function mutants, but were dominant in their genetic behavior (Ambros & Horvitz, 1987). These are gain-of-function alleles of *lin-14* in which portions of the 3′ untranslated region (UTR) of the gene's mRNA are deleted, leaving the open reading frame intact (Wightman, Burglin, Gatto, Arasu, & Ruvkun, 1991). These alleles, together with analysis of *lin-4*, were key to making the most significant discovery in the pathway: microRNAs.

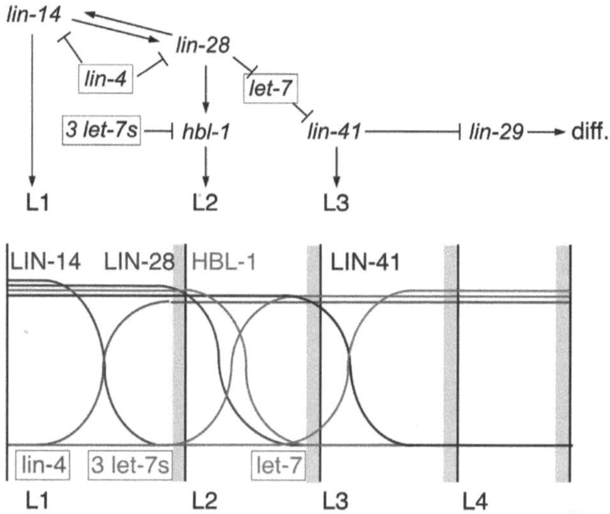

Figure 6.2 Genetic and dynamic models for the heterochronic timing mechanism. The top panel shows one model for the established genetic interactions of core players in the timing mechanism indicating approximately when in developmental time they are active. MicroRNA genes are boxed. Stage-specific fates are indicated by larval stage and terminal differentiation by "diff." The bottom panel represents the rising and falling levels of the mature products of key genes. All microRNAs (boxed) begin development in the off state and rise at specific times, and their targets are repressed accordingly, leading to falling protein levels. Not all factors are shown. (See color plate.)

4. microRNA SWITCHES

The hallmark of the developmental timing mechanism is the sequential action of microRNAs, which are 21–25 nucleotides long, the shortest functional RNAs known. These RNAs bind sequences in the 3′UTRs of target mRNAs, slowing, reducing, or blocking the accumulation of the encoded proteins (Ambros, 2011).

lin-4 encodes a microRNA which begins to accumulate from the mid-L1 and is constitutively expressed thereafter (Feinbaum & Ambros, 1999; Lee, Feinbaum, & Ambros, 1993). The LIN-14 protein is normally present early in the L1 and is downregulated as development proceeds (Ruvkun & Giusto, 1989). Sequences complementary to the *lin-4* RNA are present in the *lin-14* mRNA's 3′UTR, and these are missing in the *lin-14* dominant alleles (Wightman et al., 1991; Wightman, Ha, & Ruvkun, 1993). Thus, the loss of *lin-4* or the *lin-14* 3′UTR sequences causes LIN-14 protein to be expressed throughout larval development resulting in reiteration of L1-specific events. The transition from L1- to L2-specific events is governed by this microRNA switch.

In addition to *lin-4*, several other microRNA genes act in the pathway: *let-7* and its three relatives *mir-48*, *mir-84*, and *mir-241* (Abbott et al., 2005; Reinhart et al., 2000). The *let-7* family of microRNAs share nearly identical 5′ halves of their short 22-nucleotide lengths. Theory predicts that these microRNAs can target the same mRNAs, although whether or not they do *in vivo* is an open question. The three *let-7* relatives are expressed with nearly identical profiles and act redundantly to repress *hbl-1*, which encodes a Hunchback-like transcription factor that controls the transition from L2 to L3 (Abbott et al., 2005; Abrahante et al., 2003; Lin et al., 2003). *let-7* accumulates slightly later and its primary target is *lin-41*, which appears to encode an RNA-regulatory protein or ubiquitin ligase (Loedige, Gaidatzis, Sack, Meister, & Filipowicz, 2012; Reinhart et al., 2000; Slack et al., 2000). There is some controversy about whether it controls the L3 to L4 transition or the L4 to adult transition (Vadla, Kemper, Alaimo, Heine, & Moss, 2012).

In general, the microRNAs show an expression profile that is "off early, on late," and their targets are "on early, off late" (Fig. 6.2, bottom). What distinguishes these switches is the time of accumulation of the microRNA and the consequent time of repression of the target. Although largely still ill-defined, some mechanisms controlling microRNA accumulation have been found.

5. LIN-28: A microRNA REGULATOR AT THE PATHWAY'S CENTER

A second target of the *lin-4* microRNA is *lin-28* (Moss, Lee, & Ambros, 1997). *lin-28* encodes a small RNA-binding protein that is known to bind the precursor of the *let-7* microRNA and block its maturation, thereby supporting *lin-41* against repression (Lehrbach et al., 2009; Moss et al., 1997; Vadla et al., 2012; Van Wynsberghe et al., 2011; Viswanathan, Daley, & Gregory, 2008). It does this by recruiting a poly(U) polymerase (encoded by *pup-2*) that helps to stimulate the degradation of the *let-7* precursor RNA (Heo et al., 2008). Thus, the accumulation of *let-7* is prevented until *lin-28* is repressed by *lin-4*.

However, repressing *let-7* is not all that *lin-28* does. As mentioned above, *lin-28*'s primary role is to promote L2 events. Neither *let-7* nor *lin-41* controls the L2. Instead, *lin-28* appears to act via *hbl-1*, supporting *hbl-1* against repression (Vadla et al., 2012). The mechanism for this is independent of the *let-7* family members, so something else must be repressing *hbl-1*. What this repressor is and how the LIN-28 protein acts in this *let-7*-independent mechanism remain unclear.

lin-28 also has a supportive role in *lin-14* expression, and visa versa. Because *lin-4* targets each of these genes, when *lin-4* is removed both are constitutively expressed (Arasu, Wightman, & Ruvkun, 1991; Moss et al., 1997). Remove either one and the other is repressed (Arasu et al., 1991; Moss et al., 1997; Seggerson, Tang, & Moss, 2002). Thus, *lin-14* and *lin-28* are tied to each other's expression, but again the molecular mechanisms are unknown. Nevertheless, *lin-28* is a stage-specific positive regulator of three stage-specific effectors in the pathway: *lin-14*, *hbl-1*, and *lin-41*.

6. THE HETEROCHRONIC PATHWAY

There are three distinct phases of the heterochronic pathway, each involving a microRNA switch and a connection to the next phase via a key regulator (Fig. 6.2). *lin-4* and *lin-14* comprise the first phase, which transitions the animal from the L1 to the L2. As mentioned, *lin-4* is off early and accumulates during the L1, but how its rise is controlled is not known. *lin-14*, which appears to encode a transcription factor, seems to be all that is needed to specify L1 fates. What was clear from an early analysis of a variety of *lin-14* mutant alleles is that *lin-14* also has a role in determining what happens in the L2 (Ambros & Horvitz, 1987). It is believed that how it exerts that influence is via its positive regulation of *lin-28* (Pepper et al., 2004). Thus, *lin-14* has a direct role in specifying one stage, and influences a key regulator of the next.

The second phase is governed by *lin-28* and *hbl-1*, which are repressed by *lin-4* and the three *let-7* relatives, respectively. *hbl-1* seems to have the more direct role in specifying L2 fates (Abbott et al., 2005; Abrahante et al., 2003). But *lin-28*—like *lin-14*—has the interesting property of controlling two consecutive stages: it positively regulates *hbl-1* and then positively regulates *lin-41* (by inhibiting *let-7*). The repression of *lin-41* by *let-7*, which leads ultimately to the activation of *lin-29*, is the third phase. Recent data suggest the *let-7*/*lin-41* switch might directly regulate L3 fates (Vadla et al., 2012). It is not yet clear if *lin-41* controls two larval stages; a great deal of investigation needs to be done to fill in how this third phase works, including whether *lin-41* acts alone and how *lin-29* is repressed. It has been suggested that *lin-41*, whose mechanism is unclear, might act directly on *lin-29* post-transcriptionally (Del Rio-Albrechtsen, Kiontke, Chiou, & Fitch, 2006).

lin-29, which is required for terminal differentiation of epidermal cells, encodes a transcription factor that can directly regulate terminal differentiation genes (Rougvie & Ambros, 1995).

The genes directing these three phases form the core heterochronic pathway. In general, these genes have highly penetrant and unambiguous phenotypes when mutated. Other factors governing developmental timing in *C. elegans* are characterized by how they integrate into this core pathway. Still, there is quite a bit missing from this picture, not least of which is how the boundaries between the larval stages fit in.

7. DEVELOPMENTAL TRANSITIONS: THE MOLTS

The heterochronic pathway and molting control must be intimately linked. The very boundaries of the stages are the molts, and the heterochronic pathway confers temporal identity on cells during the intermolt period. Furthermore, precocious mutants like *lin-14* and *lin-28* cease the molting cycle at least one stage too soon, while retarded mutants like *lin-4* and *lin-29* continue molting indefinitely. One expects that the mechanisms coordinating the animal-wide process of molting influence the rise and fall of heterochronic factors that determine the stage-specific cell fates.

Molting is a defining feature of the Ecdysozoa clade of which both *C. elegans* and *Drosophila* are members (Aguinaldo et al., 1997). The mechanisms that guide molting are best understood in flies where pulses of the cholesterol-derived steroid hormone 20-hydroxyecdysone (20-E) impose temporal transitions during the life cycle, timing molting and metamorphosis, by acting through a nuclear receptor heterodimer composed of the ecdysone receptor (EcR) and its partner ultraspiracle (USP) (Koelle et al., 1991; Yao et al., 1993; Yao, Segraves, Oro, McKeown, & Evans, 1992). The signals generated by these periodic interactions induce expression of additional nuclear receptors (DHR3, βFtz-F1, and others) in a precise sequence that temporally patterns the molts and metamorphosis in a coordinated fashion throughout the organism.

C. elegans neither synthesizes nor appears to respond to ecdysteroids, and it lacks EcR and USP orthologs (Chitwood & Feldlaufer, 1990; Sluder, Mathews, Hough, Yin, & Maina, 1999), indicating that molting in this organism is launched by a different trigger. Nevertheless, some aspects of molting control appear to be more universal, with steroid mediated responses providing a unifying theme. The worm lacks cholesterol biosynthesis enzymes and instead depends upon dietary intake of this steroid. When grown on cholesterol deficient media, *C. elegans* larvae exhibit molting defects, supporting a role for steroid ligands in regulating aspects of this process (Yochem, Tuck, Greenwald, & Han, 1999). In addition, downstream components of the fly ecdysone-activated transcriptional cascade are broadly conserved and several function in molting in the nematode. The best characterized of these are NHR-23 (similar to fly DHR3 and mammalian RORα) and NHR-25 (fly βFTZ-F1; mammalian SF1), nuclear receptors whose expression levels cycle with the molts, peaking in the intermolt (Asahina et al., 2000; Gissendanner, Crossgrove, Kraus, Maina, & Sluder, 2004; Gissendanner & Sluder, 2000; Kostrouchova, Krause, Kostrouch, & Rall, 1998, 2001). Whether these conserved receptors respond to ligand, steroidal or otherwise, has not been established.

The loss of NHR-25 function by mutation or RNAi causes molting defects and an array of additional phenotypes, including embryonic and larval lethality and cuticle structural abnormalities (Asahina et al., 2000; Chen, Eastburn, & Han, 2004 Gissendanner & Sluder, 2000; Silhankova, Jindra, & Asahina, 2005). The molting defect is a manifest as an inability to shed the old cuticle and results from the loss of NHR-25 function in the epidermis (Chen et al., 2004). The loss of *nhr-23* function causes a similar suite of phenotypes, including the inability to shed cuticle at each larval stage (Kostrouchova et al., 1998, 2001). Clarifying the roles of these and potentially other nuclear receptors in molting will require the identification of their direct targets, a process now underway (Kouns et al., 2011), and the identification of their ligands (if any). Genome-wide searches have also revealed genes required to execute molts (Frand, Russel, & Ruvkun, 2005; Kang et al., 2013) and together with more directed studies are beginning to reveal the elements of a molting pathway. Although some signaling molecules are included, many of the genes identified act in the mechanics of molting, encoding proteins such as the collagens that build each cuticle and various peptidases whose functions include detaching the old cuticle from the epidermis. These types of genes are critical, providing the nuts and bolts of the molt. But perhaps more fascinating are the factors that orchestrate molting behaviors with the stage-specific developmental programs. Periodic regulation of molting is required during larval stages and must ultimately be terminated to halt the program when the adult stage is reached. How this regulation is achieved brings us back to the heterochronic gene pathway.

The failure of retarded heterochronic mutants to limit molting to four cycles reveals that a terminal output of the timing pathway is to coordinate the cessation of molting with the transition to adulthood. Here too, miRNAs play a central role. *let-7* mutants on their own, as well as animals multiply mutant for *let-7* family miRNAs, inappropriately enter lethargus as adults and partly shed the cuticle, indicating these miRNAs engage the mechanism to terminate molting upon reproductive maturity (Abbott et al., 2005; Hayes, Frand, & Ruvkun, 2006; Reinhart et al., 2000). Let-7 family miRNAs appear to terminate the molting cycles through repression of *nhr-23* and *nhr-25* (Hayes et al., 2006); however, only the *nhr-25* 3′UTR contains a predicted *let-7* family binding site, and whether it is a direct target has not been established. Ties among molting, nuclear receptors, miRNAs, and other timing pathway components, are further strengthened by complex genetic interactions between *nhr-25* and various heterochronic genes, and the identification of *apl-1* (amyloid precursor protein-like-1) as a potential *nhr-25* target (Niwa & Hada, 2010; Niwa, Zhou, Li, & Slack, 2008). *apl-1* functions include molting, it acts downstream of *let-7*, and it interacts genetically with other heterochronic pathway members (Hornsten et al., 2007; Niwa et al., 2008). Many details of these gene interactions remain unresolved.

A striking observation is that the extra molt in *let-7 mir-84* double mutants occurs synchronously in the adult population (Hayes et al., 2006), indicating a timing mechanism at work. This finding emphasizes key questions—what paces the molts, and how are the stage-specific programs integrated with the molting timer? Exciting recent work again implicates the heterochronic gene network, this time through *lin-42* (Monsalve, Van Buskirk, & Frand, 2011), a gene first identified by mutations that cause a precocious heterochronic phenotype (Jeon, Gardner, Miller, Deshler, & Rougvie, 1999).

8. LIN-42: A LINK BETWEEN MOLTING AND DEVELOPMENTAL DECISIONS

LIN-42 provides a remarkable link to biological timing mechanisms in flies and mammals: it is the nematode homolog of the Period family of circadian rhythm proteins. *lin-42* and Per genes exhibit an intriguing parallel in regulation: their mRNA and protein levels oscillate over biologically pertinent time scales. LIN-42 levels fluctuate with the molting cycle (~8–10 h at 25 °C) while PER levels are synchronized with the 24-h clock (Hardin, Hall, & Rosbash, 1990; Jeon et al., 1999; Shearman, Zylka, Weaver, Kolakowski, & Reppert, 1997; Sun et al., 1997; Tei et al., 1997; Tennessen, Gardner, Volk, & Rougvie, 2006). As a result of its phenotypes and expression patterns, *lin-42* function has been most studied in a developmental context; however, one report links *lin-42* activity to a circadian rhythm (Simonetta, Migliori, Romanowski, & Golombek, 2009). For additional consideration of circadian rhythms in *C. elegans*, the reader is directed to the recent literature (Hasegawa, Saigusa, & Tamai, 2005; Migliori, Simonetta, Romanowski, & Golombek, 2011; Olmedo et al., 2012; Temmerman et al., 2011; van der Linden et al., 2010).

The cyclic developmental profile of LIN-42 accumulation dramatically contrasts with the patterns of LIN-14, LIN-28, and HBL-1. Levels of these proteins are initially high in the L1 epidermis, then stage-specifically decrease during early larval stages and remain absent (Fig. 6.2), yet their loss-of-function also causes precocious phenotypes (Abrahante et al., 2003; Lin et al., 2003; Moss et al., 1997; Ruvkun & Giusto, 1989). Thus, *lin-42* appears qualitatively different, with an expression pattern suggestive of multiple or reiterated roles during development. Such a repeated role is not reflected in the early epidermal lineage patterns of *lin-42* mutants which appear normal until the L3 molt when an adult-type cuticle appears one stage early (Jeon et al., 1999). However, certain double mutant combinations do reveal epidermal lineage defects prior to the L3 stage, implying earlier redundant role(s) for *lin-42* in controlling these divisions (Abrahante, Miller, & Rougvie, 1998; Ren & Zhang, 2010). But, it is the conspicuous reiterative pattern of LIN-42 levels, peaking once per larval stage, that is most evocative, suggesting a possible link to the molting cycle.

Curiously, the first described *lin-42* alleles were only observed to have defects in the final molt; the first three molts appeared normal despite the cyclical pattern of *lin-42* expression (Jeon et al., 1999). We now know that early molting defects were hidden by complexities of the *lin-42* locus. LIN-42 and Period proteins share multiple regions of homology, notably a PAS domain that mediates protein–protein interactions and smaller SYQ and LT domains that reside within a C-terminal regulatory region of Per (Chang & Reppert, 2003; Tennessen et al., 2006). *lin-42* encodes four isoforms, two of which do not overlap. One of these nonoverlapping isoforms contains the PAS domain (LIN-42c), while the other contains the SYQ and LT domains (LIN-42a); the remaining two isoforms contain all of these homology domains (Fig. 6.3). The first alleles of *lin-42* left the SYQ/LT isoform intact, but an important recent study of an allele (*ok2385*) that completely deletes this isoform (and truncates the long isoforms) reveals reiterative function(s) for *lin-42* in the molting cycle (Monsalve et al., 2011), and implies that there is some division of labor among the isoforms.

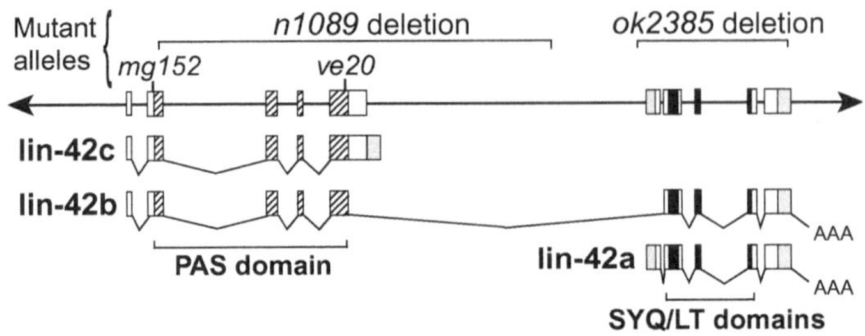

Figure 6.3 Schematic of the *lin-42* locus. The top line indicates the *lin-42* genomic locus with boxes representing exons. The regions encoding the PAS and SYQ/LT domains conserved in fly and mammalian PER proteins are indicated by hatching and black boxes, respectively. Untranslated portions of exons are gray. Positions of lesions in mutant alleles that affect subsets of isoforms are shown above the locus. The first described alleles (e.g., *mg152, n1089, ve20*) did not alter the *lin-42a* transcription unit, which plays a key role in molting regulation and is deleted in *ok2853*. A fourth transcription unit that lacks the last two exons of lin-42b has been omitted for simplicity.

lin-42a mutants differ dramatically from wild-type animals during larval development, exhibiting defects in epidermal cell dynamics and molting at each larval stage, along with the precocious phenotype in the L3 (Monsalve et al., 2011). Rather than developing in near unison as do populations of wild-type animals hatched at the same time, *lin-42a* mutants exhibit severe developmental delays, feed for variable amounts of time prior to an extended lethargus, and molt asynchronously. Animals at each larval stage can become stuck during the molt, failing to complete ecdysis, revealing that LIN-42a is repeatedly required for this process. Moreover, the epidermal seam appears to be the key site of LIN-42 action for regulating these functions. These observations indicate that LIN-42a is required for the rapid synchronous development of wild-type animals and proper execution of molts, leading to the proposal that the rise and fall of the *lin-42a* isoform directs the cyclical entry and exit from each molt, thereby pacing this process. That forced temporal mis-expression of *lin-42a* can induce a lethargus-like state and molting (albeit abnormal) supports this model. This role of *lin-42* in specifying a periodic behavior harkens back again to circadian rhythms, where oscillations in levels of a similar protein in other organisms link behaviors to the 24-h clock.

lin-42a appears to be the elusive factor that integrates successive stage-specific cell lineage programs and epidermal dynamics with the oscillatory molting behaviors, pacing this aspect of postembryonic development. If levels of this LIN-42 isoform are key to the execution and pacing of molts, then several questions need answers. How are its expression dynamics controlled? What are the targets of LIN-42a activity and do they include *let-7* family miRNAs? And what are contributions of the other isoforms? These issues are presently being intensely studied. Transcription factors required for cyclical *lin-42* expression have not been reported, but nuclear receptors, of which there are many in *C. elegans*, are good candidates and could allow for endocrine control of the process. There is also potential for regulatory interactions among the LIN-42 isoforms. Over-expression of *lin-42a* can rescue the heterochronic defects of all tested mutations in the locus, suggesting that its high dosage can compensate for lack of the other isoforms (Tennessen et al., 2006), perhaps overcoming a missing regulatory or intracellular targeting function. Also curious is the observation that over-expression of the *lin-42c* PAS-domain isoform can antagonize the *lin-42a* mutant phenotype (Monsalve et al., 2011). Although the biochemical roles of each isoform await identification, analogy to Per, together with its nuclear localization, suggests LIN-42 is likely to regulate transcription of downstream genes that control epidermal cell behaviors and integrate them temporally with the molts.

In addition to programming molts, *lin-42* exerts another important influence as it charts an efficient path to reproductive maturity—it mediates organismal responses to environmental conditions. In this capacity, *lin-42* participates in the decision of whether to interrupt the molting cycles of continuous development to enable dauer formation. Here, LIN-42 acts to antagonize the ligand-free form of the nuclear receptor DAF-12 (Tennessen, Opperman, & Rougvie, 2010).

9. REVERSIBLE INTERRUPTION OF DEVELOPMENTAL PROGRESSION: DAF-12

The DAF-12 nuclear receptor plays a unique role in nematode life history regulation by controlling the choice between developing rapidly through each larval stage to reach reproductive maturity versus executing a contingency plan, dauer larva formation, in response to unfavorable growth conditions (Antebi, Culotti, & Hedgecock, 1998; Antebi, Yeh, Tait, Hedgecock, & Riddle, 2000). The ability to interrupt reproductive development for dauer formation intertwines many regulatory pathways. Environmental signals are received and conveyed throughout the animal to interrupt the molting cycle and halt one set of stage-specific programs in order to implement another, while maintaining the ability to coordinately reverse the process and restart reproductive development from the point where it left off.

DAF-12 is distinctive among worm nuclear receptors in that its ligands, the dafachronics acids (DA) have been identified (Held et al., 2006; Motola et al., 2006) (see also Chapter 7). These cholesterol-derived steroidal ligands bind to DAF-12 during favorable growth conditions and promote reproductive development. In adverse conditions of elevated temperature and/or limited food supply, DA levels are low and ligand-free DAF-12 binds the DIN-1 corepressor and promotes dauer formation (Ludewig et al., 2004). Consequently, DAF-12 acts as a switch to modulate nematode life history. Animals lacking *daf-12* activity altogether are dauer defective and undergo reproductive development regardless of conditions. Developmental timing programs in these animals are largely normal, with only a weakly penetrant reiteration of L2 programs at the L3 stage. However, this retarded phenotype predominates in mutants where DAF-12 is truncated in the ligand binding domain, and it is likely to arise from ligand-free DAF-12 acting with DIN-1 as a constitutive repressor (Antebi et al., 1998, 2000). DAF-12 is thus absolutely required for dauer formation at the L2 molt, but plays a more modulatory role in the heterochronic pathway by helping to control the L2 to L3 transition.

LIN-42 inhibits dauer formation by antagonizing ligand-free DAF-12, thereby promoting continuous development under conditions of mild stress and adding a level of robustness to the dauer decision (Tennessen et al., 2010). At elevated temperatures, *lin-42* mutants constitutively enter the dauer pathway in a DIN-1-dependent manner. Consistent with a role in preventing dauer formation, LIN-42 levels are dramatically reduced in larvae destined to be dauers, and forced expression of LIN-42 in this stage can bypass dauer formation. *lin-42* and *daf-12* also act in opposition during continuous development; *lin-42*; *daf-12* double mutants exhibit mutual suppression of epidermal phenotypes (Tennessen et al., 2006). Together, these genes indirectly sense external stimuli and make the choice between reproduction and an extended juvenile phase.

The mechanism by which DAF-12 coordinates postembryonic developmental progression is coming into focus through the identification of its direct targets. DAF-12 targets appear diverse, as might be expected given its global roles in development and aging, but conspicuously include core heterochronic pathway components, such as *hbl-1*, *lin-28*, *lin-41*, and *lin-42*, as well as the *din-1*, *daf-12*'s partner in dauer formation (Hochbaum et al., 2011). Directed studies are now required to tease apart the relative contributions of these interactions to DAF-12 functions. Also of key relevance are miRNA targets of DAF-12, notably the *let-7* family miRNA genes. Ligand bound DAF-12 moderately activates *let-7* family miRNAs during continuous development, and, for at least *mir-84* and *mir-241*, this regulation appears direct, resulting in promotion of L3 fates through repression of HBL-1 levels (Bethke, Fielenbach, Wang, Mangelsdorf, & Antebi, 2009; Hammell, Karp, & Ambros, 2009). Ligand-free DAF-12, in contrast, is a potent repressor of these miRNAs in dauers, suspending their ability to program stage transitions upon dauer entry. Not only are these miRNA genes targets of DAF-12, but repression of DAF-12 levels during the L3 stage is achieved through feedback targeting of *daf-12* mRNAs by the *let-7* family miRs, regulation that is necessary for proper developmental progression.

10. INTERRUPTED DEVELOPMENT: REPROGRAMMING AND RESYNCHRONIZATION

An unexpected phenomenon was revealed when heterochronic mutants progressed through the dauer larval stage. Certain heterochronic mutants come out of the dauer stage appearing developmentally normal and are able to lay eggs and mate (Abrahante et al., 2003; Euling & Ambros, 1996; Liu & Ambros, 1991). However, only those mutants whose defects begin early, in the L1 or L2, are subject to this postdauer suppression. Mutants whose developmental defects appear late in development, such as *lin-29* mutants, are not suppressed by going through dauer. But regardless of whether the mutant would have ceased molting precociously after the third stage, or whether it would have continued molting after the fourth stage, all of these suppressed postdauer mutants undergo exactly two molts after emerging from the dauer (Fig. 6.4). This phenomenon implies the existence of a molecular mechanism for setting developmental time that is independent of the early-acting heterochronic genes.

This reprogramming of cell fates upon passage through the dauer is seen more vividly in the hermaphrodite vulva (Euling & Ambros, 1996). Because dauer larvae arrest midway through larval development, precocious vulva development sometimes gets caught in process when the dauer arrest occurs (Fig. 6.4). Remarkably, when the animal emerges from the dauer it develops a perfectly normal vulva, but not in the way that one would expect: it does not simply pick up where it left off. Normally, three of six VPCs divide multiple times to generate the 22 cells of the vulva. In precocious mutants that

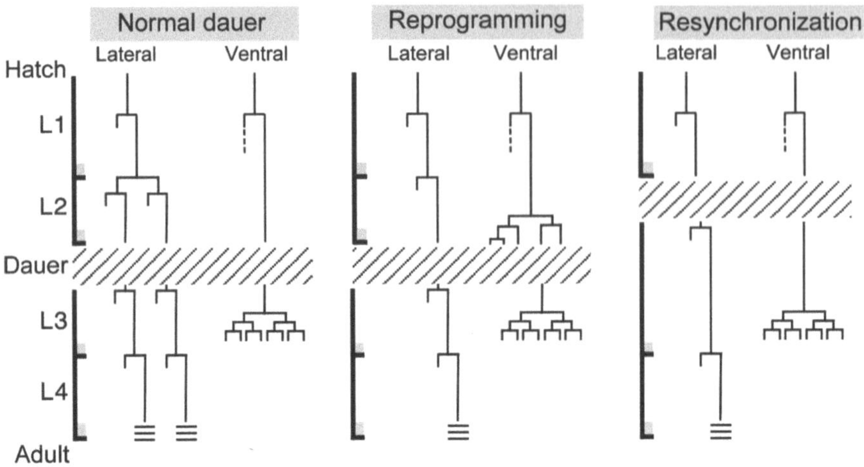

Figure 6.4 Schematic of cell lineage behaviors following dauer development. Examples of lateral and ventral epidermal lineages are shown for wild-type animals and heterochronic mutants that undergo reprogramming or resynchronization. Lineage diagrams are as in Fig. 6.1, with the variable amount of time spent in dauer arrest indicated by hatching. The vulva arises from a subset of the ventral epidermal cells, the vulva precursor cells (VPCs). Vulval reprogramming occurs in precocious mutants where VPCs have divided prior to dauer entry; three VPC progeny (only one is shown for simplicity) will be reprogrammed to divide postdauer, producing the wild-type vulva pattern. Animals that enter dauer too early, at the L1 molt (third panel), undergo a lengthened larval stage upon dauer recovery and tissues resynchronize.

pass through dauer, regardless of how many cells were generated prior to dauer arrest, vulva development is reinitiated, and only three cells divide further to generate a proper vulva. Passage through the dauer stage reprograms the cells so that they revert to their ancestral fates. The only exception is if a VPC divided three times prior to dauer arrest; such cells are incapable of being reprogrammed.

The mechanistic explanation for this reprogramming is incomplete; however, it is known that the expression of certain vulva fate markers and signaling pathways is repressed during the dauer stage, indicating that any commitment prior to arrest is therefore reversible (Karp & Greenwald, 2013). In addition, the important developmental regulator DAF-16/FOXO has a critical role in the VPCs themselves for the maintenance of multipotency (Karp & Greenwald, 2013).

One further variation on this phenomenon is that if certain mutants enter the dauer stage at the wrong time, then they can still emerge from dauer with "corrected" development. Mutants that alter *lin-14* expression can form dauer larvae either early, at the L1 stage, or late, at the L3 stage (Liu & Ambros, 1989). When these animals emerge from the dauer, they still undergo two additional larval stages; however, despite having undergone a total of three or five larval stages, they can be fully functional, egg-laying competent animals. This is something they would not have achieved if they had developed without arrest. A curiosity is that the postdauer animal seems to synchronize its somatic development, which has undergone one too few or one too many larval stages already, with its gonad development, whose progress is unaffected by the heterochronic genes. Thus, the first postdauer larval stage is sped-up or slowed-down, presumably to get in sync with the gonad. The mechanism of this resynchronization is entirely unknown.

11. LAUNCHING LARVAL DEVELOPMENT

By definition, the postembryonic phase of the worm life cycle begins when the worm hatches from the egg. With this embryo to larva transition, life for the worm changes dramatically; direct exposure to the environment brings with it the ability to forage, feed, and grow. But what launches the larval developmental programs? A growing body of work indicates that postembryonic developmental programs are genetically repressed until appropriate signals are received from the local milieu.

The key trigger is nutritional. Larval progression initiates only in the presence of a food source, typically provided by a lawn of *E. coli* in the lab. When *E. coli* is present, development begins and proceeds apace, in a stereotypic fashion with high temporal synchrony between individuals. Without food, the L1 larva arrests both somatic and germline development, entering a state known as L1 arrest or L1 diapause, and it can survive more than 2 weeks. If a food source becomes available, development resumes synchronously throughout the animal with stage-specific programs appropriately scheduled relative to each other and to the molts, suggesting endocrine involvement. As for postdauer development, what is particularly impressive here is that the same cell division pattern and sequence occurs regardless of the interruption.

Diverse developmental programs are suspended during L1 arrest, and the heterochronic gene pathway is not activated. The *lin-4* miRNA fails to accumulate, and consequently, LIN-14 levels remain high (Arasu et al., 1991; Feinbaum & Ambros, 1999). The *lin-4* response appears transcriptional because this expression pattern is recapitulated by assays of promoter fusions; the promoter remains inactive unless food is present (Baugh & Sternberg, 2006). Important for understanding how temporal development initiates then is to link nutritional cue(s) supplied by *E. coli* to signaling pathways in the worm that ultimately activate *lin-4* setting off the pathway. Some clues to the mechanism of L1 arrest are beginning to emerge. Notably, components of the insulin/IGF signaling pathway are required for the maintenance of L1 diapause (Baugh & Sternberg, 2006; Fukuyama, Rougvie, & Rothman, 2006; Gems et al., 1998). Animals mutant for DAF-18/PTEN or DAF-16/FOXO initiate postembryonic programs involving cell division, migration, and cell fusion events even under starvation conditions and have reduced survival rates during L1 diapause.

miRNAs help couple nutrient availability to the activation of larval programs. miRNAs have been identified that are required for the maintenance of L1 diapause, survival during diapause, or whose levels correlate with L1 diapause entry and/or exit (Karp, Hammell, Ow, & Ambros, 2011; Kasuga, Fukuyama, Kitazawa, Kontani, & Katada, 2013; Zhang, Zabinsky, Teng, Cui, & Han, 2011). Perhaps most intriguing among these are miR-235 and miR-71. Similar to insulin pathway mutants, miR-235 is required for the arrest of developmental programs during L1 diapause; in its absence, the activation of *lin-4* and *lin-42* is observed, blast cells divide and molting cycles initiate, even when food is lacking (Kasuga et al., 2013). Although much remains to be learned about how activity of this miRNA feeds into heterochronic gene regulation, an insulin pathway mediated reduction in miR-235 levels appears to be an important step in launching larval development.

mir-71 links L1 diapause with postembryonic developmental timing in a surprisingly different way. The loss of *mir-71* has little phenotypic consequence during continuous larval development (Miska et al., 2007) but, unexpectedly, *mir-71* mutants develop with retarded heterochronic defects in the vulva if they have experienced L1 diapause (Zhang et al., 2011). This result suggests that *mir-71* plays a role in ensuring developmental robustness when the worm enters L1 arrest. Further, it predicts that *mir-71* targets mRNAs of proteins whose over-expression would delay the resumption of vulval programs upon feeding; that is, perhaps *mir-71* interfaces with heterochronic gene regulation. Indeed, miR-71 binding sites are predicted in the 3′UTRs of two temporal regulators of vulva development, *lin-42* and *hbl-1*, and these UTRs are *mir-71*-repsonsive during L1 diapause. Given that the loss of *lin-42* or *hbl-1* activity causes precocious defects in vulva formation during continuous development (Abrahante et al., 2003; Tennessen et al., 2006), their inappropriate elevation could delay vulval cell divisions during recovery from L1 arrest. The influence of nutritional status on *mir-71*, *hbl-1*, and *lin-42* expression in the vulval lineages warrants further investigation.

12. A HETEROCHRONIC PATHWAY FOR LARGER ANIMALS?

A question from the very beginning of the heterochronic gene studies has been: "Is this pathway peculiar to nematodes, or does it recur in other animals?" Of the core pathway components, a few have clear orthologs in animals, including mammals. Others have homologs, but it is too early to say whether those are likely to be performing similar roles.

Mammalian LIN28 has an established role in pluripotency, proliferation, development, and cancer (Shyh-Chang & Daley, 2013). It is expressed in a variety of developing tissues, consistent with a role in timing development on a tissue-by-tissue basis (Yang & Moss, 2003). Its role at the cellular level was explored in neural development where it was shown to have a function remarkably like that in *C. elegans* (Balzer, Heine, Jiang, Lee, & Moss, 2010). Continuous expression of LIN28 caused a kind of retarded development where early fates (neurons) were reiterated at the expense of later fates (glia). It remains to be seen whether all of mammalian LIN28's functions can be traced to defects like those in *C. elegans*, or whether it has diversified its role over evolutionary time.

LIN41 has also been shown to control development in the central nervous system, and mouse LIN41 knockouts display a kind of precocious development in this tissue (Chen, Lai, & Niswander, 2012). LIN41 is also widely expressed and in many of the same tissues as LIN28 (Kanamoto, Terada, Yoshikawa, & Furukawa, 2006; Schulman, Esquela-Kerscher, & Slack, 2005; Yang & Moss, 2003). It is intriguing that these two genes might represent a bit of the heterochronic pathway that has been conserved in mammals. A great deal of work remains to be done to see whether they work together to regulate developmental transitions.

13. CONCLUSIONS AND FUTURE DIRECTIONS

In the interval since Ambros and Horvitz first realized that a few *C. elegans* mutants could be classified as displaying developmental timing defects, much progress has been made toward understanding how temporal patterning is achieved. It is perhaps fitting that the first two miRNAs were discovered here; miRNAs play pivotal roles in the heterochronic gene pathway, acting as a sequential switches that guide transitions from one larval stage to the next. Indeed, miRNAs now provide a common thread among diverse timing pathways in both plant and animal systems. The key miRNA-target gene associations that guide stage transitions have been identified in the worm, but much remains to be learned about the functions of the protein coding targets. The molecular activities of some remain unresolved and the identities of their direct contacts are largely unknown. In the case of the transcription factors (e.g., LIN-14, HBL-1, DAF-12), direct targets have begun to be identified, but how they time and organize cell division cycles and other events remains unclear. Similarly, with LIN-28 we have a glimpse of one way in which miRNA levels can be timed, but the complete picture of how miRNA levels are controlled in the pathway promises to be more complex with additional transcriptional and posttranscriptional inputs.

The first inkling of how molting cycles are paced and aligned with stage-specific programs has been revealed through the studies of LIN-42; this protein participates in both functions, controlling molt time and execution while also acting to specify epidermal lineage patterns. How oscillatory levels of its various isoforms are programmed and combined to achieve these regulatory outputs remains to be learned, and the identification of LIN-42's activity, targets, and interaction partners are prime goals for future studies.

Layered on top of the core heterochronic gene pathway that times larval development and paces the molts are external inputs that launch and can alter life history paths. How environmental conditions are transduced to coordinately initiate, and sometimes interrupt and subsequently restart, developmental programs are great problems to pursue. Linking these factors to environmental cues are key next steps. How does feeding control *mir-235* levels and how do they ultimately modulate the heterochronic pathway? How does the external environment modulate DAF-12 ligand levels to allow interruption of larval progression? Ultimately, these combined avenues of study will advance our understanding of the intimate connections between environmental cues and the mechanisms that temporally pattern orchestrate developmental transitions.

ACKNOWLEDGMENTS

The authors thank Kate McCulloch for helpful comments on the manuscript. We apologize to our colleagues whose work could not be cited due to space constraints.

REFERENCES

Abbott, A. L., Alvarez-Saavedra, E., Miska, E. A., Lau, N. C., Bartel, D. P., & Horvitz, H. R. (2005). The *let-7* microRNA family members *mir-48*, *mir-84*, and *mir-241* function together to regulate developmental timing in *Caenorhabditis elegans*. *Developmental Cell*, *9*(3), 403–414.

Abrahante, J. E., Daul, A. L., Li, M., Volk, M. L., Tennessen, J. M., & Miller, E. A. (2003). The *Caenorhabditis elegans hunchback*-like gene *lin-57/hbl-1* controls developmental time and is regulated by microRNAs. *Developmental Cell*, *4*(5), 625–637.

Abrahante, J. E., Miller, E. A., & Rougvie, A. E. (1998). Identification of heterochronic mutants in *Caenorhabditis elegans*: Temporal misexpression of a collagen::green fluorescent protein fusion gene. *Genetics*, *149*(3), 1335–1351.

Aguinaldo, A. M., Turbeville, J. M., Linford, L. S., Rivera, M. C., Garey, J. R., & Raff, R. A. (1997). Evidence for a clade of nematodes, arthropods and other moulting animals. *Nature*, *387*(6632), 489–493. http://dx.doi.org/10.1038/387489a0.

Ambros, V. (1989). A hierarchy of regulatory genes controls a larva-to-adult developmental switch in *C. elegans*. *Cell*, *57*(1), 49–57.

Ambros, V. (2011). MicroRNAs and developmental timing. *Current Opinion in Genetics and Development*, *21*(4), 511–517. http://dx.doi.org/10.1016/j.gde.2011.04.003.

Ambros, V., & Horvitz, H. R. (1984). Heterochronic mutants of the nematode *Caenorhabditis elegans*. *Science*, *226*(4673), 409–416.

Ambros, V., & Horvitz, H. R. (1987). The *lin-14* locus of *Caenorhabditis elegans* controls the time of expression of specific postembryonic developmental events. *Genes & Development*, *1*(4), 398–414.

Antebi, A., Culotti, J. G., & Hedgecock, E. M. (1998). *daf-12* regulates developmental age and the dauer alternative in *Caenorhabditis elegans*. *Development*, *125*(7), 1191–1205.

Antebi, A., Yeh, W. H., Tait, D., Hedgecock, E. M., & Riddle, D. L. (2000). *daf-12* encodes a nuclear receptor that regulates the dauer diapause and developmental age in *C. elegans*. *Genes & Development*, *14*(12), 1512–1527.

Arasu, P. A., Wightman, B., & Ruvkun, G. (1991). Temporal regulation of *lin-14* by the antagonistic action of two other heterochronic genes, *lin-4* and *lin-28*. *Genes & Development*, *5*, 1825–1833.

Asahina, M., Ishihara, T., Jindra, M., Kohara, Y., Katsura, I., & Hirose, S. (2000). The conserved nuclear receptor Ftz-F1 is required for embryogenesis, moulting and reproduction in *Caenorhabditis elegans*. *Genes to Cells*, *5*(9), 711–723.

Balzer, E., Heine, C., Jiang, Q., Lee, V. M., & Moss, E. G. (2010). LIN28 alters cell fate succession and acts independently of the let-7 microRNA during neurogliogenesis in vitro. *Development*, *137*(6), 891–900. http://dx.doi.org/10.1242/dev.042895.

Baugh, L. R., & Sternberg, P. W. (2006). DAF-16/FOXO regulates transcription of cki-1/Cip/Kip and repression of lin-4 during *C. elegans* L1 arrest. *Current Biology*, *16*(8), 780–785.

Bethke, A., Fielenbach, N., Wang, Z., Mangelsdorf, D. J., & Antebi, A. (2009). Nuclear hormone receptor regulation of microRNAs controls developmental progression. *Science*, *324*(5923), 95–98. http://dx.doi.org/10.1126/science.1164899.

Chang, D. C., & Reppert, S. M. (2003). A novel C-terminal domain of *Drosophila* PERIOD inhibits dCLOCK:CYCLE-mediated transcription. *Current Biology*, *13*(9), 758–762.

Chen, Z., Eastburn, D. J., & Han, M. (2004). The *Caenorhabditis elegans* nuclear receptor gene nhr-25 regulates epidermal cell development. *Molecular and Cellular Biology*, *24*(17), 7345–7358. http://dx.doi.org/10.1128/MCB.24.17.7345-7358.2004.

Chen, J., Lai, F., & Niswander, L. (2012). The ubiquitin ligase mLin41 temporally promotes neural progenitor cell maintenance through FGF signaling. *Genes & Development*, *26*(8), 803–815. http://dx.doi.org/10.1101/gad.187641.112.

Chitwood, D. J., & Feldlaufer, M. F. (1990). Ecdysteroids in axenically propagated *Caenorhabditis elegans* and culture medium. *Journal of Nematology*, *22*(4), 598–607.

Del Rio-Albrechtsen, T., Kiontke, K., Chiou, S. Y., & Fitch, D. H. (2006). Novel gain-of-function alleles demonstrate a role for the heterochronic gene lin-41 in *C. elegans* male tail tip morphogenesis. *Developmental Biology*, *297*(1), 74–86.

Euling, S., & Ambros, V. (1996). Reversal of cell fate determination in *Caenorhabditis elegans* vulval development. *Development*, *122*, 2507–2515.

Feinbaum, R., & Ambros, V. (1999). The timing of *lin-4* RNA accumulation controls the timing of postembryonic developmental events in *Caenorhabditis elegans*. *Developmental Biology*, *210*(1), 87–95.

Ferguson, E. L., & Horvitz, H. R. (1985). Identification and characterization of 22 genes that affect the vulval cell lineages of the nematode *Caenorhabditis elegans*. *Genetics*, *110*(1), 17–72.

Frand, A. R., Russel, S., & Ruvkun, G. (2005). Functional genomic analysis of *C. elegans* molting. *PLoS Biology*, *3*(10), e312.

Fukuyama, M., Rougvie, A. E., & Rothman, J. H. (2006). *C. elegans* DAF-18/PTEN mediates nutrient-dependent arrest of cell cycle and growth in the germline. *Current Biology*, *16*(8), 773–779.

Gems, D., Sutton, A. J., Sundermeyer, M. L., Albert, P. S., King, K. V., & Edgley, M. L. (1998). Two pleiotropic classes of daf-2 mutation affect larval arrest, adult behavior, reproduction and longevity in *Caenorhabditis elegans*. *Genetics*, *150*(1), 129–155.

Gissendanner, C. R., Crossgrove, K., Kraus, K. A., Maina, C. V., & Sluder, A. E. (2004). Expression and function of conserved nuclear receptor genes in *Caenorhabditis elegans*. *Developmental Biology*, *266*(2), 399–416.

Gissendanner, C. R., & Sluder, A. E. (2000). nhr-25, the *Caenorhabditis elegans* ortholog of ftz-f1, is required for epidermal and somatic gonad development. *Developmental Biology*, *221*(1), 259–272. http://dx.doi.org/10.1006/dbio.2000.9679.

Hammell, C. M., Karp, X., & Ambros, V. (2009). A feedback circuit involving let-7-family miRNAs and DAF-12 integrates environmental signals and developmental timing in *Caenorhabditis elegans*. *Proceedings of the National Academy of Sciences of the United States of America*, *106*(44), 18668–18673. http://dx.doi.org/10.1073/pnas.0908131106.

Hardin, P. E., Hall, J. C., & Rosbash, M. (1990). Feedback of the *Drosophila* period gene product on circadian cycling of its messenger RNA levels. *Nature*, *343*, 536–540.

Hasegawa, K., Saigusa, T., & Tamai, Y. (2005). *Caenorhabditis elegans* opens up new insights into circadian clock mechanisms. *Chronobiology International*, *22*(1), 1–19.

Hayes, G. D., Frand, A. R., & Ruvkun, G. (2006). The mir-84 and let-7 paralogous micro-RNA genes of *Caenorhabditis elegans* direct the cessation of molting via the conserved nuclear hormone receptors NHR-23 and NHR-25. *Development*, *133*(23), 4631–4641. http://dx.doi.org/10.1242/dev.02655.

Held, J. M., White, M. P., Fisher, A. L., Gibson, B. W., Lithgow, G. J., & Gill, M. S. (2006). DAF-12-dependent rescue of dauer formation in *Caenorhabditis elegans* by (25S)-cholestenoic acid. *Aging Cell*, *5*(4), 283–291.

Heo, I., Joo, C., Cho, J., Ha, M., Han, J., & Kim, V. N. (2008). Lin28 mediates the terminal uridylation of let-7 precursor MicroRNA. *Molecular Cell*, *32*(2), 276–284. http://dx.doi.org/10.1016/j.molcel.2008.09.014.

Hochbaum, D., Zhang, Y., Stuckenholz, C., Labhart, P., Alexiadis, V., & Martin, R. (2011). DAF-12 regulates a connected network of genes to ensure robust developmental decisions. *PLoS Genetics*, 7(7), e1002179. http://dx.doi.org/10.1371/journal.pgen.1002179.

Hornsten, A., Lieberthal, J., Fadia, S., Malins, R., Ha, L., & Xu, X. (2007). APL-1, a *Caenorhabditis elegans* protein related to the human beta-amyloid precursor protein, is essential for viability. *Proceedings of the National Academy of Sciences of the United States of America*, *104*(6), 1971–1976. http://dx.doi.org/10.1073/pnas.0603997104.

Jeon, M., Gardner, H. F., Miller, E. A., Deshler, J., & Rougvie, A. E. (1999). Similarity of the *C. elegans* developmental timing protein LIN-42 to circadian rhythm proteins. *Science, 286*(5442), 1141–1146.

Kanamoto, T., Terada, K., Yoshikawa, H., & Furukawa, T. (2006). Cloning and regulation of the vertebrate homologue of lin-41 that functions as a heterochronic gene in *Caenorhabditis elegans. Developmental Dynamics, 235*(4), 1142–1149. http://dx.doi.org/10.1002/dvdy.20712.

Kang, Y. L., Yochem, J., Bell, L., Sorensen, E. B., Chen, L., & Conner, S. D. (2013). *Caenorhabditis elegans* reveals a FxNPxY-independent low-density lipoprotein receptor internalization mechanism mediated by epsin1. *Molecular Biology of the Cell, 24*(3), 308–318. http://dx.doi.org/10.1091/mbc.E12-02-0163.

Karp, X., & Greenwald, I. (2013). Control of cell-fate plasticity and maintenance of multipotency by DAF-16/FoxO in quiescent *Caenorhabditis elegans. Proceedings of the National Academy of Sciences of the United States of America, 110*(6), 2181–2186. http://dx.doi.org/10.1073/pnas.1222377110.

Karp, X., Hammell, M., Ow, M. C., & Ambros, V. (2011). Effect of life history on microRNA expression during *C. elegans* development. *RNA, 17*(4), 639–651. http://dx.doi.org/10.1261/rna.2310111.

Kasuga, H., Fukuyama, M., Kitazawa, A., Kontani, K., & Katada, T. (2013). The microRNA miR-235 couples blast-cell quiescence to the nutritional state. *Nature, 497*, 503–506. http://dx.doi.org/10.1038/nature12117.

Koelle, M. R., Talbot, W. S., Segraves, W. A., Bender, M. T., Cherbas, P., & Hogness, D. S. (1991). The Drosophila EcR gene encodes an ecdysone receptor, a new member of the steroid receptor superfamily. *Cell, 67*(1), 59–77. http://dx.doi.org/10.1016/0092-8674(91)90572-G.

Kostrouchova, M., Krause, M., Kostrouch, Z., & Rall, J. E. (1998). CHR3: A *Caenorhabditis elegans* orphan nuclear hormone receptor required for proper epidermal development and molting. *Development, 125*(9), 1617–1626.

Kostrouchova, M., Krause, M., Kostrouch, Z., & Rall, J. E. (2001). Nuclear hormone receptor CHR3 is a critical regulator of all four larval molts of the nematode *Caenorhabditis elegans. Proceedings of the National Academy of Sciences of the United States of America, 98*(13), 7360–7365. http://dx.doi.org/10.1073/pnas.131171898.

Kouns, N. A., Nakielna, J., Behensky, F., Krause, M. W., Kostrouch, Z., & Kostrouchova, M. (2011). NHR-23 dependent collagen and hedgehog-related genes required for molting. *Biochemical and Biophysical Research Communications, 413*(4), 515–520. http://dx.doi.org/10.1016/j.bbrc.2011.08.124.

Lee, R. C., Feinbaum, R. L., & Ambros, V. (1993). The *C. elegans* heterochronic gene *lin-4* encodes small RNAs with antisense complementarity to *lin-14. Cell, 75*, 843–854.

Lehrbach, N. J., Armisen, J., Lightfoot, H. L., Murfitt, K. J., Bugaut, A., & Balasubramanian, S. (2009). LIN-28 and the poly(U) polymerase PUP-2 regulate let-7 microRNA processing in *Caenorhabditis elegans. Nature Structural and Molecular Biology, 16*(10), 1016–1020. http://dx.doi.org/10.1038/nsmb.1675.

Lin, S. Y., Johnson, S. M., Abraham, M., Vella, M. C., Pasquinelli, A., & Gamberi, C. (2003). The *C. elegans hunchback* homolog, *hbl-1*, controls temporal patterning and is a probable microRNA target. *Developmental Cell, 4*(5), 639–650.

Liu, Z., & Ambros, V. (1989). Heterochronic genes control the stage-specific initiation and expression of the dauer larva developmental program in *Caenorhabditis elegans. Genes & Development, 3*(12B), 2039–2049.

Liu, Z., & Ambros, V. (1991). Alternative temporal control systems for hypodermal cell differentiation in *Caenorhabditis elegans. Nature, 350*, 162–165.

Loedige, I., Gaidatzis, D., Sack, R., Meister, G., & Filipowicz, W. (2012). The mammalian TRIM-NHL protein TRIM71/LIN-41 is a repressor of mRNA function. *Nucleic Acids Research, 41*(1), 518–532. http://dx.doi.org/10.1093/nar/gks1032.

Ludewig, A. H., Kober-Eisermann, C., Weitzel, C., Bethke, A., Neubert, K., & Gerisch, B. (2004). A novel nuclear receptor/coregulator complex controls *C. elegans* lipid metabolism, larval development, and aging. *Genes & Development*, *18*(17), 2120–2133.

Migliori, M. L., Simonetta, S. H., Romanowski, A., & Golombek, D. A. (2011). Circadian rhythms in metabolic variables in *Caenorhabditis elegans*. *Physiology and Behavior*, *103*(3–4), 315–320. http://dx.doi.org/10.1016/j.physbeh.2011.01.026.

Miska, E. A., Alvarez-Saavedra, E., Abbott, A. L., Lau, N. C., Hellman, A. B., & McGonagle, S. M. (2007). Most *Caenorhabditis elegans* microRNAs are individually not essential for development or viability. *PLoS Genetics*, *3*(12), e215. http://dx.doi.org/10.1371/journal.pgen.0030215.

Monsalve, G. C., Van Buskirk, C., & Frand, A. R. (2011). LIN-42/PERIOD controls cyclical and developmental progression of *C. elegans* molts. *Current Biology*, *21*(24), 2033–2045. http://dx.doi.org/10.1016/j.cub.2011.10.054.

Moss, E. G., Lee, R. C., & Ambros, V. (1997). The cold shock domain protein LIN-28 controls developmental timing in *C. elegans* and is regulated by the *lin-4* RNA. *Cell*, *88*, 637–646.

Motola, D. L., Cummins, C. L., Rottiers, V., Sharma, K. K., Li, T., & Li, Y. (2006). Identification of ligands for DAF-12 that govern dauer formation and reproduction in *C. elegans*. *Cell*, *124*(6), 1209–1223.

Niwa, R., & Hada, K. (2010). Identification of a spatio-temporal enhancer element for the Alzheimer's amyloid precursor protein-like-1 gene in the nematode *Caenorhabditis elegans*. *Bioscience, Biotechnology, and Biochemistry*, *74*(12), 2497–2500. http://dx.doi.org/10.1271/bbb.100450.

Niwa, R., Zhou, F., Li, C., & Slack, F. J. (2008). The expression of the Alzheimer's amyloid precursor protein-like gene is regulated by developmental timing microRNAs and their targets in *Caenorhabditis elegans*. *Developmental Biology*, *315*(2), 418–425. http://dx.doi.org/10.1016/j.ydbio.2007.12.044.

Olmedo, M., O'Neill, J. S., Edgar, R. S., Valekunja, U. K., Reddy, A. B., & Merrow, M. (2012). Circadian regulation of olfaction and an evolutionarily conserved, non-transcriptional marker in *Caenorhabditis elegans*. *Proceedings of the National Academy of Sciences of the United States of America*, *109*(50), 20479–20484. http://dx.doi.org/10.1073/pnas.1211705109.

Pepper, A. S., McCane, J. E., Kemper, K., Yeung, D. A., Lee, R. C., & Ambros, V. (2004). The *C. elegans* heterochronic gene *lin-46* affects developmental timing at two larval stages and encodes a relative of the scaffolding protein gephyrin. *Development*, *131*(9), 2049–2059.

Reinhart, B. J., Slack, F. J., Basson, M., Pasquinelli, A. E., Bettinger, J. C., & Rougvie, A. E. (2000). The 21-nucleotide *let-7* RNA regulates developmental timing in *Caenorhabditis elegans*. *Nature*, *403*(6772), 901–906.

Ren, H., & Zhang, H. (2010). Wnt signaling controls temporal identities of seam cells in *Caenorhabditis elegans*. *Developmental Biology*, *345*(2), 144–155. http://dx.doi.org/10.1016/j.ydbio.2010.07.002.

Rougvie, A. E., & Ambros, V. (1995). The heterochronic gene *lin-29* encodes a zinc finger protein that controls a terminal differentiation event in *C. elegans*. *Development*, *121*, 2491–2500.

Ruvkun, G., & Giusto, J. (1989). The *Caenorhabditis elegans* heterochronic gene *lin-14* encodes a nuclear protein that forms a temporal developmental switch. *Nature*, *338*(6213), 313–319.

Schulman, B. R., Esquela-Kerscher, A., & Slack, F. J. (2005). Reciprocal expression of lin-41 and the microRNAs let-7 and mir-125 during mouse embryogenesis. *Developmental Dynamics*, *234*(4), 1046–1054. http://dx.doi.org/10.1002/dvdy.20599.

Seggerson, K., Tang, L., & Moss, E. G. (2002). Two genetic circuits repress the *Caenorhabditis elegans* heterochronic gene *lin-28* after translation initiation. *Developmental Biology*, *243*(2), 215–225.

Shearman, L. P., Zylka, M. J., Weaver, D. R., Kolakowski, L. F., Jr., & Reppert, S. M. (1997). Two period homologs: Circadian expression and photic regulation in the suprachiasmatic nuclei. *Neuron, 19*(6), 1261–1269.

Shyh-Chang, N., & Daley, G. Q. (2013). Lin28: Primal regulator of growth and metabolism in stem cells. *Cell Stem Cell, 12*(4), 395–406. http://dx.doi.org/10.1016/j.stem.2013.03.005.

Silhankova, M., Jindra, M., & Asahina, M. (2005). Nuclear receptor NHR-25 is required for cell-shape dynamics during epidermal differentiation in *Caenorhabditis elegans*. *Journal of Cell Science, 118*(Pt 1), 223–232. http://dx.doi.org/10.1242/jcs.01609.

Simonetta, S. H., Migliori, M. L., Romanowski, A., & Golombek, D. A. (2009). Timing of locomotor activity circadian rhythms in *Caenorhabditis elegans*. *PLoS One, 4*(10), e7571. http://dx.doi.org/10.1371/journal.pone.0007571.

Slack, F. J., Basson, M., Liu, Z., Ambros, V., Horvitz, H. R., & Ruvkun, G. (2000). The *lin-41* RBCC gene acts in the *C. elegans* heterochronic pathway between the *let-7* regulatory RNA and the LIN-29 transcription factor. *Molecular Cell, 5*(4), 659–669.

Sluder, A. E., Mathews, S. W., Hough, D., Yin, V. P., & Maina, C. V. (1999). The nuclear receptor superfamily has undergone extensive proliferation and diversification in nematodes. *Genome Research, 9*(2), 103–120.

Sun, Z. S., Albrecht, U., Zhuchenko, O., Bailey, J., Eichele, G., & Lee, C. C. (1997). *RIGUI*, a putative mammalian ortholog of the *Drosophila period* gene. *Cell, 90*, 1003–1011.

Tei, H., Okamura, H., Shigeyoshi, Y., Fukuhara, C., Ozawa, R., & Hirose, M. (1997). Circadian oscillation of a mammalian homologue of the *Drosophila period* gene. *Nature, 389*, 512–516.

Temmerman, L., Meelkop, E., Janssen, T., Bogaerts, A., Lindemans, M., & Husson, S. J. (2011). *C. elegans* homologs of insect clock proteins: A tale of many stories. *Annals of the New York Academy of Sciences, 1220*, 137–148. http://dx.doi.org/10.1111/j.1749-6632.2010.05927.x.

Tennessen, J. M., Gardner, H. F., Volk, M. L., & Rougvie, A. E. (2006). Novel heterochronic functions of the *Caenorhabditis elegans* period-related protein LIN-42. *Developmental Biology, 289*(1), 30–43.

Tennessen, J. M., Opperman, K. J., & Rougvie, A. E. (2010). The *C. elegans* developmental timing protein LIN-42 regulates diapause in response to environmental cues. *Development, 137*(20), 3501–3511. http://dx.doi.org/10.1242/dev.048850.

Thummel, C. S. (2001). Molecular mechanisms of developmental timing in *C. elegans* and *Drosophila*. *Developmental Cell, 1*(4), 453–465.

Vadla, B., Kemper, K., Alaimo, J., Heine, C., & Moss, E. G. (2012). lin-28 controls the succession of cell fate choices via two distinct activities. *PLoS Genetics, 8*(3), e1002588. http://dx.doi.org/10.1371/journal.pgen.1002588.

van der Linden, A. M., Beverly, M., Kadener, S., Rodriguez, J., Wasserman, S., & Rosbash, M. (2010). Genome-wide analysis of light- and temperature-entrained circadian transcripts in *Caenorhabditis elegans*. *PLoS Biology, 8*(10), e1000503. http://dx.doi.org/10.1371/journal.pbio.1000503.

Van Wynsberghe, P. M., Kai, Z. S., Massirer, K. B., Burton, V. H., Yeo, G. W., & Pasquinelli, A. E. (2011). LIN-28 co-transcriptionally binds primary let-7 to regulate miRNA maturation in *Caenorhabditis elegans*. *Nature Structural and Molecular Biology, 18*, 302–308. http://dx.doi.org/10.1038/nsmb.1986.

Viswanathan, S. R., Daley, G. Q., & Gregory, R. I. (2008). Selective blockade of microRNA processing by Lin28. *Science, 320*(5872), 97–100. http://dx.doi.org/10.1126/science.1154040.

Wightman, B., Burglin, T. R., Gatto, J., Arasu, P., & Ruvkun, G. (1991). Negative regulatory sequences in the *lin-14* 3′-untranslated region are necessary to generate a temporal switch during *Caenorhabditis elegans* development. *Genes & Development, 5*, 1813–1824.

Wightman, B., Ha, I., & Ruvkun, G. (1993). Posttranscriptional regulation of the heterochronic gene *lin-14* by *lin-4* mediates temporal pattern formation in *C. elegans*. *Cell*, *75*, 855–862.
Yang, D. H., & Moss, E. G. (2003). Temporally regulated expression of Lin-28 in diverse tissues of the developing mouse. *Gene Expression Patterns*, *3*(6), 719–726.
Yao, T. P., Forman, B. M., Jiang, Z., Cherbas, L., Chen, J. D., & McKeown, M. (1993). Functional ecdysone receptor is the product of EcR and Ultraspiracle genes. *Nature*, *366*(6454), 476–479. http://dx.doi.org/10.1038/366476a0.
Yao, T. P., Segraves, W. A., Oro, A. E., McKeown, M., & Evans, R. M. (1992). Drosophila ultraspiracle modulates ecdysone receptor function via heterodimer formation. *Cell*, *71*(1), 63–72. http://dx.doi.org/10.1016/0092-8674(92)90266-F.
Yochem, J., Tuck, S., Greenwald, I., & Han, M. (1999). A gp330/megalin-related protein is required in the major epidermis of *Caenorhabditis elegans* for completion of molting. *Development*, *126*(3), 597–606.
Zhang, X., Zabinsky, R., Teng, Y., Cui, M., & Han, M. (2011). microRNAs play critical roles in the survival and recovery of *Caenorhabditis elegans* from starvation-induced L1 diapause. *Proceedings of the National Academy of Sciences of the United States of America*, *108*(44), 17997–18002. http://dx.doi.org/10.1073/pnas.1105982108.

CHAPTER SEVEN

Steroid Regulation of *C. elegans* Diapause, Developmental Timing, and Longevity

Adam Antebi[*,†,‡,1]

[*]Max Planck Institute for Biology of Ageing, Cologne, Germany
[†]Cologne Excellence Cluster on Cellular Stress Responses in Aging Associated Diseases (CECAD), University of Cologne, Cologne, Germany
[‡]Department of Molecular and Cellular Biology, Huffington Center on Aging, Baylor College of Medicine, Houston, Texas, USA
[1]Corresponding author: e-mail address: AAntebi@age.mpg.de

Contents

1. Introduction 182
   1.1 Hormonal regulation of maturation 182
2. Endocrine Regulation of the Dauer Diapause 183
   2.1 *C. elegans* life history 183
   2.2 A hormone-regulated switch governs the dauer decision 183
   2.3 DA biosynthetic pathway 186
   2.4 DA production is distributed 189
   2.5 DA signal amplification enforces reproductive commitments 190
3. DAF-12 Regulates Developmental Timing 192
   3.1 Heterochronic genes 192
   3.2 The microRNA paradigm 192
   3.3 DAF-12 regulates the *let-7* microRNAs 193
   3.4 DAF-12 targets include heterochronic genes 195
   3.5 *daf-12* is feedback regulated by the *let-7* microRNAs 196
   3.6 *daf-12* acts in opposition to *lin-42*/period 196
   3.7 Dauer reprogramming 198
   3.8 Heterochronic genes and the molt cycle 198
4. Hormonal Control of Longevity 201
   4.1 Gonadal longevity pathway 201
5. Perspectives 203
Acknowledgments 205
References 205

Current Topics in Developmental Biology, Volume 105
ISSN 0070-2153
http://dx.doi.org/10.1016/B978-0-12-396968-2.00007-5

Abstract

Hormones play a critical role in driving major stage transitions and developmental timing events in many species. In the nematode *C. elegans* the steroid hormone receptor, DAF-12, works at the confluence of pathways regulating developmental timing, stage specification, and longevity. DAF-12 couples environmental and physiologic signals to life history regulation, and it is embedded in a rich architecture governing diverse processes. Here, we highlight the molecular insights, extraordinary circuitry, and signaling pathways governing life stage transitions in the worm and how they have yielded fundamental insights into steroid regulation of biological time.

1. INTRODUCTION

1.1. Hormonal regulation of maturation

Hormones play a key role in coordinating metabolism, reproduction, and homeostasis in all metazoans. Some of the earliest seminal discoveries on endocrine mechanisms revealed how small molecular metabolites govern major maturational transitions. These include estrogens/androgens in mammalian puberty, thyroid hormone in amphibian metamorphosis, and ecdysteroids during insect molting and metamorphosis (Mangelsdorf & Evans, 1995; Mangelsdorf et al., 1995). These small molecules bind to cognate nuclear hormone receptor transcription factors androgen, estrogen, thyroid hormone, and the ecdysone receptors, respectively and thereby work as switches to direct broad changes in gene expression that drive developmental transitions.

Nuclear receptors have a conserved architecture, which underlie their signaling capacity. The N-terminus contains paired zinc fingers that constitute the DNA-binding domain, while the C-terminus harbors the ligand-binding domain (LBD), which also serves to recruit coactivator and corepressor complexes. Typically, nuclear receptors activate transcription in the presence of ligand, and in some cases they can repress the same genes in the absence of ligand (Mangelsdorf & Evans, 1995; Mangelsdorf et al., 1995).

The molecular mechanism of steroid receptor signal transduction is conserved in *C. elegans* and exemplified in the nuclear receptor DAF-12. DAF-12 is most homologous to vertebrate farnesoid-X (FXR), liver-X, and vitamin-D receptors, and it binds to bile acid-like steroids called the dafachronic acids (DAs), which regulate its transcriptional activity (Antebi, Yeh, Tait, Hedgecock, & Riddle, 2000; Motola et al., 2006). Because of the worm's relative simplicity and powerful genetics, dissection of steroid signaling has illuminated important and novel facets of metazoan life history regulation. DAF-12 controls broad aspects of *C. elegans* maturation, including regulation of developmental arrest at the long-lived dauer stage, progression from second-to third-larval stage programs in developmental timing circuits, and organismal longevity. DAF-12's essential role in these processes is to couple environmental and physiologic information to the precise timing of reproductive development. In each context, it is embedded in a remarkably rich circuitry, whose elucidation has yielded major insights into endocrine regulation of metazoan reproductive commitments, developmental timing, and longevity. Here, the author reviews the role of steroid receptor DAF-12 in the framework of these circuits.

2. ENDOCRINE REGULATION OF THE DAUER DIAPAUSE

2.1. *C. elegans* life history

All animals have the ability to sense environmental quality and nutrient availability, and adjust rates of maturation accordingly. In favorable environments, *C. elegans* develops rapidly from embryo through four larval stages (L1–L4) marked by molts to adulthood in about 3.5 days, termed reproductive development. They produce progeny for a week, and then typically live another two weeks. In unfavorable environments marked by food scarcity, elevated temperatures, and overcrowding indicated by ascaroside pheromones (Butcher, Fujita, Schroeder, & Clardy, 2007; Golden & Riddle, 1984), they arrest development at an alternate third larval stage called the dauer diapause, which is specialized for survival and dispersal (Cassada & Russell, 1975). Dauer larvae are arrested at a "prepubertal" quiescent state, morphologically and metabolically distinct, highly stress resistant and long lived, surviving over 4 months without food. Yet, when returned to favorable conditions they will mature into reproductive adults with a normal life span. Although mammals do not have explicit states of diapause, they can enter alternate metabolic states of hibernation, torpor, caloric restriction, or maturational delay, which mimic aspects of dauer.

2.2. A hormone-regulated switch governs the dauer decision

What are the molecular pathways that govern dauer formation? Early genetic screens identified abnormal *da*uer *f*ormation (Daf) mutants that either enter dauer constitutively (Daf-c) or are defective in dauer formation (Daf-d) (Riddle, Swanson, & Albert, 1981). A molecular genetic analysis of the

Daf genes reveals that they comprise an endocrine network converging on DAF-12, which works as a molecular switch to mediate the choice between arrest at diapause and progression to reproductive development (Fig. 7.1A). These studies have led to a model in which environmental and physiologic cues detected by ciliated sensory neurons in the head, are transduced through guanylyl cyclase and various signal transduction pathways (Fielenbach & Antebi, 2008). These pathways in turn impinge on two major endocrine pathways, TGF-β and insulin/IGF signaling (IIS), which regulate DAF-12 steroidal signal transduction. In favorable environments, TGF-β and insulin/IGF hormones are produced and secreted from sensory neurons (Li, Kennedy, & Ruvkun, 2003; Ren et al., 1996), activating their respective signal transduction pathways in target tissues. Specifically TGF-β binds to its receptors, which regulate Smad/Co-Smad transcriptional complexes (da Graca et al., 2004; Estevez et al., 1993; Georgi, Albert, & Riddle, 1990; Patterson, Koweek, Wong, Liu, & Ruvkun, 1997; Ren et al., 1996). Insulin-like peptides bind to the insulin receptor, stimulating a PI3/AKT kinase cascade that inhibits the DAF-16/FOXO transcription factor (Kimura, Tissenbaum, Liu, & Ruvkun, 1997; Li et al., 2003; Lin, Dorman, Rodan, & Kenyon, 1997; Morris, Tissenbaum, & Ruvkun, 1996; Ogg et al., 1997; Paradis & Ruvkun, 1998). These two pathways converge within steroidogenic tissues, to promote production of the cholesterol derived DAF-12 ligands the DAs (Motola et al., 2006). Thereafter, DA binds to nuclear DAF-12 expressed throughout the body to prevent dauer formation and promote reproductive development. It is postulated that ligand bound DAF-12 associates in coactivator complexes, but these remain to be identified. Conversely in unfavorable environments, TGF-β and insulin-like peptides are downregulated, stimulating transcriptional complexes of DAF-3/Co-Smad and DAF-5/SKI as well as DAF-16/FOXO (da Graca et al., 2004; Patterson et al., 1997), which initiate specific changes toward thrifty metabolism and increased stress response. Biosynthesis of the DAs is repressed, unliganded DAF-12 forms a potent repression complex with DIN-1/SHARP corepressor, and thereby specifies the long-lived dauer stage (Ludewig et al., 2004). Thus DA/DAF-12 serves as a molecular switch between reproductive and survival modes, and is instructive in this choice: null mutations that contain dual lesions affecting both DNA and LBDs result in animals unable to enter the dauer stage (Daf-d), while LBD missense mutations cause animals to constitutively enter the dauer stage (Daf-c) (Antebi et al., 2000). While DAF-12 responsibilities during dauer formation are essential, its functions in reproductive development are modulatory and only partly required.

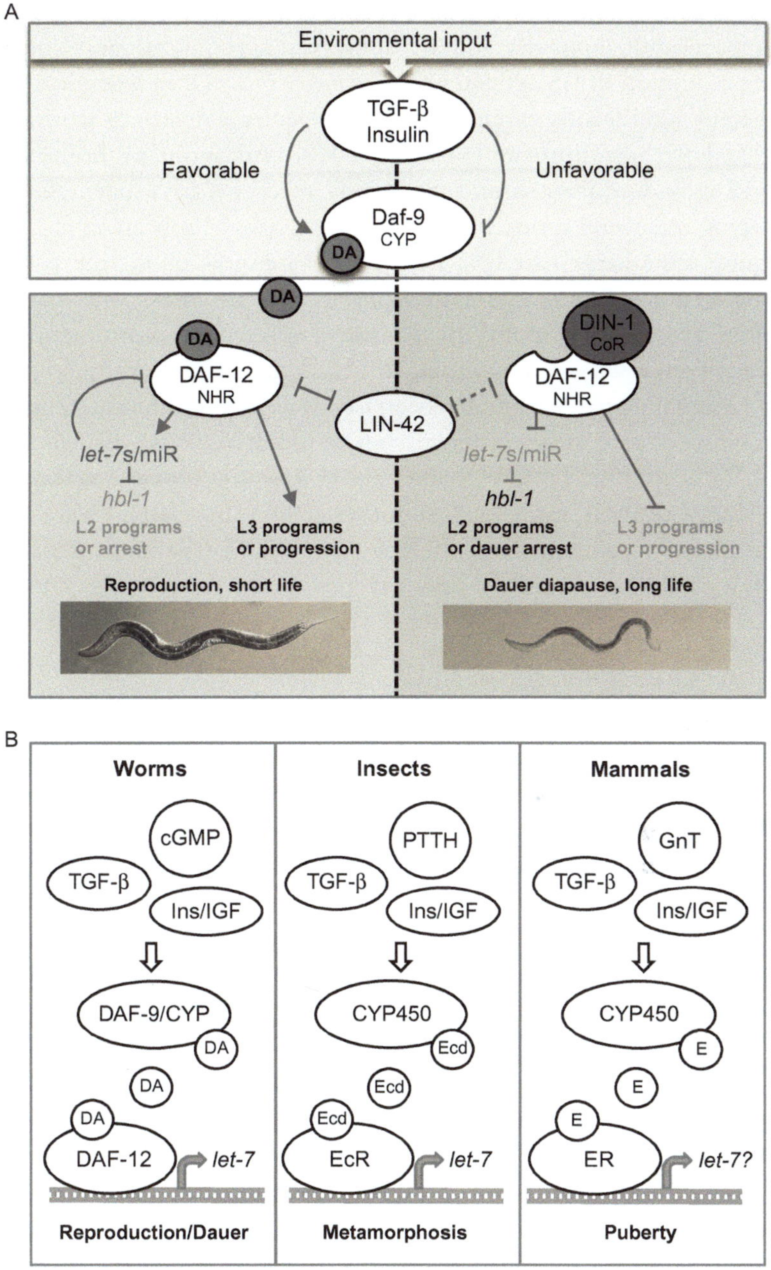

Figure 7.1 A. Model for regulation of dauer and reproduction development. *C. elegans* dauer signaling pathways work as a molecular switch (see text). In favorable conditions, TGF-β and IIS positively affect production of the DAs. DA binds to DAF-12 activating its transcriptional activity, and DA production is amplified through feedback on *daf-9*

(Continued)

Significantly many of the major endocrine pathways regulating dauer also impact adult longevity and are conserved across taxa. Best known, a modest reduction of IIS activates the DAF-16/FOXO transcription factor, and thereby significantly extends life span in species as diverse as worms, flies, mice, and perhaps humans (Kenyon, 2010). Additionally, homologs of DAF-12 including liver-X and vitamin-D receptors have been associated with aging in several models (Keisala et al., 2009; Mooijaart et al., 2007). It is quite remarkable that IIS, TGF-β, and steroidal endocrine pathways regulate major maturational transitions in worms, flies, and mammals, revealing an ancient origin for pathways governing reproduction and survival (Tennessen & Thummel, 2011) (Fig. 7.1B). Finally, the DA/DAF-12 signaling mechanism has been highly conserved among nematodes, and used to couple environmental cues to dauer diapause, and to specify mouth dimorphisms in free living species (Bento, Ogawa, & Sommer, 2010; Ogawa, Streit, Antebi, & Sommer, 2009). Similarly, DA/DAF-12 regulates entry and exit from the infective stage of parasitic nematodes, and thus could serve as a therapeutic entry point for treating pathologic diseases (Ogawa et al., 2009; Wang et al., 2009).

2.3. DA biosynthetic pathway

Given the importance of the DAs in regulating DAF-12 activity and nematode life history, many questions arise: how are they made, how is synthesis regulated, and how do they govern organismal fate choice? An important clue came from the discovery of *daf-9*, a cytochrome P450 critical for DA production (Gerisch, Weitzel, Kober-Eisermann, Rottiers, & Antebi, 2001; Jia, Albert, & Riddle, 2002; Motola et al., 2006). *daf-9* mutants form

Figure 7.1—Cont'd (not shown). Upon commitment to reproductive development DAF-12 turns on the *let-7* family microRNAs, downregulating HBL-1/ZnF to catalyze L2/L3 transitions. *let-7* microRNAs also feedback regulate *daf-12* to diminish expression at later stages. In unfavorable conditions, endocrine pathways are down, the unliganded receptor forms a corepressor complex with DIN-1/SHARP, shutting down microRNA expression and the developmental timer. LIN-42 and DAF-12 antagonize one another. LIN-42/PERIOD inhibits DAF-12's dauer promoting activity under mild stress, and antagonizes its heterochronic functions in seam and gonad during reproductive development. It is unknown if this is direct or indirect. B. Conserved regulation of reproductive signaling. In worms, cGMP, TGF-β, and IIS regulate dafachronic acid (DA) production. In insects IIS, TGF-β, and PTTH (prothoracicoctropic hormone) regulate ecdysteroid (Ecd) production. Similarly in mammals, gonadotrophs (GnT), insulin/IGF, and TGF-β/activin signaling affect production of estrogens (E) and androgens. Downstream steroid receptors promote expression of *let-7* and related microRNAs to promote maturation in various tissues. (For color version of this figure, the reader is referred to the online version of this chapter.)

dauer larvae constitutively and have gonadal outgrowth defects, similar to *daf-12* LBD mutants. Nematodes do not synthesize cholesterol and depriving wild type animals of cholesterol phenocopies *daf-9* mutants (Gerisch et al., 2001; Matyash et al., 2004). Similar phenotypes are seen in mutants of the Niemann–Pick type C1 homologs (Li, Brown, Ailion, Lee, & Thomas, 2004), which affect cholesterol transport, suggesting that the DAF-12 ligands are derived from dietary cholesterol. Additional genes involved in hormone biosynthesis were found based on genetic screens for similar phenotypes (Patel, Fang, Svy, Ruvkun, & Li, 2008; Rottiers et al., 2006; Wollam et al., 2012).

Of course a critical understanding of steroidogenic pathways came with the discovery of the DAs and their structure (Motola et al., 2006). They are 3-keto bile acid-like steroids with a 25S-carboxylic acid moiety at the end of the cholesterol sidechain (Fig. 7.2). X-ray crystal structures reveal that they bind within the DAF-12 ligand-binding pocket much like the bile acids bind to mammalian FXR (Zhi et al., 2012). Of the two known DAs, Δ-4 and Δ-7-DA, the latter is the more prevalent and potent form. Both ligands activate the receptor in the nM range, are absent in *daf-9* mutants, and supplementation rescues all known *daf-9* phenotypes. Whether the different DAs have distinct transcriptomes or functions, remains unknown.

With these structures in mind, a model for DA biosynthesis was deduced through a combination of genetics, sterol feeding, and biochemical approaches. It has been proposed that cholesterol is converted through branched biosynthetic pathways, dubbed the Δ-7 and the Δ-4 branches, into the two DAs. The first committed step in the Δ-7 biosynthetic branch is introduction of the Δ-7 double bond by the DAF-36/Rieske oxygenase, converting cholesterol to 7-dehydrocholesterol (Wollam et al., 2011) (Fig. 7.2). Similarly, the *Drosophila* ortholog, *neverland*, carries out the same first step in ecdysteroid biosynthesis (Yoshiyama-Yanagawa et al., 2011), while the mammalian cytochrome P450, CYP7A1, carries out analogous chemistry hydroxylating the 7-position of the cholesterol backbone at the first step of bile acid synthesis (Russell, 2003). An unknown Δ-5 reductase is surmised to convert 7-dehydrocholesterol to lathosterol. Thereafter the 3β-hydrosteroid dehydrogenase, DHS-16, converts lathosterol to the 3-keto steroid lathosterone (Wollam et al., 2012). Finally, the last step is carried out by DAF-9/CYP27A1, which oxidizes the cholesterol sidechain to a carboxylic acid moiety, with chemistry orthologous to the mammalian CYP27A1, which is also involved in bile acid biosynthetic pathways (Motola et al., 2006). HSD-1 was suggested to work in the production of Δ-4-DA by catalyzing oxidation of cholesterol to 4-cholesten-3-one, but

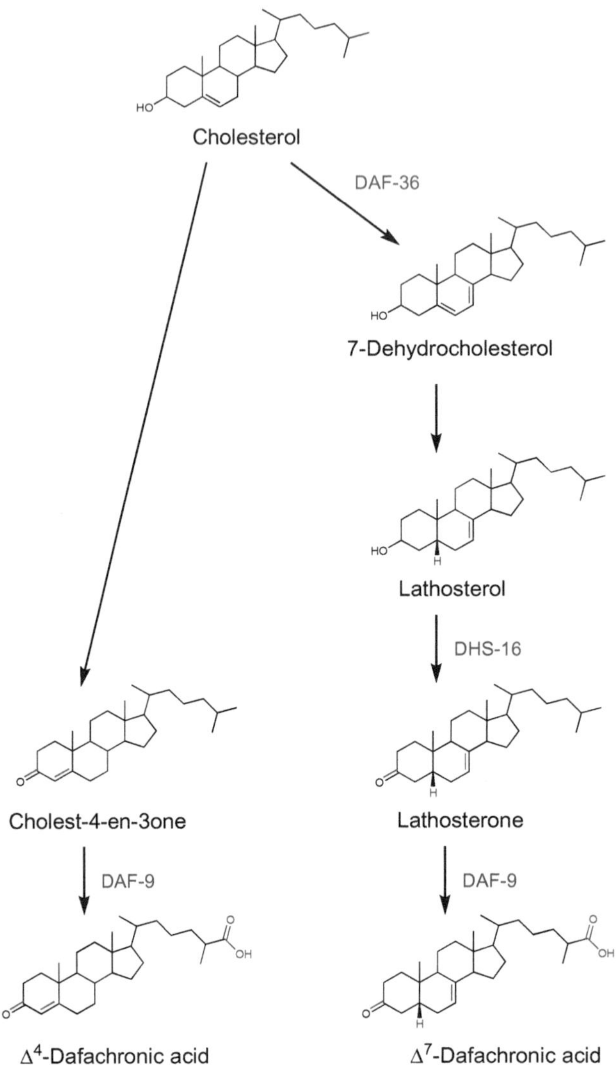

Figure 7.2 DA biosynthesis. Proposed pathways for Δ-7 and Δ-4 DA production, with identified activities DAF-36/Rieske, DHS-16/3HSD, and DAF-9/CYP27A1 shown (see text). (For color version of this figure, the reader is referred to the online version of this chapter.)

evidence for this has remained elusive (Patel et al., 2008) possibly because of vanishingly small quantities of this ligand or that HSD-1 affects other aspects of biosynthesis. In fact, recent evidence suggests there may be other DAs and that the biosynthetic pathways are more complex than previously thought (Wollam et al., 2012). Importantly, despite gaps in our knowledge, the emergent view is that bile acid-like steroidal biosynthetic pathways in nematodes strikingly resemble those found in mammals. Interestingly long-lived dwarf mice have alternations in bile acid levels (Amador-Noguez et al., 2007). Knowledge of these biosynthetic pathways may provide novel avenues for manipulating signaling pathways and enhancing animal longevity.

2.4. DA production is distributed

DAF-12 is found in the nucleus of cells throughout the body, but the hormone biosynthetic genes are expressed in discrete steroidogenic tissues (Fig. 7.3). The most important enzyme in the pathway DAF-9/CYP27A1 is found in a pair of neuroendocrine cells called the XXX, the hypodermal syncytium, and the hermaphrodite spermatheca (Gerisch et al., 2001; Jia et al., 2002). By contrast, DAF-36/Rieske oxygenase is found primarily in the gut and does not overlap with DAF-9, while DHS-16 resides in sensory neurons, pharynx, and hypodermis (Rottiers et al., 2006; Wollam et al.,

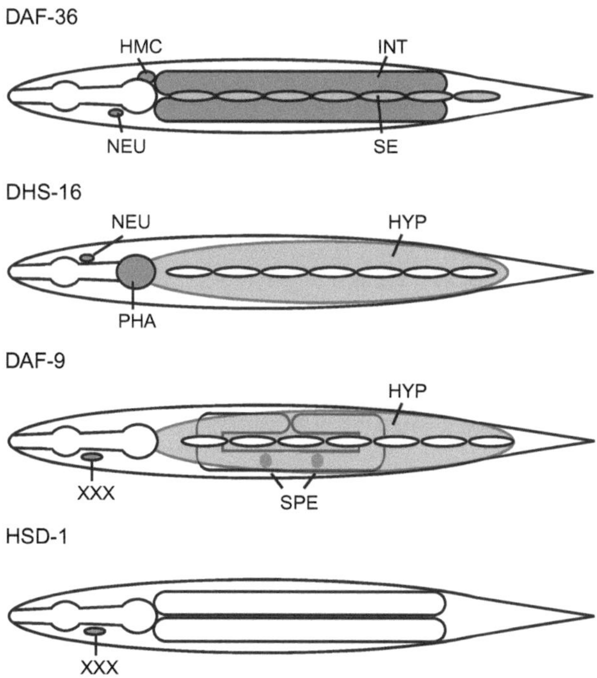

Figure 7.3 Expression patterns of the DA hormone biosynthetic genes. Expression patterns are based on GFP fusion proteins. Intestine (int), unidentified head neurons (neu), head mesodermal cell (hmc), seam cells (se), pharynx (pha), hypodermis (hyp) XXX neuroendocrine cells (XXX), spermatheca (spe). (For color version of this figure, the reader is referred to the online version of this chapter.)

2012). Finally, HSD-1 resides in the XXX cells (Patel et al., 2008) and possibly at low levels in the intestine (Antebi, unpublished).

The disparate expression patterns and partial overlap of the biosynthetic genes are surprising, and suggest that DA production is distributed and requires transport mechanisms between the tissues. Perhaps the Niemann–Pick proteins or ABC transporters function to do so. The intriguing question is, why has the nematode configured the biosynthetic pathway itself to be cell nonautonomous? Conceivably this could provide a way for each tissue to communicate to the other and imprint its physiologic state. In this view, each tissue can cast a vote to regulate developmental progression. Alternately, it may be a way to concoct tissue-specific ligands at various points in development with distinct functions. Understanding the roles of the different tissues in DA hormone transactions and transport will be an important area to map out in the future. While classical steroidogenic enzymes of mammals reside mainly in adrenal and ovary, they are also found in other tissues such as the skin (Slominski et al., 2013). Moreover, vitamin D production requires specific ring opening steps to be catalyzed in the skin upon exposure to UV light.

2.5. DA signal amplification enforces reproductive commitments

Like other life fate choices, the dauer decision entails integrating variable, complex, and competing environmental cues, as well as diverse physiologic and metabolic states over time, which must be transformed into a binary all or none commitment to reproductive growth or arrest. How is this decision achieved and what are the underlying principles?

Key insights into these questions have come from a study of tissue-specific *daf-9*/CYP27A1 regulation and the timing of proreproductive (DA-induced) versus prodauer (ascaroside-induced) activities. *daf-9::gfp* is expressed at near constant levels in the XXX neuroendocrine cells from late embryogenesis into adulthood, and surprisingly, it is not obviously regulated by inputs from IIS and TGF-β pathways. Evidence indicates, nevertheless, that *daf-9* expressed in these cells helps to prevent dauer formation (Gerisch & Antebi, 2004; Mak & Ruvkun, 2004). By contrast, hypodermal *daf-9* is expressed from mid-L2–L4 around the time of reproductive commitments, and is dramatically regulated by environmental cues as well as dauer signaling pathways, and thus appears to mediate major regulatory inputs into DA production (Gerisch & Antebi, 2004). Accordingly, hypodermal *daf-9* shuts off entirely under dauer-inducing conditions, and is low but visible under reproductive growth conditions, strictly correlating with developmental choice. Perplexingly, under conditions of mild stress, modestly-reduced TGF-β or IIS, or limiting DA production, hypodermal *daf-9* is massively upregulated precisely from mid-L2 onwards. It is hypothesized that this upregulation increases DA production and thus ensures reproductive development in the face of mild adversity. By inference, continuous tonic levels of DA from the XXX are augmented by regulated production in the hypodermis.

Schaedel and coworkers further addressed organismal fate locking mechanisms by examining the antagonistic actions of dauer pheromone and DAs on the dauer decision during the course of early larval development (Schaedel, Gerisch, Antebi, & Sternberg, 2012). By performing shift experiments they determined that a critical time window (~12–18 h) exists during the L1/L2 stage in which dauer pheromone exposure was sufficient to drive dauer formation in wild type. Conversely, coincident exposure of *daf-9* null mutants to DA during this same window could drive reproductive development. Through a series of intricate laser ablation studies, they were able to discern a two step mechanism in which tonic levels of DA emanating from the XXX cells during the L1 must reach a critical threshold to bypass dauer (~5–10 nM), which is then amplified within the L2 stage in the hypodermis to lock in this fate choice, and drive reproductive programs in the somatic tissues (Schaedel et al., 2012). In particular in *daf-9* null mutants, which are unable to amplify this signal, limiting amounts of supplemented DA were sufficient to bypass dauer but insufficient to promote normal reproductive programs of gonadal outgrowth. Higher and sustained DA levels were required to promote normal reproductive development and maximal growth rates. Moreover, the threshold for dauer bypass could be modulated by environmental conditions; treatment with ascarosides raised the threshold for dauer bypass by about 30 nM. Thus, commitment to reproductive development is set by plastic thresholds that are modulated by environmental inputs.

Correlatively, titrating DA into *daf-9* mutant strains harboring a *pdaf-9::gfp* transcriptional reporter revealed that the reporter is off in the absence of DA, on strongly with limiting amounts of DA, but diminished by DA excess. These dynamics in hypodermal *daf-9* expression depend entirely on *daf-12* as well as several other regulators (Barna et al., 2012; Gerisch & Antebi, 2004; Mak & Ruvkun, 2004; Monje, Brokate-Llanos, Perez-Jimenez, Fidalgo, & Munoz, 2011). These observations imply that both feedforward and feedback regulation are critical to ensure binary organismal fate choices. They also give a striking visible readout of commitments to reproductive development. Critical tests of this model await measuring DA production in each tissue during the course of development. Conceivably analogous amplification mechanisms are deployed to drive maturational processes in insects and mammals.

3. DAF-12 REGULATES DEVELOPMENTAL TIMING

3.1. Heterochronic genes

In the course of animal development, cells throughout the body acquire both temporal and positional identities. As *C. elegans* develops from embryo through four larval stages L1–L4 and on to adult, each stage elaborates a characteristic array of cellular programs of division, differentiation, morphogenesis/migration, and death. In particular, certain tissues are explicitly temporally patterned, including division patterns of epidermal seam cells and intestinal cells, or morphogenesis of the vulva and gonad during larval development. Temporal selector genes called the heterochronic loci dictate the stage-specific temporal patterning of cell fate. Their mutation can lead to either precocious or retarded execution of cellular temporal patterns, which in some tissues manifests as a deletion or repetition of stage-specific programs. They often work in a stage- and tissue-specific manner and cooperate in a regulatory hierarchy, called the heterochronic circuit, which controls the proper succession of developmental events.

Many of the heterochronic loci encode conserved regulators including the very first discovered microRNAs, *lin-4* and *let-7*, as well as genes involved in their activity, biogenesis, and maturation (Lee, Feinbaum, & Ambros, 1993; Reinhart et al., 2000). For example, the conserved *lin-28* gene encodes an RNA-binding protein that inhibits the maturation of *let-7* microRNA (Moss, Lee, & Ambros, 1997; Rybak et al., 2008; Viswanathan, Daley, & Gregory, 2008). Extraordinarily, the mammalian homologs of LIN-28 regulate the timing of puberty (He et al., 2009; Ong et al., 2009; Sulem et al., 2009; Zhu et al., 2010). Moreover, mammalian LIN28 works as a pluripotency factor essentially collaborating with other such factors to reprogram terminally differentiated cells to a primordial temporal fate (Yu et al., 2007). These and other studies highlight the original and groundbreaking work first uncovered by this tiny corner of *C. elegans* biology.

3.2. The microRNA paradigm

An emergent paradigm for temporal control of development is that microRNAs work in succession as switches that catalyze transitions from earlier stage programs to the next by downregulating key regulatory factors. Within *C. elegans* epidermal stem cells, microRNA *lin-4* successively downregulates expression of the LIN-14 transcription factor and the LIN-28 RNA-binding protein to promote L1/L2 and L2/L3 transitions, respectively (Lee et al., 1993; Moss et al., 1997; Wightman, Ha, & Ruvkun, 1993). Similarly, the *let-7* related microRNAs *mir-48*, *mir-84*, and *mir-241* collaborate to downregulate the kruppel-like zinc finger protein HBL-1/hunchback thereby catalyzing L2/L3 transitions (Abbott et al., 2005). Finally, the *let-7* microRNA downregulates the LIN-41/TRIM71 homolog and other factors to trigger L4/adult transitions (Slack et al., 2000). Remarkably, similar regulatory interactions are seen driving differentiation events and stem cell dynamics in higher animals. For example, mammalian *let-7* downregulates *lin-41* to promote neural tube closure, cell cycle exit and differentiation, and inhibits self renewal in embryonic stem cells (Chang et al., 2012; Maller Schulman et al., 2008).

The microRNAs appear to work somewhat cell autonomously, that is, within the tissues in which they are expressed (Zhang & Fire, 2010). The question arises, how are the microRNAs themselves regulated? How are the different temporally patterned events coordinated across tissues? How are environmental, nutritional, and growth signals integrated into developmental timing circuits? And how are such events coordinated with global transitions such as the molt cycle or dauer arrest?

3.3. DAF-12 regulates the *let-7* microRNAs

Initial answers to some of these questions came with the discovery that DAF-12 functions within the heterochronic circuit to modulate L2/L3 transitions (Antebi, Culotti, & Hedgecock, 1998). Mutants exhibit canonical retarded heterochronic phenotypes in which they inappropriately repeat L2 programs at the L3 stage. Such phenotypes are visible in epidermal seam cells, intestinal nuclei, migratory cells of the gonad, and other tissues, suggesting that DAF-12 broadly coordinates temporal programs throughout the body. Intriguingly, the most penetrant phenotypes are provoked by *daf-12(rh61)* mutations that truncate the receptor within the LBD, while less penetrant phenotypes are seen in null mutants that disrupt DNA and LBDs (Antebi et al., 2000). Apparently, these LBD mutations abrogate ligand binding and transcriptional activation, and instead assemble corepressor complexes that inhibit target gene expression, thus exacerbating phenotypes. Accordingly, removal of the corepressor *din-1* restores phenotypes to that seen in the null mutants (Ludewig et al., 2004). This behavior is reminiscent of various syndromes in which nuclear receptors behave as constitutive repressors because of ligand insufficiency. Impenetrant heterochronic phenotypes for null mutants are not unique to *daf-12*. Most other heterochronic mutants that affect L2/L3 temporal programs also display impenetrant phenotypes, reflecting functional redundancy or multiple layers of regulation of microRNA transcription, maturation, and activity (Ding, Spencer, Morita, & Han, 2005; Grishok et al., 2001; Hammell, Lubin, Boag, Blackwell, & Ambros, 2009; Morita & Han, 2006; Pepper et al., 2004; Xia, Huang, & Zhang, 2009). In support of this idea, the heterochronic phenotypes of *daf-12* nulls are strongly enhanced in double mutant combinations with these loci.

An important insight into how DAF-12 regulates these transitions came with the observation that mutants showed reduced RNA levels of a number of microRNAs of the *let-7* family (Bethke, Fielenbach, Wang, Mangelsdorf, & Antebi, 2009; Esquela-Kerscher et al., 2005; Hammell, Karp, & Ambros, 2009). Additionally, a triple deletion of the *let-7* related microRNAs, *mir-48*, *mir-84*, *mir-241* gave phenotypes similar to *daf-12* mutants, in which seam cells repeat L2 programs at the L3 stage giving rise to extra cells (Abbott et al., 2005). This led to the hypothesis that DAF-12 directly regulates these microRNAs. Indeed, DA/DAF-12 were shown to potently activate *mir-84* and *mir-241* in human cell culture and bind directly to the microRNA promoters. Similarly in worms, *mir-84* and *mir-241* expression showed DA and DAF-12 dependence in specific tissues (Bethke et al., 2009). Notably, *mir-84* expression in seam cells was abrogated in *daf-12* null mutants, suggesting that heterochronic phenotypes in the epidermis arise in part from a failure to activate this microRNA. Consistent with a downstream role, the microRNA target, HBL-1/hunchback, was inappropriately elevated in *daf-12* mutants, particularly in *daf-12(rh61)* LBD truncation mutants and to a lesser extent in null mutants (Bethke et al., 2009; Hammell, Karp, & Ambros, 2009). Moreover loss of *hbl-1* completely suppressed the extra seam cell phenotype seen in *daf-12* mutants, showing that *daf-12* works through the *let-7* related microRNAs and *hbl-1* (Fig. 7.1A).

The interactions described above reveal DAF-12's influence on microRNA expression during reproductive development, but what about during dauer arrest? In DA-deficient *daf-9* mutants, DAF-12 switches to repressor mode, animals enter the dauer stage, and overall expression of the *let-7* family microRNAs in most tissues is dramatically repressed (Bethke et al., 2009; Hammell, Karp, & Ambros, 2009). In fact, DAF-12's function as a repressor in these circuits is more potent than its role as activator. Similarly, repression of the microRNAs is seen in *daf-12(rh61)* alleles, dependent on corepressor *din-1/SHARP*.

A simplifying model is that, in favorable environments global production of the DAs results in activation of DAF-12 in target tissues, where it turns on the microRNAs. They in turn downregulate HBL-1, inhibiting earlier L2 programs to allow for later L3 programs (Abbott et al., 2005). DAF-12 may also directly promote L3 programs by turning on genes that specify this temporal fate. In unfavorable environments, animals enter the dauer stage, generally repress microRNA expression, and shut down the developmental timer (Fig. 7.1A). Thus, DAF-12-*let-7s* work as hormone-regulated switch governing developmental progression in response to the environment.

Evidence suggests that steroid regulation of *let-7* microRNAs is a conserved ancestral pathway. In *Drosophila*, 20-hydroxyecdysone and its receptor rapidly transactivate pri-*let-7* complex and regulate larval to adult developmental events in various tissues (Chawla & Sokol, 2012; Wu, Chen, Mercer, & Sokol, 2012). In mammals, the estrogen and vitamin-D receptors reportedly regulate *let-7* and *let-7*-related microRNAs (Bhat-Nakshatri et al., 2009; Ting, Messing, Yasmin-Karim, & Lee, 2013). Estrogen also regulates genes involved in microRNA biogenesis (Gupta, Caffrey, Callagy, & Gupta, 2012). These observations hint that mammalian steroidal receptors too may regulate stage transitions by influencing microRNA transcription, biogenesis, and activity.

3.4. DAF-12 targets include heterochronic genes

let-7 microRNAs are not the only DAF-12 targets responsible for influencing developmental timing. Identification of DAF-12 target genes by chromatin immunoprecipitation and transcriptome analysis reveals that DAF-12 resides at the promoters of multiple heterochronic genes including those involved in microRNA biogenesis and activity (*alg-2/argonaut, nhl-2*), genes involved in the L2/L3 switch (*lin-46/gephryin-like, lin-28*), and the larval to adult transition (*lin-41*/TRIM71, *dre-1*/FBXO11, *lin-42*/PER) (Hochbaum et al., 2011). In particular, DAF-12 may have a role in downregulating *lin-28* expression at both the mRNA and protein level (Hochbaum et al., 2011; Morita & Han, 2006; Seggerson, Tang, & Moss, 2002), although some observations contradict these results (Hammell, Karp, & Ambros, 2009). DAF-12 also affects several dauer transcription factors (*daf-3/Smad, daf-16/FOXO*), as well as its own corepressor (*din-1/SHARP*). These cross-regulatory relationships suggest that DAF-12 activity is important to establish and reinforce robust developmental decisions (Hochbaum et al., 2011). Although expression of these genes in human cell culture shows clear DA and DAF-12 dependence, the requirement *in vivo* is more modest or even absent. Conceivably other transcription factors can come in to do the job, highlighting functional redundancy, or DAF-12 regulation might operate only under specific conditions, such as dauer entry or exit.

3.5. *daf-12* is feedback regulated by the *let-7* microRNAs

Transcriptional feedback loops play an important role in timing devices as well as homeostatic regulation. Interestingly, the *daf-12* 3′ untranslated region (UTR) contains multiple *let-7*-binding sites, which mediate feedback regulation by *let-7* family microRNAs. In wild type, DAF-12 protein peaks around early L3 and then diminishes into L4. By contrast, in the *mir-48, mir-84, mir-241* triple mutant, DAF-12 expression, as measured with full-length *daf-12::gfp* or 3′UTR reporter constructs, is sustained to later stages (Hammell, Karp, & Ambros, 2009). Moreover, microRNA deletion results in increased DAF-12 repressor activity under limiting hormone conditions, exacerbating gonadal cell migration defects, and biasing dauer formation. Such feedback regulation might be important for temporal control of DAF-12 activity as well as modulating activity in the face of fluctuating environments.

Similarly, *let-7* regulates DAF-12 levels through its 3′UTR during the L4/adult transition (Grosshans, Johnson, Reinert, Gerstein, & Slack, 2005). *let-7* mediates terminal differentiation events at the L4/adult switch and loss-of-function mutations result in retarded phenotypes in which seam cells fail to express adult fates on schedule (Reinhart et al., 2000). Consistent with a downstream functional role, *daf-12* mutation suppresses *let-7* phenotypes. Moreover, *daf-12* mutants on their own exhibit mild precocious seam cell fusion phenotypes, and enhance the precocious phenotypes of *lin-41*/Trim71 mutants. These observations are somewhat puzzling since they suggest that *daf-12(+)* retards development at the L4/adult switch, but advances development at the L2/L3 transition. Speculatively, this janus-like behavior in both retarding and advancing development could reflect its switch like capabilities.

3.6. *daf-12* acts in opposition to *lin-42*/period

One of the key global regulators of *C. elegans* life history is *lin-42*, a homolog of the circadian oscillator regulator *period* (Jeon, Gardner, Miller, Deshler, &

Rougvie, 1999). *lin-42* controls developmental timing, molting, and dauer formation, and it works in opposition to *daf-12* for some of these processes. In developmental timing circuits, *lin-42* loss-of-function mutants display precocious phenotypes, most conspicuously around the L3 molt, when they exhibit precocious seam cell fusion and adult alae formation. These studies suggest that *lin-42* prevents early expression of adult fates. Additionally, mutants show earlier defects in the L2 stage in certain genetic backgrounds, suggesting that it works at multiple steps (Abrahante, Miller, & Rougvie, 1998).

lin-42 and *daf-12* display intriguing genetic interactions, which suggest they antagonize one another's activity. As mentioned earlier, *daf-12* mutants repeat L2-proliferative divisions of the seam cells at the L3 stage: *lin-42* mutation reverts this phenotype, restoring seam cell number back to normal (Tennessen, Gardner, Volk, & Rougvie, 2006). Likewise, *daf-12* mutation counteracts the *lin-42* precocious expression of adult seam fates. Similar countervailing interactions are seen in gonadal developmental timing circuits. Whereas *lin-42* mutants cause precocious dorsal migration of the distal tip cells in the L2, one full stage earlier than normal, *daf-12(rh61)* mutants cause delayed gonadal distal tip cell migration. In *lin-42;daf-12* double mutants this migration is nearly normal or delayed, depending on the allele (Fielenbach et al., 2007; Tennessen et al., 2006).

lin-42 also opposes *daf-12* activity with respect to dauer formation (Tennessen, Opperman, & Rougvie, 2010). *lin-42* mutants inappropriately enter the dauer stage at moderately elevated temperatures, suggesting that *lin-42(+)* normally prevents dauer entry and promotes reproductive development under mild stress. Consistent with this notion, *lin-42* expression drops as animals enter the dauer stage. Genetic epistasis experiments suggest *lin-42* works downstream of IIS and TGF-β, coincident with *daf-12*. Specifically *daf-12* Daf-d null mutants suppress the *lin-42* Daf-c defect. Moreover, *lin-42(+)* overexpression prevails over the Daf-c phenotypes of *daf-9*/CYP27A1 mutants as well as *daf-12* LBD mutants, placing it at the same step as steroidal signaling. This latter finding is quite striking because it implies that *lin-42* overexpression can drive reproductive development in the absence of the DAF-12 ligands, rendering the receptor ligand independent.

Currently it is unknown how *lin-42* and *daf-12* affect one another's activity. The molecular basis of the yin-yang behavior could result from a convergence of parallel antagonistic pathways on the same target genes. Another possibility is that there exists a more intimate physical interaction between *daf-12* and *lin-42* in which they regulate one another's activity. Notably, yeast two-hybrid experiments reveal that LIN-42 interacts with a portion of the DAF-12 LBD and hinge (Tennessen et al., 2010). A simple model is that LIN-42 inhibits DAF-12 dauer promoting activity, perhaps by displacing DIN-1 corepressor or other coregulators. Interestingly, a similar antagonism of nuclear receptor activity by Period operates in mammals. Mammalian Period2 protein interacts with the adipogenic nuclear receptor PPARγ, and inhibits its recruitment to target promoters. Deletion of Period2 results in upregulation of PPARγ target gene expression and adipogenesis (Grimaldi et al., 2010). Similarly, another regulator in the circadian clock, the cryptochromes, interact with the glucocorticoid receptor and inhibit its activity (Lamia et al., 2011). However, it remains to be seen whether DAF-12 and LIN-42 work through a similar biochemical mechanism.

3.7. Dauer reprogramming

An extraordinary finding over the last few years is that differentiated cells in metazoans can be reprogrammed to earlier pluripotent fates (Takahashi & Yamanaka, 2006). Interestingly, *C. elegans* can also reprogram temporal fates when development is channeled through the dauer stage or by starvation. Basically, mutants of multiple heterochronic loci, precocious or retarded, exhibit near wild type like development when worms traverse the dauer stage, revealing that mid-larval temporal fate transformations can be prevented or even remodeled (Euling & Ambros, 1996; Liu & Ambros, 1991). The molecular basis of this reset is still obscure, but likely involves *daf-12* and dauer signaling pathways shutting down the heterochronic timer during the dauer diapause (Bethke et al., 2009; Hochbaum et al., 2011). Presumably upon dauer exit, development is reinitiated using a different spectrum of regulators. Indeed, Karp and Ambros recently showed that a remodeling of the regulatory activity of the *lin-4* and *mir-84* microRNAs and associated factors affecting the *hbl-1* target gene occurs upon dauer exit, resulting in a restoration of more normal temporal programming (Karp & Ambros, 2012). Further studies in this area are bound to yield important insights into this fascinating phenomenon.

3.8. Heterochronic genes and the molt cycle

Molting entails synthesis of the new skin and shedding of the old, and it is important for the growth and maturation of many animals. The molt cycle constitutes a cyclical molecular clock that coordinates many aspects of development and takes about 8–10 h in *C. elegans*. Strikingly, several cellular programs such as seam cell division and intestinal endoreplication are closely coordinated with cuticle synthesis and molting, and are iterated with each larval stage (Hedgecock & White, 1985; Sulston & Horvitz, 1977). Even migratory transitions in gonadal morphogenesis may be closely tied to this clock (Antebi et al., 1998).

The close coordination of cellular events across tissues during the molt cycle is suggestive of hormonal regulation. In other molting organisms such as *Drosophila*, molting is driven by ecdysteroids working through the ecdysone receptor and its obligate partner ultraspiracle/retinoid-X-receptor (RXR) (Tennessen & Thummel, 2011). The ecdysone receptor activates downstream transcriptional cascades including the nuclear receptors HR3, FTZ-F1, and others. *C. elegans* lacks ecdysteroids and the ecdysteroid receptor. Although steroid receptor *daf-12* mRNA may be expressed in a cyclical fashion (Merris, Wang, Soteropoulos, & Lenard, 2007), neither the receptor nor its ligands have an overt affect on the molt cycle except for the dauer molt. However, two other nuclear receptors, NHR-23 and NHR-25, homologs of HR3 and FTZ-F1, respectively, oscillate with the molt cycle and are responsible for molting, since their mutation results in ecdysis defects (Asahina et al., 2000; Gissendanner, Crossgrove, Kraus, Maina, & Sluder, 2004; Gissendanner & Sluder, 2000; Kostrouchova, Krause, Kostrouch, & Rall, 2001). Their ligands remain unknown, but could be sterol derivatives as cholesterol deprivation results in molting defects (Yochem, Tuck, Greenwald, & Han, 1999), and the mammalian homolog of NHR-23, RORα, binds to sterols (Wang et al., 2010). Alternately, they could be phospho, or sphingolipids, since NHR-25 and homologs appear to bind these molecules (Krylova et al., 2005; Lee et al., 2011; Mullaney et al., 2010; Urs et al., 2007). Another possibility is that NHR-23 and NHR-25 are not ligand gated at all, but instead bind lipids as competency factors; in this view oscillations of other components such as LIN-42 (below) could regulate their activity.

Remarkably the period homolog LIN-42 is also a central component of the molt cycle. *Lin-42* mRNA and protein strikingly oscillate in concert with the molt cycle with peak protein levels around the molts (Jeon et al., 1999; Monsalve, Van Buskirk, & Frand, 2011). Null mutants exhibit molting defects at all four molts, and have erratic arrhythmic molts and extended molting quiescence, a sleep-like state during which the old cuticle is shed (Monsalve et al., 2011). The *lin-42* genomic locus is complex, as it encodes four different isoforms of diverse structure and function. The shortest isoform appears to be mainly responsible for molting, since its expression can restore molt defects in the background of *lin-42* loss-of-function mutants, and forced overexpression can induce erratic molts and quiescence. The cyclical pattern of expression and function of *lin-42* resembles that of circadian timers, with an 8–10 h cycle instead of 24 h. Interestingly, in mammals RORα is an integral component of the circadian clock and is regulated by the *lin-42* homolog period (Ranhotra, 2012).

There are other intriguing parallels between the molt cycle and circadian cycles: the molt cycle begins with feeding behavior, anabolic processes and activity, comparable to the day period. Thereafter worms enter lethargus, a sleep-like quiescent state during which there is a cessation of feeding and movement, and reserves are drawn upon for energy. Concurrently, the old cuticle is shed and the new one laid down, coupled to concerted growth. Both molt and circadian cycles likely entail negative feedback loops although the functional relationships between the *C. elegans* molting regulators *lin-42*, *nhr-23*, and *nhr-25* are poorly understood. All three gene products are regulated in a cyclical fashion, but whether they cross regulate one another's activity is unknown. Generally speaking the molecular mechanisms underlying molting and circadian clocks might inform one another, and both may deploy nuclear receptor cascades organized in feedback loops that are regulated by *lin-42*/period.

Although the function of heterochronic and molting timers are somewhat separable, there is also considerable overlap as already described for *lin-42*. Other examples include NHR-25 whose major function is to drive the molt cycle, but mutants also display complex heterochronic phenotypes suggestive of defects in the larval to adult switch (Hada et al., 2010). Another gene implicated in molting, the amyloid precursor protein homolog *apl-1*, can suppress some of the lethality and heterochronic phenotypes of *let-7* (Hornsten et al., 2007; Niwa, Zhou, Li, & Slack, 2008; Wiese, Antebi, & Zheng, 2010). A component of SCF ubiquitin ligases, the FBXO11 homolog *dre-1*, works in both developmental timing and molting pathways (Fielenbach et al., 2007). Conversely, several genes in the heterochronic pathway oscillate with the molt cycle including the pri-microRNA form of *let-7* and possibly *daf-12* (Merris et al., 2007; Van Wynsberghe et al., 2011). *let-7* and related microRNAs as well as many other heterochronic genes affect the terminal molt (Ambros & Horvitz, 1984; Reinhart et al., 2000). In particular retarded heterochronic mutants often have supernumerary molts, while precocious mutants cease molting one or two stages earlier than normal. Evidence suggests that the *let-7* microRNAs may play a role in downregulating *nhr-25* to bring about a cessation of molting (Hayes, Frand, & Ruvkun, 2006). Further exploring the relationship between the molting clock and the heterochronic timer should help us understand how iterative cyclical timing devices interface with switches to promote the proper succession of temporal fates.

4. HORMONAL CONTROL OF LONGEVITY

Pioneering studies in the worm have illuminated conserved signaling pathways that regulate longevity. Several pathways have emerged that regulate longevity across taxa including reduced IIS, mitochondrial function, dietary-restriction-mediated longevity, and signaling from the gonad (Kenyon, 2010). Below, we highlight recent discoveries linking heterochronic functions and hormonal signaling to the gonadal longevity pathway.

4.1. Gonadal longevity pathway

Reproductive tissues are not only involved in propagation of the species but can also affect the life span. Removal of the gonad or parts thereof can extend life span in a variety of species (Hansen, Flatt, & Aguilaniu, 2013). Kenyon and coworkers originally showed that removal of *C. elegans* germline precursor cells by laser microsurgery or genetic mutation in germlineless *glp-1* mutants extends life span by 60% (Arantes-Oliveira, Apfeld, Dillin, & Kenyon, 2002; Hsin & Kenyon, 1999). This longevity depends upon the presence of the somatic gonad, since removal of somatic gonadoblasts abrogates the longevity. That the reproductive system controls the life span of the whole organism implies that hormonal mechanisms might be involved.

An analysis of genes required for life span extension of germlineless animals revealed that a number of transcription factors including *daf-12*, the FOXO homolog *daf-16* (Hsin & Kenyon, 1999), the HNF4α-like nuclear hormone receptor *nhr-80* (Goudeau et al., 2011), and the FOXA homolog *pha-4* are required (Lapierre, Gelino, Melendez, & Hansen, 2011). These transcription factors stimulate proteasomal activity, lipolysis, fatty acid desaturation, and autophagy, demonstrating that germline removal entails a remodeling of the regulatory and metabolic state of the whole organism to somehow promote longevity (Vilchez et al., 2012; Hansen et al., 2013). For example, the liberation of fatty acids is central to the longevity phenotype in the gonadal pathway (Wang, O'Rourke, & Ruvkun, 2008). One way in which fatty acids contribute to the longevity is through stimulation of autophagy (Lapierre et al., 2011), which is important for energy homeostasis and protein quality control. But it also remains possible that fatty acids themselves serve as hormonal signals that coordinate other events. Exploring the interface between metabolism and longevity will certainly be an exciting area of future research.

Importantly, gonadal longevity also requires the DAs produced in the somatic gonad and other tissues, which work through DAF-12 to promote longevity in at least two ways (Gerisch et al., 2001; Rottiers et al., 2006; Yamawaki et al., 2010). First, DA and DAF-12 facilitate DAF-16/FOXO nuclear localization and activity within the intestine, a major fat depot and metabolic organ (Berman & Kenyon, 2006; Gerisch et al., 2007). DAF-16/FOXO is a powerful regulator of longevity well known to be activated by reduced IIS, but appears to be independently regulated in gonadal longevity. Second, DAF-12 and DAF-16 share responsibilities for transcribing a set of life span-enhancing genes (McCormick, Chen, Ramaswamy, & Kenyon, 2012). Yet the mechanism of crosstalk between DAF-16/FOXO and DA/DAF-12 and how DA/DAF-12 regulates longevity has remained obscure until recently.

If germline removal requires DAF-12 signaling, then it could be predicted to activate this signaling axis. Indeed, Shen and coworkers recently showed that germline removal results in increased *daf-36*/Rieske oxygenase expression, heightened production of the DAs and augmented DAF-12 transcriptional activity (Shen, Wollam, Magner, Karalay, & Antebi, 2012). In particular, DA/DAF-12 signaling turns on expression of its *let-7* family microRNA targets including *mir-84* and *mir-241*, previously implicated in developmental timing circuits. Importantly, just as DA/DAF-12 are required for long life, removal of these microRNAs also abrogrates gonadal longevity. This requirement is specific as microRNA deletion has no effect on longevity due to reduced IIS or mitochondrial function. Evidence suggests that *mir-84* and *mir-241* downregulate at least two inhibitors of DAF-16/FOXO, namely the AKT-1 kinase and the LIN-14 nuclear factor. Consequently, both DAF-12 and DAF-16 become activated by DA to promote longevity programs. Interestingly, LIN-14 had previously been implicated in life span regulation; knockdown of the gene extends life, while *lin-4* mutation, which leads to excess LIN-14, shortens life span (Boehm & Slack, 2005). Longevity was shown to be *daf-16* dependent, but additive with longevity from decreased *daf-2*/InsR signaling, suggesting a parallel pathway. Additionally, AKT-1 had been previously shown to prevent longevity through IIS by phosphorylating and inhibiting DAF-16/FOXO (Paradis & Ruvkun, 1998).

Collectively these data suggest a model whereby germline removal stimulates DA/DAF-12 signaling and expression of the *let-7* microRNAs (Fig. 7.4). These in turn diminish expression of AKT-1 and LIN-14, and thereby activate DAF-16/FOXO. Together DAF-16 and DAF-12, as well as other transcriptional regulators, promote longevity genes. What is remarkable about these findings is that multiple components of the gonadal longevity network DA, *daf-12*, *mir-84*, *mir-241*, and *lin-14* are also key players in the heterochronic circuitry, revealing an intimate connection between developmental timing and life span regulation. Notably, in the heterochronic circuit, DAF-12 turns on these microRNAs, which then turn off earlier L2 programs to make way for L3 programs. Conceivably in the context of the gonadal pathway, DAF-12 turns on these microRNAs to switch between a reproductive mode and a survival mode of adulthood in response to environmental and physiologic conditions (Fig. 7.4).

Two important questions come to mind. What is the ecological significance of germline loss and why could the heterochronic genes be involved? It is quite possible that under environmental stress, nutrient limitation, or infection, the germline arrests proliferation. For example, it is known that in worms that are nutrient limited, the germline and gonad retract and go into a state of quiescence (Angelo & Van Gilst, 2009). Heterochronic activities may be invoked as a mechanism to coordinate the relative timing of events between gonad and soma, so that they mature concurrently. From an evolutionary perspective, this would be critical to future reproductive success, but with secondary consequences for the life span.

Taken together the finding that first *daf-12* steroidal signaling (Gerisch et al., 2001; Hsin & Kenyon, 1999), then *lin-14*, *lin-4*, and the *let-7* microRNAs all affect life span (Boehm & Slack, 2005; Shen et al., 2012) reveals an important connection between developmental timing components, maturation, and longevity. In mammals, the *let-7* microRNAs have been implicated in regulating insulin signaling and glucose homeostasis (Zhu et al., 2011), raising the possibility that such microRNAs could also affect longevity circuits in higher animals.

5. PERSPECTIVES

Pioneering studies in *C. elegans* reveal that environmental cues are integrated by major growth factor pathways, including IIS and TGF-β signaling, which converge on steroidal receptor DAF-12. Depending on DA availability DAF-12 either triggers reproductive maturation by activating

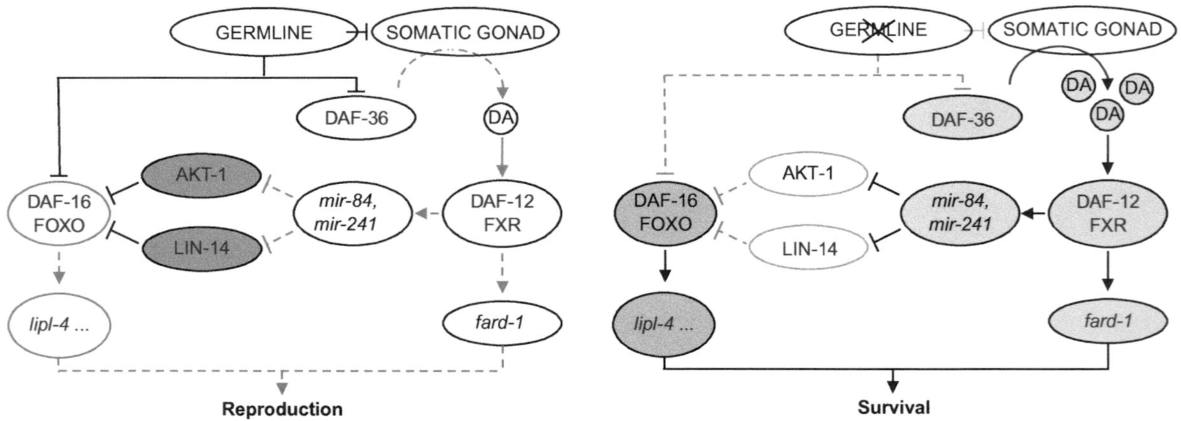

Figure 7.4 Gonadal longevity pathway. In reproductive mode, proliferating germline stem cells inhibit the activity of DAF-12 and DAF-16, resulting in normal life span. In survival mode, the germline is quiescent, resulting in upregulation of *daf-36* in intestine and DA production in the somatic gonad. Activated DAF-12 upregulates *mir-84*, *mir-241* expression, which downregulate LIN-14 and AKT-1, thereby stimulating DAF-16/FOXO. Together DAF-16/FOXO and DAF-12 turn on longevity genes. (For color version of this figure, the reader is referred to the online version of this chapter.)

let-7 family microRNAs, catalyzing stage transitions through the body, or quiescence by shutting down the heterochronic and molting timers. Some of the same components that regulate developmental timing and metabolism during maturation are co-opted to regulate life span in response to signals from the reproductive system. Steroid receptor DAF-12 acts at the convergence of these pathways conveying environmental information into these circuits. The studies in the worm point to many new directions to be explored in the future. How are environmental cues integrated to produce hormonal signals? What is the molecular architecture of feedback loops governing commitments to reproductive development and how are they integrated into developmental timing? What is the relationship between the molt cycle and the heterochronic timer? How do these circuits regulate the life span? The foremost question is to what extent are these physiologic functions conserved? Answers to these questions in the future are certain to help us understand the fundamental mechanisms underlying specification of life stages, longevity, and the nature of biological time.

ACKNOWLEDGMENTS

The author would like to thank members of the Antebi lab for reading the manuscript, and Dr. Birgit Gerisch for help with the figures. The author apologizes to those colleagues whom he could not cite due to length restrictions.

REFERENCES

Abbott, A. L., Alvarez-Saavedra, E., Miska, E. A., Lau, N. C., Bartel, D. P., Horvitz, H. R., et al. (2005). The let-7 MicroRNA family members mir-48, mir-84, and mir-241 function together to regulate developmental timing in *Caenorhabditis elegans*. *Developmental Cell*, *9*, 403–414.

Abrahante, J. E., Miller, E. A., & Rougvie, A. E. (1998). Identification of heterochronic mutants in *Caenorhabditis elegans*: Temporal misexpression of a collagen::green fluorescent protein fusion gene. *Genetics*, *149*, 1335–1351.

Amador-Noguez, D., Dean, A., Huang, W., Setchell, K., Moore, D., & Darlington, G. (2007). Alterations in xenobiotic metabolism in the long-lived Little mice. *Aging cell*, *6*, 453–470.

Ambros, V., & Horvitz, H. R. (1984). Heterochronic mutants of the nematode *C. elegans*. *Science*, *226*, 409–416.

Angelo, G., & Van Gilst, M. R. (2009). Starvation protects germline stem cells and extends reproductive longevity in *C. elegans*. *Science*, *326*, 954–958.

Antebi, A., Culotti, J. G., & Hedgecock, E. M. (1998). daf-12 regulates developmental age and the dauer alternative in *Caenorhabditis elegans*. *Development*, *125*, 1191–1205.

Antebi, A., Yeh, W. H., Tait, D., Hedgecock, E. M., & Riddle, D. L. (2000). daf-12 encodes a nuclear receptor that regulates the dauer diapause and developmental age in *C. elegans*. *Genes & Development*, *14*, 1512–1527.

Arantes-Oliveira, N., Apfeld, J., Dillin, A., & Kenyon, C. (2002). Regulation of life-span by germ-line stem cells in *Caenorhabditis elegans*. *Science*, *295*, 502–505.

Asahina, M., Ishihara, T., Jindra, M., Kohara, Y., Katsura, I., & Hirose, S. (2000). The conserved nuclear receptor Ftz-F1 is required for embryogenesis, moulting and reproduction in *Caenorhabditis elegans*. *Genes to Cells*, *5*(9), 711–723.

Barna, J., Princz, A., Kosztelnik, M., Hargitai, B., Takacs-Vellai, K., & Vellai, T. (2012). Heat shock factor-1 intertwines insulin/IGF-1, TGF-beta and cGMP signaling to control development and aging. *BMC Developmental Biology*, *12*, 32.

Bento, G., Ogawa, A., & Sommer, R. J. (2010). Co-option of the hormone-signalling module dafachronic acid-DAF-12 in nematode evolution. *Nature*, *466*, 494–497.

Berman, J. R., & Kenyon, C. (2006). Germ-cell loss extends *C. elegans* life span through regulation of DAF-16 by kri-1 and lipophilic-hormone signaling. *Cell*, *124*, 1055–1068.

Bethke, A., Fielenbach, N., Wang, Z., Mangelsdorf, D. J., & Antebi, A. (2009). Nuclear hormone receptor regulation of microRNAs controls developmental progression. *Science*, *324*, 95–98.

Bhat-Nakshatri, P., Wang, G., Collins, N. R., Thomson, M. J., Geistlinger, T. R., Carroll, J. S., et al. (2009). Estradiol-regulated microRNAs control estradiol response in breast cancer cells. *Nucleic Acids Research*, *37*, 4850–4861.

Boehm, M., & Slack, F. (2005). A developmental timing microRNA and its target regulate life span in *C. elegans*. *Science*, *310*, 1954–1957.

Butcher, R. A., Fujita, M., Schroeder, F. C., & Clardy, J. (2007). Small-molecule pheromones that control dauer development in *Caenorhabditis elegans*. *Nature Chemical Biology*, *3*, 420–422.

Cassada, R., & Russell, R. (1975). The dauer-larva: A post-embryonic developmental variant of the nematode *C. elegans*. *Developmental Biology*, *46*, 326–342.

Chang, H. M., Martinez, N. J., Thornton, J. E., Hagan, J. P., Nguyen, K. D., & Gregory, R. I. (2012). Trim71 cooperates with microRNAs to repress Cdkn1a expression and promote embryonic stem cell proliferation. *Nature Communications*, *3*, 923.

Chawla, G., & Sokol, N. S. (2012). Hormonal activation of let-7-C microRNAs via EcR is required for adult Drosophila melanogaster morphology and function. *Development*, *139*, 1788–1797.

da Graca, L. S., Zimmerman, K. K., Mitchell, M. C., Kozhan-Gorodetska, M., Sekiewicz, K., Morales, Y., et al. (2004). DAF-5 is a Ski oncoprotein homolog that functions in a neuronal TGF beta pathway to regulate *C. elegans* dauer development. *Development*, *131*, 435–446.

Ding, L., Spencer, A., Morita, K., & Han, M. (2005). The developmental timing regulator AIN-1 interacts with miRISCs and may target the argonaute protein ALG-1 to cytoplasmic P bodies in *C. elegans*. *Molecular Cell*, *19*, 437–447.

Esquela-Kerscher, A., Johnson, S. M., Bai, L., Saito, K., Partridge, J., Reinert, K. L., et al. (2005). Post-embryonic expression of *C. elegans* microRNAs belonging to the lin-4 and let-7 families in the hypodermis and the reproductive system. *Developmental Dynamics*, *234*, 868–877.

Estevez, M., Attisano, L., Wrana, J. L., Albert, P. S., Massague, J., & Riddle, D. L. (1993). The daf-4 gene encodes a bone morphogenetic protein receptor controlling *C. elegans* dauer larva development. *Nature*, *365*, 644–649.

Euling, S., & Ambros, V. (1996). Reversal of cell fate determination in *Caenorhabditis elegans* vulval development. *Development*, *122*, 2507–2515.

Fielenbach, N., & Antebi, A. (2008). *C. elegans* dauer formation and the molecular basis of plasticity. *Genes & Development*, *22*, 2149–2165.

Fielenbach, N., Guardavaccaro, D., Neubert, K., Chan, T., Li, D., Feng, Q., et al. (2007). DRE-1: An evolutionarily conserved F box protein that regulates *C. elegans* developmental age. *Developmental Cell*, *12*, 443–455.

Georgi, L. L., Albert, P. S., & Riddle, D. L. (1990). daf-1, a *C. elegans* gene controlling dauer larva development, encodes a novel receptor protein kinase. *Cell*, *61*, 635–645.

Gerisch, B., & Antebi, A. (2004). Hormonal signals produced by DAF-9/cytochrome P450 regulate *C. elegans* dauer diapause in response to environmental cues. *Development, 131*, 1765–1776.

Gerisch, B., Rottiers, V., Li, D., Motola, D. L., Cummins, C. L., Lehrach, H., et al. (2007). A bile acid-like steroid modulates *Caenorhabditis elegans* lifespan through nuclear receptor signaling. *Proceedings of the National Academy of Sciences of the United States of America, 104*, 5014–5019.

Gerisch, B., Weitzel, C., Kober-Eisermann, C., Rottiers, V., & Antebi, A. (2001). A hormonal signaling pathway influencing *C. elegans* metabolism, reproductive development, and life span. *Developmental Cell, 1*, 841–851.

Gissendanner, C. R., Crossgrove, K., Kraus, K. A., Maina, C. V., & Sluder, A. E. (2004). Expression and function of conserved nuclear receptor genes in *Caenorhabditis elegans*. *Developmental Biology, 266*, 399–416.

Gissendanner, C. R., & Sluder, A. E. (2000). nhr-25, the *Caenorhabditis elegans* ortholog of ftz-f1, is required for epidermal and somatic gonad development. *Developmental Biology, 221*, 259–272.

Golden, J. W., & Riddle, D. L. (1984). The *C. elegans* dauer larva: Developmental effects of pheromone, food, and temperature. *Developmental Biology, 102*, 368–378.

Goudeau, J., Bellemin, S., Toselli-Mollereau, E., Shamalnasab, M., Chen, Y., & Aguilaniu, H. (2011). Fatty acid desaturation links germ cell loss to longevity through NHR-80/HNF4 in *C. elegans*. *PLoS Biology, 9*, e1000599.

Grimaldi, B., Bellet, M. M., Katada, S., Astarita, G., Hirayama, J., Amin, R. H., et al. (2010). PER2 controls lipid metabolism by direct regulation of PPARgamma. *Cell Metabolism, 12*, 509–520.

Grishok, A., Pasquinelli, A. E., Conte, D., Li, N., Parrish, S., Ha, I., et al. (2001). Genes and mechanisms related to RNA interference regulate expression of the small temporal RNAs that control *C. elegans* developmental timing. *Cell, 106*, 23–34.

Grosshans, H., Johnson, T., Reinert, K. L., Gerstein, M., & Slack, F. J. (2005). The temporal patterning microRNA let-7 regulates several transcription factors at the larval to adult transition in *C. elegans*. *Developmental Cell, 8*, 321–330.

Gupta, A., Caffrey, E., Callagy, G., & Gupta, S. (2012). Oestrogen-dependent regulation of miRNA biogenesis: many ways to skin the cat. *Biochemical Society Transactions, 40*, 752–758.

Hada, K., Asahina, M., Hasegawa, H., Kanaho, Y., Slack, F. J., & Niwa, R. (2010). The nuclear receptor gene nhr-25 plays multiple roles in the *Caenorhabditis elegans* heterochronic gene network to control the larva-to-adult transition. *Developmental Biology, 344*(2), 1100–1109.

Hammell, C. M., Karp, X., & Ambros, V. (2009). A feedback circuit involving let-7-family miRNAs and DAF-12 integrates environmental signals and developmental timing in *Caenorhabditis elegans*. *Proceedings of the National Academy of Sciences of the United States of America, 106*, 18668–18673.

Hammell, C. M., Lubin, I., Boag, P. R., Blackwell, T. K., & Ambros, V. (2009). nhl-2 Modulates microRNA activity in *Caenorhabditis elegans*. *Cell, 136*, 926–938.

Hansen, M., Flatt, T., & Aguilaniu, H. (2013). Reproduction, fat metabolism, and life span: What is the connection? *Cell Metabolism, 17*(1), 10–19.

Hayes, G. D., Frand, A. R., & Ruvkun, G. (2006). The mir-84 and let-7 paralogous microRNA genes of *Caenorhabditis elegans* direct the cessation of molting via the conserved nuclear hormone receptors NHR-23 and NHR-25. *Development, 133*(23), 4631–4641.

He, C., Kraft, P., Chen, C., Buring, J. E., Pare, G., Hankinson, S. E., et al. (2009). Genome-wide association studies identify loci associated with age at menarche and age at natural menopause. *Nature Genetics, 41*, 724–728.

Hedgecock, E. M., & White, J. G. (1985). Polyploid tissues in the nematode *C. elegans*. *Developmental Biology, 107*, 128–133.

Hochbaum, D., Zhang, Y., Stuckenholz, C., Labhart, P., Alexiadis, V., Martin, R., et al. (2011). DAF-12 regulates a connected network of genes to ensure robust developmental decisions. *PLoS Genetics*, 7(7), e1002179.
Hornsten, A., Lieberthal, J., Fadia, S., Malins, R., Ha, L., Xu, X., et al. (2007). APL-1, a *Caenorhabditis elegans* protein related to the human beta-amyloid precursor protein, is essential for viability. *Proceedings of the National Academy of Sciences of the United States of America*, *104*(6), 1971–1976.
Hsin, H., & Kenyon, C. (1999). Signals from the reproductive system regulate the lifespan of *C. elegans*. *Nature*, *399*, 362–366.
Jeon, M., Gardner, H. F., Miller, E. A., Deshler, J., & Rougvie, A. E. (1999). Similarity of the *C. elegans* developmental timing protein LIN-42 to circadian rhythm proteins. *Science*, *286*, 1141–1146.
Jia, K., Albert, P. S., & Riddle, D. L. (2002). DAF-9, a cytochrome P450 regulating *C. elegans* larval development and adult longevity. *Development*, *129*(1), 221–231.
Karp, X., & Ambros, V. (2012). Dauer larva quiescence alters the circuitry of microRNA pathways regulating cell fate progression in *C. elegans*. *Development*, *139*, 2177–2186.
Keisala, T., Minasyan, A., Lou, Y. R., Zou, J., Kalueff, A. V., Pyykko, I., et al. (2009). Premature aging in vitamin D receptor mutant mice. *In J Steroid Biochem Mol Biol*, *115*(3–5), 91–97.
Kenyon, C. J. (2010). The genetics of ageing. *Nature*, *464*, 504–512.
Kimura, K. D., Tissenbaum, H. A., Liu, Y., & Ruvkun, G. (1997). daf-2, an insulin receptor-like gene that regulates longevity and diapause in *Caenorhabditis elegans*. *Science*, *277*, 942–946.
Kostrouchova, M., Krause, M., Kostrouch, Z., & Rall, J. E. (2001). Nuclear hormone receptor CHR3 is a critical regulator of all four larval molts of the nematode *Caenorhabditis elegans*. *Proceedings of the National Academy of Sciences of the United States of America*, *98*, 7360–7365.
Krylova, I. N., Sablin, E. P., Moore, J., Xu, R. X., Waitt, G. M., MacKay, J. A., et al. (2005). Structural analyses reveal phosphatidyl inositols as ligands for the NR5 orphan receptors SF-1 and LRH-1. *Cell*, *120*, 343–355.
Lamia, K. A., Papp, S. J., Yu, R. T., Barish, G. D., Uhlenhaut, N. H., Jonker, J. W., et al. (2011). Cryptochromes mediate rhythmic repression of the glucocorticoid receptor. *Nature*, *480*, 552–556.
Lapierre, L. R., Gelino, S., Melendez, A., & Hansen, M. (2011). Autophagy and lipid metabolism coordinately modulate life span in germline-less *C. elegans*. *Current Biology*, *21*, 1507–1514.
Lee, R. C., Feinbaum, R. L., & Ambros, V. (1993). The *C. elegans* heterochronic gene lin-4 encodes small RNAs with antisense complementarity to lin-14. *Cell*, *75*, 843–854.
Lee, J. M., Lee, Y. K., Mamrosh, J. L., Busby, S. A., Griffin, P. R., Pathak, M. C., et al. (2011). A nuclear-receptor-dependent phosphatidylcholine pathway with antidiabetic effects. *Nature*, *474*, 506–510.
Li, J., Brown, G., Ailion, M., Lee, S., & Thomas, J. H. (2004). NCR-1 and NCR-2, the *C. elegans* homologs of the human Niemann-Pick type C1 disease protein, function upstream of DAF-9 in the dauer formation pathways. *Development*, *131*, 5741–5752.
Li, W., Kennedy, S. G., & Ruvkun, G. (2003). daf-28 encodes a *C. elegans* insulin superfamily member that is regulated by environmental cues and acts in the DAF-2 signaling pathway. *Genes & Development*, *17*, 844–858.
Lin, K., Dorman, J. B., Rodan, A., & Kenyon, C. (1997). daf-16: An HNF-3/forkhead family member that can function to double the life-span of *Caenorhabditis elegans*. *Science*, *278*, 1319–1322.
Liu, Z., & Ambros, V. (1991). Alternative temporal control systems for hypodermal cell differentiation in *C. elegans*. *Nature*, *350*, 162–165.

Ludewig, A. H., Kober-Eisermann, C., Weitzel, C., Bethke, A., Neubert, K., Gerisch, B., et al. (2004). A novel nuclear receptor/coregulator complex controls *C. elegans* lipid metabolism, larval development, and aging. *Genes & Development, 18*, 2120–2133.

Mak, H. Y., & Ruvkun, G. (2004). Intercellular signaling of reproductive development by the *C. elegans* DAF-9 cytochrome P450. *Development, 131*, 1777–1786.

Maller Schulman, B. R., Liang, X., Stahlhut, C., DelConte, C., Stefani, G., & Slack, F. J. (2008). The let-7 microRNA target gene, Mlin41/Trim71 is required for mouse embryonic survival and neural tube closure. *Cell Cycle, 7*, 3935–3942.

Mangelsdorf, D. J., & Evans, R. M. (1995). The RXR heterodimers and orphan receptors. *Cell, 83*, 841–850.

Mangelsdorf, D. J., Thummel, C., Beato, M., Herrlich, P., Schutz, G., Umesono, K., et al. (1995). The nuclear receptor superfamily: The second decade. *Cell, 83*, 835–839.

Matyash, V., Entchev, E. V., Mende, F., Wilsch-Brauninger, M., Thiele, C., Schmidt, A. W., et al. (2004). Sterol-derived hormone(s) controls entry into diapause in *Caenorhabditis elegans* by consecutive activation of DAF-12 and DAF-16. *PLoS Biology, 2*, e280.

McCormick, M., Chen, K., Ramaswamy, P., & Kenyon, C. (2012). New genes that extend *Caenorhabditis elegans*' lifespan in response to reproductive signals. *Aging Cell, 11*, 192–202.

Merris, M., Wang, T., Soteropoulos, P., & Lenard, J. (2007). Differential gene expression of *Caenorhabditis elegans* grown on unmethylated sterols or 4alpha-methylsterols. *Journal of Lipid Research, 48*, 1159–1166.

Monje, J. M., Brokate-Llanos, A. M., Perez-Jimenez, M. M., Fidalgo, M. A., & Munoz, M. J. (2011). pkc-1 regulates daf-2 insulin/IGF signalling-dependent control of dauer formation in *Caenorhabditis elegans*. *Aging Cell, 10*, 1021–1031.

Monsalve, G. C., Van Buskirk, C., & Frand, A. R. (2011). LIN-42/PERIOD controls cyclical and developmental progression of *C. elegans* molts. *Current Biology, 21*, 2033–2045.

Mooijaart, S. P., Kuningas, M., Westendorp, R. G., Houwing-Duistermaat, J. J., Slagboom, P. E., Rensen, P. C., et al. (2007). Liver X receptor alpha associates with human life span. *The Journals of Gerontology. Series A, Biological Sciences and Medical Sciences, 62*, 343–349.

Morita, K., & Han, M. (2006). Multiple mechanisms are involved in regulating the expression of the developmental timing regulator lin-28 in *Caenorhabditis elegans*. *EMBO Journal, 25*, 5794–5804.

Morris, J. Z., Tissenbaum, H. A., & Ruvkun, G. (1996). A phosphatidylinositol-3-OH kinase family member regulating longevity and diapause in *Caenorhabditis elegans*. *Nature, 382*, 536–539.

Moss, E. G., Lee, R. C., & Ambros, V. (1997). The cold shock domain protein LIN-28 controls developmental timing in *C. elegans* and is regulated by the lin-4 RNA. *Cell, 88*, 637–646.

Motola, D. L., Cummins, C. L., Rottiers, V., Sharma, K. K., Li, T., Li, Y., et al. (2006). Identification of ligands for DAF-12 that govern dauer formation and reproduction in *C. elegans*. *Cell, 124*, 1209–1223.

Mullaney, B. C., Blind, R. D., Lemieux, G. A., Perez, C. L., Elle, I. C., Faergeman, N. J., et al. (2010). Regulation of *C. elegans* fat uptake and storage by acyl-CoA synthase-3 is dependent on NR5A family nuclear hormone receptor nhr-25. *Cell Metabolism, 12*, 398–410.

Niwa, R., Zhou, F., Li, C., & Slack, F. J. (2008). The expression of the Alzheimer's amyloid precursor protein-like gene is regulated by developmental timing microRNAs and their targets in *Caenorhabditis elegans*. *Developmental Biology, 315*, 418–425.

Ogawa, A., Streit, A., Antebi, A., & Sommer, R. J. (2009). A conserved endocrine mechanism controls the formation of dauer and infective larvae in nematodes. *Current Biology, 19*, 67–71.

Ogg, S., Paradis, S., Gottlieb, S., Patterson, G. I., Lee, L., Tissenbaum, H. A., et al. (1997). The Fork head transcription factor DAF-16 transduces insulin-like metabolic and longevity signals in *C. elegans*. *Nature*, *389*, 994–999.

Ong, K. K., Elks, C. E., Li, S., Zhao, J. H., Luan, J., Andersen, L. B., et al. (2009). Genetic variation in LIN28B is associated with the timing of puberty. *Nature Genetics*, *41*, 729–733.

Paradis, S., & Ruvkun, G. (1998). *C. elegans* Akt/PKB transduces insulin receptor-like signals from AGE-1 PI3 kinase to the DAF-16 transcription factor. *Genes & Development*, *12*, 2488–2498.

Patel, D. S., Fang, L. L., Svy, D. K., Ruvkun, G., & Li, W. (2008). Genetic identification of HSD-1, a conserved steroidogenic enzyme that directs larval development in *Caenorhabditis elegans*. *Development*, *135*, 2239–2249.

Patterson, G. I., Koweek, A., Wong, A., Liu, Y., & Ruvkun, G. (1997). The DAF-3 Smad protein antagonizes TGF-beta-related receptor signaling in the *Caenorhabditis elegans* dauer pathway. *Genes & Development*, *11*, 2679–2690.

Pepper, A. S., McCane, J. E., Kemper, K., Yeung, D. A., Lee, R. C., Ambros, V., et al. (2004). The *C. elegans* heterochronic gene lin-46 affects developmental timing at two larval stages and encodes a relative of the scaffolding protein gephyrin. *Development*, *131*, 2049–2059.

Ranhotra, H. S. (2012). The interplay between retinoic acid receptor-related orphan receptors and human diseases. *Journal of Receptor and Signal Transduction Research*, *32*, 181–189.

Reinhart, B. J., Slack, F. J., Basson, M., Pasquinelli, A. E., Bettinger, J. C., Rougvie, A. E., et al. (2000). The 21-nucleotide let-7 RNA regulates developmental timing in *Caenorhabditis elegans*. *Nature*, *403*, 901–906.

Ren, P., Lim, C. S., Johnsen, R., Albert, P. S., Pilgrim, D., & Riddle, D. L. (1996). Control of *C. elegans* larval development by neuronal expression of a TGF-beta homolog. *Science*, *274*, 1389–1391.

Riddle, D. L., Swanson, M. M., & Albert, P. S. (1981). Interacting genes in nematode dauer larva formation. *Nature*, *290*, 668–671.

Rottiers, V., Motola, D. L., Gerisch, B., Cummins, C. L., Nishiwaki, K., Mangelsdorf, D. J., et al. (2006). Hormonal control of *C. elegans* dauer formation and life span by a Rieske-like oxygenase. *Developmental Cell*, *10*, 473–482.

Russell, D. W. (2003). The enzymes, regulation, and genetics of bile acid synthesis. *Annual Review of Biochemistry*, *72*, 137–174.

Rybak, A., Fuchs, H., Smirnova, L., Brandt, C., Pohl, E. E., Nitsch, R., et al. (2008). A feedback loop comprising lin-28 and let-7 controls pre-let-7 maturation during neural stem-cell commitment. *Nature Cell Biology*, *10*, 987–993.

Schaedel, O. N., Gerisch, B., Antebi, A., & Sternberg, P. W. (2012). Hormonal signal amplification mediates environmental conditions during development and controls an irreversible commitment to adulthood. *PLoS Biology*, *10*, e1001306.

Seggerson, K., Tang, L., & Moss, E. G. (2002). Two genetic circuits repress the *Caenorhabditis elegans* heterochronic gene lin-28 after translation initiation. *Developmental Biology*, *243*, 215–225.

Shen, Y., Wollam, J., Magner, D., Karalay, O., & Antebi, A. (2012). A steroid receptor-microRNA switch regulates life span in response to signals from the gonad. *Science*, *338*, 1472–1476.

Slack, F. J., Basson, M., Liu, Z., Ambros, V., Horvitz, H. R., & Ruvkun, G. (2000). The lin-41 RBCC gene acts in the *C. elegans* heterochronic pathway between the let-7 regulatory RNA and the LIN-29 transcription factor. *Molecular Cell*, *5*, 659–669.

Slominski, A., Zbytek, B., Nikolakis, G., Manna, P. R., Skobowiat, C., Zmijewski, M., et al. (2013). Steroidogenesis in the skin: Implications for local immune functions. *The Journal of Steroid Biochemistry and Molecular Biology*, *67*, 123–128.

Sulem, P., Gudbjartsson, D. F., Rafnar, T., Holm, H., Olafsdottir, E. J., Olafsdottir, G. H., et al. (2009). Genome-wide association study identifies sequence variants on 6q21 associated with age at menarche. *Nature Genetics*, *41*, 734–738.

Sulston, J. E., & Horvitz, H. R. (1977). Post-embryonic cell lineages of the nematode, *C. elegans*. *Developmental Biology*, *56*, 110–156.

Takahashi, K., & Yamanaka, S. (2006). Induction of pluripotent stem cells from mouse embryonic and adult fibroblast cultures by defined factors. *Cell*, *126*, 663–676.

Tennessen, J. M., Gardner, H. F., Volk, M. L., & Rougvie, A. E. (2006). Novel heterochronic functions of the *Caenorhabditis elegans* period-related protein LIN-42. *Developmental Biology*, *289*, 30–43.

Tennessen, J. M., Opperman, K. J., & Rougvie, A. E. (2010). The *C. elegans* developmental timing protein LIN-42 regulates diapause in response to environmental cues. *Development*, *137*, 3501–3511.

Tennessen, J. M., & Thummel, C. S. (2011). Coordinating growth and maturation—Insights from Drosophila. *Current Biology*, *21*, R750–R757.

Ting, H. J., Messing, J., Yasmin-Karim, S., & Lee, Y. F. (2013). Identification of microRNA-98 as a therapeutic target inhibiting prostate cancer growth and a biomarker induced by vitamin D. *Journal of Biological Chemistry*, *288*, 1–9.

Urs, A. N., Dammer, E., Kelly, S., Wang, E., Merrill, A. H., Jr., & Sewer, M. B. (2007). Steroidogenic factor-1 is a sphingolipid binding protein. *Molecular and Cellular Endocrinology*, *265–266*, 174–178.

Van Wynsberghe, P. M., Kai, Z. S., Massirer, K. B., Burton, V. H., Yeo, G. W., & Pasquinelli, A. E. (2011). LIN-28 co-transcriptionally binds primary let-7 to regulate miRNA maturation in *Caenorhabditis elegans*. *Nature Structural & Molecular Biology*, *18*, 302–308.

Vilchez, D., Morantte, I., Liu, Z., Douglas, P. M., Merkwirth, C., Rodrigues, A. P., et al. (2012). RPN-6 determines *C. elegans* longevity under proteotoxic stress conditions. *Nature*, *489*, 263–268.

Viswanathan, S. R., Daley, G. Q., & Gregory, R. I. (2008). Selective blockade of microRNA processing by Lin28. *Science*, *320*, 97–100.

Wang, Y., Kumar, N., Solt, L. A., Richardson, T. I., Helvering, L. M., Crumbley, C., et al. (2010). Modulation of retinoic acid receptor-related orphan receptor alpha and gamma activity by 7-oxygenated sterol ligands. *Journal of Biological Chemistry*, *285*, 5013–5025.

Wang, M. C., O'Rourke, E. J., & Ruvkun, G. (2008). Fat metabolism links germline stem cells and longevity in *C. elegans*. *Science*, *322*, 957–960.

Wang, Z., Zhou, X. E., Motola, D. L., Gao, X., Suino-Powell, K., Conneely, A., et al. (2009). Identification of the nuclear receptor DAF-12 as a therapeutic target in parasitic nematodes. *Proceedings of the National Academy of Sciences of the United States of America*, *106*, 9138–9143.

Wiese, M., Antebi, A., & Zheng, H. (2010). Intracellular trafficking and synaptic function of APL-1 in *Caenorhabditis elegans*. *PLoS One*, *5*(9), e12790.

Wightman, B., Ha, I., & Ruvkun, G. (1993). Posttranscriptional regulation of the heterochronic gene lin-14 by lin-4 mediates temporal pattern formation in *C. elegans*. *Cell*, *75*, 855–862.

Wollam, J., Magner, D. B., Magomedova, L., Rass, E., Shen, Y., Rottiers, V., et al. (2012). A novel 3-hydroxysteroid dehydrogenase that regulates reproductive development and longevity. *PLoS Biology*, *10*, e1001305.

Wollam, J., Magomedova, L., Magner, D. B., Shen, Y., Rottiers, V., Motola, D. L., et al. (2011). The Rieske oxygenase DAF-36 functions as a cholesterol 7-desaturase in steroidogenic pathways governing longevity. *Aging Cell*, *10*, 879–884.

Wu, Y. C., Chen, C. H., Mercer, A., & Sokol, N. S. (2012). Let-7-complex microRNAs regulate the temporal identity of Drosophila mushroom body neurons via chinmo. *Developmental Cell*, *23*, 202–209.

Xia, D., Huang, X., & Zhang, H. (2009). The temporally regulated transcription factor sel-7 controls developmental timing in *C. elegans*. *Developmental Biology*, *332*, 246–257.

Yamawaki, T. M., Berman, J. R., Suchanek-Kavipurapu, M., McCormick, M., Maria Gaglia, M., Lee, S. J., et al. (2010). The somatic reproductive tissues of *C. elegans* promote longevity through steroid hormone signaling. *PLoS Biology*, *8*(8), e1000468.

Yochem, J., Tuck, S., Greenwald, I., & Han, M. (1999). A gp330/megalin-related protein is required in the major epidermis of *Caenorhabditis elegans* for completion of molting. *Development*, *126*, 597–606.

Yoshiyama-Yanagawa, T., Enya, S., Shimada-Niwa, Y., Yaguchi, S., Haramoto, Y., Matsuya, T., et al. (2011). The conserved Rieske oxygenase DAF-36/Neverland is a novel cholesterol-metabolizing enzyme. *Journal of Biological Chemistry*, *286*, 25756–25762.

Yu, J., Vodyanik, M. A., Smuga-Otto, K., Antosiewicz-Bourget, J., Frane, J. L., Tian, S., et al. (2007). Induced pluripotent stem cell lines derived from human somatic cells. *Science*, *318*, 1917–1920.

Zhang, H., & Fire, A. Z. (2010). Cell autonomous specification of temporal identity by *Caenorhabditis elegans* microRNA lin-4. *Developmental Biology*, *344*, 603–610.

Zhi, X., Zhou, X. E., Melcher, K., Motola, D. L., Gelmedin, V., Hawdon, J., et al. (2012). Structural conservation of ligand binding reveals a bile acid-like signaling pathway in nematodes. *Journal of Biological Chemistry*, *287*, 4894–4903.

Zhu, H., Shah, S., Shyh-Chang, N., Shinoda, G., Einhorn, W. S., Viswanathan, S. R., et al. (2010). Lin28a transgenic mice manifest size and puberty phenotypes identified in human genetic association studies. *Nature Genetics*, *42*, 626–630.

Zhu, H., Shyh-Chang, N., Segre, A. V., Shinoda, G., Shah, S. P., Einhorn, W. S., et al. (2011). The Lin28/let-7 axis regulates glucose metabolism. *Cell*, *147*, 81–94.

CHAPTER EIGHT

Diapause: Delaying the Developmental Clock in Response to a Changing Environment

Luca Schiesari*, Michael B. O'Connor†,[1]

*Department of Biology, University of Padova, Padova, Italy
†Department of Genetics, Cell Biology and Development, University of Minnesota, Minneapolis, Minnesota, USA
[1]Corresponding author: e-mail address: moconnor@umn.edu

Contents

1. Introduction 214
2. Hormonal Control of Embryonic Diapause in Insects: The "Hatch-Ready" Example 215
3. Hormonal Control of Insect Diapause: The Mid-Embryonic Arrest Example 216
   3.1 Production of diapause hormone in pupa determines the developmental fate of the future egg 217
   3.2 Control of DH expression 219
4. Developmental Control of Larval Diapause 219
5. Pupal Diapause 221
6. Diapause Can Modify Larval/Pupal Growth 226
7. Seasonal Polyphenisms Correlate with Diapause 227
   7.1 Adaptive advantages of seasonal polyphenisms 231
   7.2 Molecular mechanism for polyphenism induction 231
8. The Killifish Diapausing Embryo: A Developmental Analogy in a Vertebrate 233
   8.1 Do insulin/IGFs play an evolutionary conserved function in diapause? 235
9. Summary and Perspective 236
Acknowledgments 237
References 237

Abstract

Seasonal changes can induce organisms to modify their developmental growth. Many holometabolous insects, especially Lepidoptera, trigger diapause, an "actively induced" dormancy, for overwintering. Diapause is an alternative developmental pathway that reversibly blocks developmental growth during specific transitions and enhances the hibernating potential of the organism. Changes in environmental cues, such as light and temperature, trigger modifications in the levels, or in the timing, of developmental hormones. These in turn switch the developmental trajectory (diapause or direct development), strongly altering larval/pupal growth and inducing the appearance of diapause-bound seasonal morphs (polyphenism). We also discuss an example of

Current Topics in Developmental Biology, Volume 105
ISSN 0070-2153
http://dx.doi.org/10.1016/B978-0-12-396968-2.00008-7

vertebrate diapause using the killifish embryo as an example where diapause is an environmentally determined developmental switch analogous to that observed in lepidopteran dormancy. Based on the examples discussed here, we propose that the complex physiological responses leading to diapause might evolve quickly by relatively limited genetic changes in the regulation of hormonal signals that program normal developmental transitions.

1. INTRODUCTION

Diapause is an "actively induced" dormancy that blocks developmental growth of an organism in anticipation of a major harsh seasonal change, such as winter. The diapausing phase is genetically specified, but it is elicited not in response to developmental cues, but by reliable environmental signals, such as photoperiod and temperature, that are perceived during an earlier developmental stage (Saunders, Steel, Vafopoulou, & Lewis, 2002). Hence, diapause profoundly influences the trajectory of holometabolous growth without causing any functional developmental aberration in the final adult form, although as we will describe, there is some developmental plasticity associated with diapause that is frequently manifested as seasonal polyphenisms in adult morphologies (Saunders et al., 2002).

While induction of diapause is genetically encoded and requires input from the endocrine system, exiting the dormant stage also requires a "genetically established" period of chilling to enable reactivation of development when a more optimal environment (spring) is encountered (Denlinger, Yocum, & Rinehart, 2012; Lee & Denlinger, 1991). For instance, diapausing pupae of *Samia cynthia* must spend 3–5 months below 4 °C prior to breaking diapause, while dormant embryos of *Bombyx mori* require about 2 months below 5 °C (Horie, Kanda, & Mochida, 2000; Nakamura et al., 2011; Suzuki, Fujita, & Miya, 1983) before reactivating the maturation process. Without the cold period, dormancy cannot be broken (Denlinger et al., 2012).

Diapausing organisms are extremely resistant to low temperatures by either undergoing supercooling or freezing, often in the presence of biologically produced cyroprotectorants (Denlinger et al., 2012; Hahn & Denlinger, 2010; Lee & Denlinger, 1991). As an example, pupae of the papilionid *Papilio machaon* can resist freezing up to −25 °C but even after freezing at −30 °C, they still remain alive for months (Shimada, 1980). Likewise, larvae of the lymantrid *Gymnaephora groenlandica* undergo supercooling until −7 °C, but they can survive freezing down to −70 °C (Kukal, Serianni, &

Duman, 1988; Lee & Denlinger, 1991). Similarly, diapausing embryos of *B. mori* can resist freezing down to −32 °C for several months using only supercooling (Lee & Denlinger, 1991; Suzuki et al., 1983).

In this chapter, we will focus primarily on a discussion of diapause in Lepidoptera species largely because of their rich diversity in the types of diapause and polyphenism responses they produce in the face of similar or unique environmental challenges. This plasticity, together with their long history of study by entomologists and naturalists, make them an ideal model system to determine at the molecular genetic level, how adaptive developmental pathways evolve. Although understanding the mechanisms behind lepidopteran diapause induction has been a long standing issue, the use of modern genomic and physiological methods now offers new opportunities to answer important outstanding issues about adaptive change, a process of increasing relevance in light of recent rapid global climatic changes.

In the first section of this chapter, we will describe the examples of lepidopteran species that arrest development at several different stages, and then discuss what is presently know about the molecular mechanisms behind diapause induction. Lastly, we will briefly describe one vertebrate example in which diapause is induced not by seasonal temperature and photoperiod changes, but instead responds to seasonal changes in hydration levels of the local environment.

2. HORMONAL CONTROL OF EMBRYONIC DIAPAUSE IN INSECTS: THE "HATCH-READY" EXAMPLE

Diapause can be triggered in embryos by blocking developmental growth. Some species undergo diapause as early embryos, while others enter diapause as pharate first instar larvae ("hatch-ready" larvae) within the eggshell. Embryonic diapauses can be very different in regulation, although they are all under the control of the steroid hormone ecdysone (ECD). However, the role of ECD as a diapause inhibitor or promoter depends on the diapausing stage (Denlinger et al., 2012; Saunders et al., 2002). For example, consider the gypsy moth, *Lymantria dispar*. It diapauses in autumn as a pharate first instar larvae and does not hatch until it has spent a genetically established period in the cold. As described, the cold phase is needed to reestablish developmental competence for growth, once more optimal conditions are encountered. In this example, diapause is induced and sustained by the production of high ECD titers, whereas under normal developmental conditions the ECD titer would be low in fully developed prehatch first instar larva. The requirement for a high ECD titer to induce and maintain diapause in this species can be further demonstrated by the observation that "chilled" diapausing embryos remain dormant, even when exposed to optimal conditions, if they are simultaneously treated with exogenous ECD. Conversely, embryos fail to induce diapause when injected with KK-42, an ECD antagonist, but this effect can be reversed by topical applications of ECD. Interestingly, "nondiapausing" Lymantria mutants that have genetically lost the ability to induce diapause can enter dormancy when injected with ECD, revealing that a latent, and inducible, diapause response is still genetically encoded in these mutants (Lee & Denlinger, 1996, 1997; Lee, Horodyski, Valaitis, & Denlinger, 2002; Lee, Valaitis, & Denlinger, 1997; Suzuki, Nakamura, Yanbe, Kurihara, & Kuwano, 1993).

A similar hormonal mechanism is thought to control diapause entry in the hesperiid *Thymelicus lineola* (McNeil & Fields, 1985) and the saturniid *Antheraea yamamai* (Suzuki et al., 1990). In the latter species, prolonged pulses of ECD induce diapause (as pharate first instar larva) and the inhibition of ECD signaling in diapausing embryos breaks dormancy. The main source of ECD in this, as well as other insects, is the embryonic/larval prothoracic gland (PG) which maintains high levels of endocrine activity until diapause termination, at which time the ECD titer drops and the embryo hatches (Denlinger et al., 2012; Lee & Denlinger, 1997; Suzuki et al., 1990, 1993). Consistent with the role of the PG in regulating diapause, excision of the thorax/head complex (containing the PG) from diapausing embryos induces the isolated abdomens to resume growth (Suzuki et al., 1990). From these studies one theme that emerges is that in certain insects, a prolonged and intense ECD pulse can induce diapause by overriding the normal developmental program. Diapause is then broken after a predetermined cool period and the restoration of a warmer environment and longer photoperiod.

3. HORMONAL CONTROL OF INSECT DIAPAUSE: THE MID-EMBRYONIC ARREST EXAMPLE

Although prolonged ECD levels can induce the formation of "hatch ready," pre-eclosion first instar larva in some insect species, in other species it is the failure of inducing the mid-embryonic ECD pulse that triggers diapause. In these cases, however, the embryos arrest at a much earlier stage of embryonic development. For example, in autumn, embryos of the silk moth, *B. mori* (Bombycidae), enter diapause just after embryonic mesoderm segmentation, at which point they become extremely resistant to cold (able to tolerate −32 °C). The cold tolerance is enhanced by the production of glycerol and sorbitol that act as cryoprotecting agents (Horie et al., 2000; Lee & Denlinger, 1991). Interestingly, sorbitol also appears to play a regulatory function since when exogenously applied to "nondiapausing embryos," it actively inhibits developmental growth and stimulates the embryos to enter dormancy. Conversely, dormant embryos break diapause when sorbitol is removed from the embryos themselves (Horie et al., 2000). Similar to the "hatch-ready" case, mid-embryonic-arrested Bombyx embryos also require a defined period of time in the cold (2–3 months at 5 °C) before they can resume normal development (Horie et al., 2000). It appears that one function of the cold period is to reduce sorbitol levels below a threshold that then enables the embryos to exit from diapause and resume developmental growth (Horie et al., 2000). Part of the exit process involves ERK signaling which promotes the conversion of the remaining sorbitol into glycogen (restoring the energy resources needed for developmental growth) and the induction of the required mid-embryonic ECD pulse. These two processes rely on ERK dependant activation of two key enzymes: sorbitol dehydrogenase-2 and ecdysteroid-phosphate phosphatase (ECDPPase). The later enzyme is crucial during the postdiapause exit stage, since embryos require an ECD pulse at this stage of development and they depend entirely on the ability of ECDPPase to covert maternal stores of inactive ecdysteroid-phosphates into ECD to reactivate normal developmental growth and differentiation (Fujiwara, Shindome, Takeda, & Shiomi, 2006; Fujiwara & Shiomi, 2006; Fujiwara, Tanaka, et al., 2006; Horie et al., 2000; Iwata et al., 2005).

3.1. Production of diapause hormone in pupa determines the developmental fate of the future egg

While sorbitol appears to play a role in the induction of Bombyx embryonic diapause, it is downstream of the actual initiating event, which is thought to be the environmentally controlled production of diapause hormone (DH), a 24-amino acid peptide, that acts inside the pupa of future mothers (Sato, Ikeda, & Yamashita, 1994; Sato et al., 1993; Sato, Shiomi, Saito, Imai, & Yamashita, 1998; Shiomi et al., 2007; Suwan, Isobe, Yamashita, Minakata, & Imai, 1994; Yamashita, 1996; Yamashita, Shiomi, Ishida, Katagiri, & Niimi, 2001). When the future mother moth grows as a larva in the summer months (long days and warm temperatures), the emerged adult female lays "autumnal" embryos that are programmed for entering diapause through the production of high levels of DH in the mother (Fig. 8.1) (Fukuda, 1951; Hasegawa, 1951; Nakagaki, Takei, Nagashima, & Yaginuma, 1991;

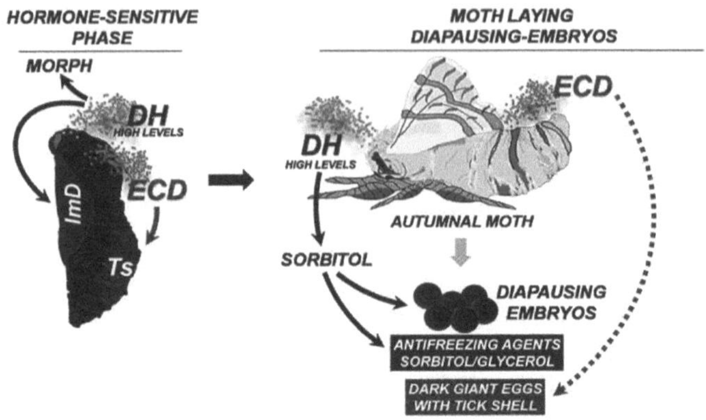

Figure 8.1 Diapause induction in *Bombyx mori*. Long summer days induce fifth last instar larvae to develop as moth laying diapausing embryos (diapause specification) in early autumn. During pupal life ("hormone-sensitive" stage), DH signals over-threshold either in target tissue (Ts, in red) and, presumably, in imaginal wing discs (ImD) committing the development of autumnal morph (brown scales on the wings). High levels of ecdysone (ECD, in red) and the shifted timing of its signaling since pupal life induce the appearance of the typical diapause traits in embryos, such as dark pigmentation, gigantism, and egg-shell thickness. Autumnal moths have enhanced DH titers that commit embryos to enter dormancy and enhance the levels of sorbitol, which acts either as cryoprotector and developmental inhibitor. (For interpretation of the references to color in this figure legend, the reader is referred to the online version of this chapter.)

Sato et al., 1994, 1993; Shiomi et al., 2007). DH is produced in the suboesophageal ganglion (SOG) of the moth nervous system and released into hemolymph via the neurohemal glands. Circulating DH signals to the developing oocytes via a G-coupled receptor, diapause hormone receptor (DHR), to enhance glycogen levels (the source of sorbitol and glycerol) (Homma et al., 2006; Horie et al., 2000). Experimental manipulations that change the strength of DH signaling in the pupa can completely alter the developmental trajectory of the future embryos. For example, a "nondiapause-conditioned" pupae (low DH levels) will develop into a moth that lays diapausing eggs if it is injected with DH (Uehara, Senoh, Yoneda, Kato, & Shiomi, 2011). Conversely, when the SOG of a "diapause-conditioned pupa" is removed, the resulting adult female lays nondiapausing eggs (Fukuda, 1951). Taken together, these observations suggest that a "DH sensitive" period occurs during early pupal development that dictates which type of embryo will be laid by the emerged mother moth (Uehara et al., 2011).

3.2. Control of DH expression

The Bombyx DH gene (BmDH) exhibits a different expression profile depending on the developmental trajectory of the larva. In "diapause-conditioned" animals (those that will eventually lay diapausing eggs), large peaks of *BmDH* expression occur during both the mid-late larval instars (4°/5°) and the pupal stage (Morita, Niimi, & Yamashita, 2004; Xu, Sato, Ikeda, & Yamashita, 1995a). One key question in the field is how environmental signals affect DH expression. The product of the *Ptix1* gene, a bicoid-like homeobox transcription factor binds cis-regulatory elements in the promoter region of *BmDH*, as do members of the POU transcription factor family (Shiomi et al., 2007; Xu, Sato, Ikeda, & Yamashita, 1995b; Zhang, Kang, Zhang, & Xu, 2004). It is not clear however, what inputs control Ptix1 expression. One possibility is dopamine (DA). Bombyx moths laying diapausing embryos produce a large amount of DA and they upregulate the *BmDdc* gene (encoding DA decarboxylase, a key enzyme in DA synthesis) during the late larval and early pupal life. Consistent with a causal role of DA in triggering diapause, "nondiapause" larvae fed with L-DOPA (a DA precursor) or injected with DA, develop as moths that express high levels of DH and lay embryos primed for diapause. This suggests that DA acts upstream of DH signaling in inducing diapause trajectory (Noguchi & Hayakawa, 2001). However, it is not clear whether DA signaling directly results in the upregulation of *Ptix1* nor is it yet clear whether DA producing neurons respond directly to environmental cues.

4. DEVELOPMENTAL CONTROL OF LARVAL DIAPAUSE

Unlike the embryonic diapause described above, in some species, the primary point of diapause control is within the larval stages. Last instar larvae of the pyralid *Diatraea grandiosella* trigger diapause by molting from the normal "spotted" morph (developing) to an "immaculate" (diapausing) one (Fig. 8.2). Some of these diapausing larvae undergo extra "stationary" molts where approximately 50% undergo one extra molt while 14% undergo two extra molts. Immaculate larvae remain unchanged in size through each molt, but they modify completely their physiology: feeding stops, respiration decreases, and fat accumulation, dehydratation, cold hardiness and fat storage are enhanced (Chippendale, 1977, 1984).

In this species as well as several others that undergo larval diapause, developmental arrest is induced by juvenile hormone (JH) which constrains ECD

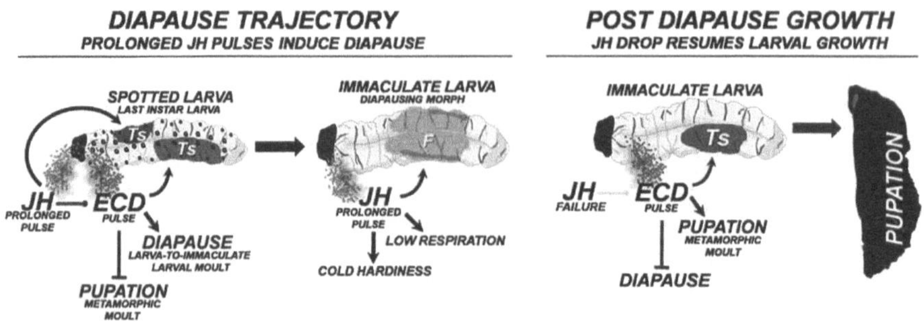

Figure 8.2 Larval diapause in *Diatraea grandiosella*. The spotted larva at the last instar enters dormancy under high levels of JH which override the molting action of ECD to induce the appearance of the dormant immaculate morph instead of the pupation. Immaculate larva fails to initiate metamorphosis. After a period of chilling, the drop of JH arrests diapause and larvae initiate normally the pupal life. Ts, target organs; F, fat accumulation. (For color version of this figure, the reader is referred to the online version of this chapter.)

signaling similar to the way it does for normal larval–larval molts. In fact, injections of ECD into immaculate larvae elicit only further immaculate-to-immaculate molts failing to induce metamorphosis. Moreover, injections of ECD into the body of neck-legated diapausing larvae (in which the hemolymphatic diffusion of JH from the neurohemal glands producing it is blocked) induce premature pupation of the segments posterior to the ligature. Consistent with the pivotal role of JH, topical application of JH to "spotted" (nondiapausing) larvae induces them to develop as immaculate dormant morphs and continued exposure to exogenous JH maintains diapause and enhances the number of "stationary" molts (Chippendale, 1977, 1984). Hemolymphatic levels of JH remain at high levels during diapause after an immaculate molt. Only when JH levels decay does diapause end and larva begin to undergo normal metamorphosis (Chippendale, 1977, 1984; Chippendale & Yin, 1973; Saunders et al., 2002; Yin & Chippendale, 1973, 1974, 1975, 1976a, 1976b)

Diapausing larval morphs or larval polyphenisms have also been reported in other species (Scott, 1979 and references therein). Asterocampa larvae (Nymphalidae) diapause as third or fourth instar and, once diapause is ended, they reach pupation in six larval instars rather than normal five. Larvae of two nymphalids, *Chlosyne gorgone* and *Chlosyne nycteis*, diapause as dark larval morphs, while *Euphydryas editha* (Nymphalidae) has a larval diapause morph with thicker integument. The green larvae of *Plebejus icarioides* (Lycaenidae) diapause as dark second larval instar, but they resume the green morph color after diapause.

Another interesting variation of larval diapause is the occurrence of stationary molts during the diapause stage. For example, larvae of the diapausing noctuid *Sesamia nonagriodes* feed slowly and undergo up to 12 (although usually 3–4) stationary larval molts without increasing the body size. In such larvae, ECD rises irregularly during the extra larval molts but high JH levels prevent pupation. Topical applications of an ECD mimic to diapausing larvae induce them to accelerate larval molts rather than induce pupation and nondiapausing fifth instar larvae make extra larval molts when fed with a JH analogue resembling the induction of diapause. Normally, diapause is only specified prior to the third instar and results in an increasing JH titer that keeps them in a juvenile form (Eizaguirre et al., 1998; Eizaguirre, Schafellner, Lopez, & Sehnal, 2005). Similarly, high levels of JH maintain larval diapause in the yellow-spotted longicorn beetle, *Psacothea hilaris* (Munyiri & Ishikawa, 2004). Many Lepidoptera also exhibit stationary molts during diapause (Kevan, 1944; Sugiki & Masaki, 1972; Usua, 1973; Yagi & Fukaya, 1974).

While stationary molts are relatively common in many species, some just arrest with no extra molting. In such cases, diapause is thought not to be induced by prolonged JH pulses, but rather, by a failure of ECD signaling (Denlinger et al., 2012; Saunders et al., 2002). For example, injection of JH into nondiapausing *Ostrinia nubilalis* larvae fails to induce dormancy, but if ECD is injected into diapause-induced larvae, they will undergo a larval to pupal molt instead of a stationary larval molt (Denlinger et al., 2012; Gadenne, Varjas, & Mauchamp, 1990; Gelman & Brents, 1984; Gelman et al., 1992; Gelman & Woods, 1983; Peypelut, Beydon, & Lavenseau, 1990). In larvae of both *Laspeyresia polmonella* and *Ostrinia nubilalis* the loss of ECD signaling appears to be a result from a block in the brain-derived prothoracicotropic hormone (PTTH) signaling cascade that is necessary for high level ECD induction in most insects (Bean & Beck, 1980, 1983; Denlinger et al., 2012; Saunders et al., 2002; Sieber and Benz, 1977, 1980; Smith & Rybczynski, 2012).

5. PUPAL DIAPAUSE

The final developmental stage in which diapause can be induced occurs just after the larval to pupal molt, but before significant metamorphosis occurs. As described above for some larval arrest examples, pupal diapause comes about through an inhibition of ECD synthesis (reviewed in Denlinger et al., 2012; Saunders et al., 2002). These arresting pupae can be extremely cold tolerant (Lee & Denlinger, 1991) as described previously for hatch-ready dormant embryos. For example, dormant pupae of *Papilio machaon* can supercool to a temperature of −25 °C, but even after freezing at −30 °C, they can remain alive for months (Shimada, 1980). Also just like diapausing embryos, diapausing pupae have to remain in the cold for several months to acquire developmental competence. Once the cold period ends, dormant pupae can break diapause and reactivate metamorphosis if exposed to an optimal environment (Denlinger et al., 2012; Saunders et al., 2002).

Pupal diapause is specified during a "sensitive period" of larval development. In *Araschnia levana*, *Pararge aegeria* (Nymphalidae), and *Pieris Napi* (Pieridae) the pathway decision is primarily made in the middle larval instars and most larvae shifted from long days (nondiapause trajectory) to short days (diapause-inducing condition) after the "sensitive stage" are irreversibly committed to the direct development and lose diapause inducibility (Friberg, Haugen, Dahlerus, Gotthard, & Wiklund, 2011; Wiklund & Friberg, 2011). While the potential for inducing diapause of "nondiapause" larvae is restricted during development by the hormonal programming, diapause-fated larvae maintain a potential of developmental reversibility (Friberg et al., 2011). The hornworm moth, *Manduca sexta*, triggers pupal diapause when, as larva, it is exposed to short days. This results in a failure of the pupal ECD pulse. Interestingly, these larvae that are exposed to short days exhibit no differences in larval ECD levels compared to "developing" ones. However, ECD levels drop in diapausing pupae blocking metamorphic progression before the pharate stage (Fig. 8.3) (Bowen, Bollenbacher, & Gilbert, 1984; Bowen et al., 1985; Smith, Bowen, Bollenbacher, & Gilbert, 1986). That the low pupal ECD level is causative for inducing diapause is suggested by the fact that dormant pupae break diapause when injected with an ECD agonist (RH5849) (Sielezniew & Cymborowski, 1997). Moreover, developing pupae of *Pieris brassicae* enter into a state of permanent diapause if their PGs (source of ECD) are excised (Calvez, 1976; Pullin & Bale, 1989). A similar ECD-based regulatory mechanism also occurs in many other lepidopteran species (Agui, 1975; Bodnaryk, 1985; Highnam, 1958; Islam et al., 2005; Loeb, 1982; Mishra et al., 2008; Roxström-Lindquist, Assefaw-Redda, Rosinska, & Faye, 2005; Williams, 1946, 1952, 1968; Zhang & Denlinger, 2012).

Since PTTH is the major neuropeptide hormone that sets the timing of ECD pulses (Smith & Rybczynski, 2012), it is not surprising that its failure of expression appears to be pivotal in eliciting pupal diapause (Denlinger et al., 2012; Saunders et al., 2002; Smith & Rybczynski, 2012). For example, in *Manduca sexta*, the *MsPTTH* gene is expressed normally during larval life,

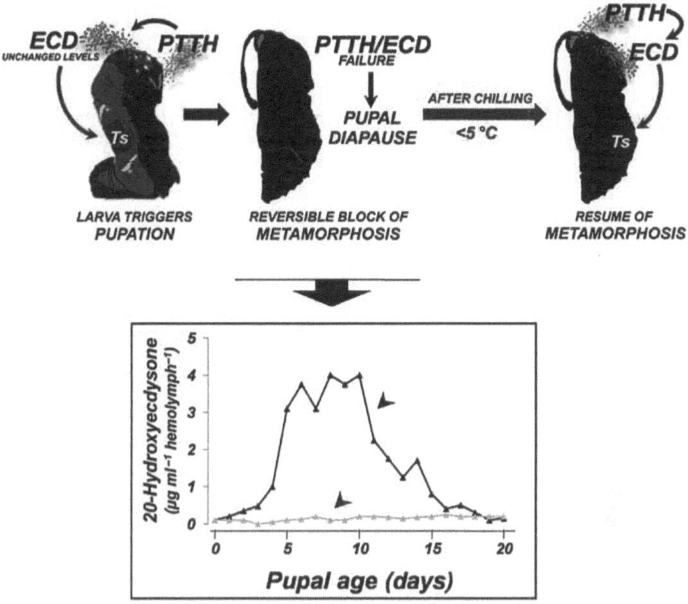

Figure 8.3 Model of pupal diapause in *Manduca sexta*. Under short days of autumn, diapausing last fifth instar larvae exhibit normal temporal profiles of both ECD (in red) and prothoracicotropic hormone (PTTH, in blue), but they metamorphose into diapausing pupae. In diapausing pupae, PTTH falls down and the ECD pulse is not elicited any more blocking the metamorphic progression and triggering dormancy. After an established period of chilling, PTTH and ECD pulses are resumed and the metamorphosis progresses without reporting growth defects (Ts, target organs). During the pupal stage, diapausing animals exhibit low levels of 20-hydroxy-ecdysone (20ECD) compared to nondiapausing ones. Red and blue arrowheads indicate diapausing and "nondiapausing" animals, respectively. *Graph redrawn from Saunders et al. (2002).* (For interpretation of the references to color in this figure legend, the reader is referred to the online version of this chapter.)

but its expression is strongly downregulated in diapausing pupae compared to developing ones (in Xu & Denlinger, 2004). Similarly, *ptth* expression is downregulated in diapausing pupae of *Heliothis virescens* (Xu & Denlinger, 2003) and *Helicoverpa armigera* (Wei, Zhang, Kang, Xu, & Denlinger, 2005). Consistent with the idea that low PTTH levels are responsible for the induction of diapause, injections of PTTH induce the termination of pupal dormancy in noctuids of Heliothis/Helicoverpa complex (Wei et al., 2005). Similarly, injections of PTTH in both *Manduca sexta* and *Anthaerea pernyi* stimulate the resumption of metamorphic growth in both normal diapausing pupae as well as in brain-less (brain surgically removed) dormant ones that are arrested in permanent diapause (Sauman & Reppert,

1996, Shionoya et al., 2003). Interestingly, the implantation of a "cold-activated" brain into "brainless" diapausing pupae of the Hyalophora giant-silkworm causes the termination of diapause (Williams, 1946, 1947, 1952). This effect is consistent with a model in which the pupal brain, when chilled, becomes competent to release PTTH which, in turn, restores the ECD pulse needed to antagonize pupal diapause (Denlinger et al., 2012).

PTTH release from the brain, however, does not appear to be the only way to break dormancy. In *Pieris rapae*, *Pieris brassicae*, and *Antheraea polyphemus* removal of the brain only induces a "permanent diapause" if it occurs within the first month of diapause. After this time, diapausing pupae of these lepidopteran species, as well as several others, can break dormancy and complete metamorphic development even in the absence of their brains (Judy, 1972; Kind, 1978; Maslennikova, 1970; McDaniel & Berry, 1967; Wilson & Larsen, 1974). Similarly, diapausing pupae of the *Helicoverpa zea* are independent from the brain and they have all the potential to resume metamorphosis after the chilling period as long as they are de-brained within 24 h of pupariation (Denlinger et al., 2012; Meola & Adkisson, 1977).

The capacity to reactivate development in the absence of a brain might depend on an autonomous process in the PG that become independent of neural secretions in an age and cold sensitive manner (Denlinger et al., 2012; Saunders et al., 2002). Several pieces of evidence support this model. The PG of de-brained diapausing pupae of *Papilio xuthus* is directly activated by cold (Ozeki, 1954) while the PG of *Samia cynthia* larvae retains a high degree of independence from the brain (Mizoguchi & Ishizaki, 1982). In *Manduca sexta*, the PG of diapausing pupae becomes refractory to PTTH signaling from the day of the larva-to-pupa molt, perhaps as result of inductive events occurring in the late larval stages, and it becomes autonomously responsive to environmental stimuli in breaking dormancy (Bowen et al., 1984, 1985). A similar refractoriness of PG to PTTH signaling also occurs in *Mamestra brassicae* (Agui, 1975) and *Pieris brassicae* (Calvez, 1976).

At present, the mechanism behind PTTH independent reactivation is unknown. One possibility is that an unknown inhibitor might block PTTH/ECD signaling in the PG and it might be gradually removed by the cold so that developmental competence is resumed. DA might be this putative inhibitor since it promotes diapause in several Lepidoptera. For example, in *Mamestra brassicae* or *Pieris brassicae*, diapausing pupae exhibit higher levels of hemolymphatic DA than those set for direct development and last instar larvae set for direct development develop into diapausing pupae when they are fed with L-DOPA (Isabel, Gourdoux, & Moreau,

2001; Noguchi & Hayakawa, 1997; Puiroux, Moreau, & Gourdoux, 1990). Interestingly, dormant pupae of *Antheraea pernyi* decrease the hemolymphatic levels of DA when they are chilled (Matsumoto & Takeda, 2002), suggesting that chilling is pivotal in removing the DA signal. This facilitates PTTH signaling to inhibit dormancy and stimulate its prometamorphic function. Consistent with this model, brainless diapausing pupae of *Hyalophora cecropia* break the "permanently induced" dormancy when they receive the brain of "chilled-activated" pupae (Smith & Rybczynski, 2012; Williams, 1946, 1947, 1952).

Although PTTH and DA signaling appear import, the full regulatory mechanism underlying pupal diapause is likely to be more complex, since DH is also required to reactivate postdiapause growth in synergism with PTTH. Such DH functions are well documented for the pupal diapause of noctuids belonging to the Heliothis/Helicoverpa complex (Sun, Zhang, Zhang, & Xu, 2003; Sun et al., 2005; Zhang, Sun, Zhang, Shen, & Xu, 2004; Zhang, Sun, Zhang, Xu, et al., 2004). In these moths, DH injections into "nonchilled" diapausing pupae induce them to break dormancy (Xu & Denlinger, 2003; Zhang, Nachman, Kaczmarek, Zabrocki, & Denlinger, 2011; Zhang, Sun, Zhang, Shen, et al., 2004; Zhang, Sun, Zhang, Xu, et al., 2004; Zhang, Zdarek, Nachman, & Denlinger, 2008) in a temperature-dependent way: it is unable to break diapause at 20 °C, but does so in dormant pupae shifted to 25 °C (Zhang, Sun, Zhang, Shen, et al., 2004; Zhang, Sun, Zhang, Xu, et al., 2004).

Consistent with these observations, DH gene expression declines during of the mid-early pupal stages in diapausing pupae of *Helicoverpa zea* and *Manduca sexta*, whereas it is upregulated in nondiapausing pupae (Xu & Denlinger, 2004; Zhang & Denlinger, 2012). Moreover, 1-year-old "chilled" pupae (activated) of *Helicoverpa zea* break dormancy when they perceive optimal temperatures ($\approx$25 °C) in coincidence of an upregulation of DH gene expression (Zhang & Denlinger, 2012).

DH also seems to control the modifications of larval development that are coincident to diapause trajectory. Larvae of *Helicoverpa armigera* set for the direct development delay larval development when injected with DH (Sun et al., 2005). Similarly, larvae of *B. mori* set to develop into "diapause" moths prolong their feeding period coincidentally to an upregulation of *BmDH* gene (Xu et al., 1995a). Interestingly, a conserved role of DH on diapause response in a variety of Lepidoptera suggests that this hormone might be pivotal in evolution of the diapause response in synergism with the hormones PTTH and ECD.

6. DIAPAUSE CAN MODIFY LARVAL/PUPAL GROWTH

Modulations of larval/pupal developmental rate are also programmed to adjust the growth on diapause trajectory. As an example, the swallowtail butterfly *Sericinus montelus* triggers a pupal diapause linked to the modifications of larval/pupal development. In fact, larvae slow down development for more than 20 days upon perception of short days ("diapause-inducing"), while they accelerate growth (<19 days) under long days (direct development). Pupal growth is also modified: diapausing pupae are fatter and heavier than developing ones (Wang, Yang, Zhou, Zhao, & Lei, 2007). Similarly, the geometrid moths *Cabera exanthemata* and *Chiasmia clathrata* trigger pupal diapause after a prolonged larval life in which the larval growth rates is lower than those of larvae set for direct development (Kivelä, Välimäki, & Mäenpää, 2012; Välimäki et al., 2013).

The nymphalid *Pararge aegeria* diapauses either as a third (penultimate) instar larva or as a pupa, depending by photoperiod (Nylin, Wickman, & Wiklund, 1989; Wiklund & Friberg, 2011). Larvae exposed to long days (>17 h of light) develop directly without diapause, those reared under intermediate days (14–16 h) induce pupal diapause and those perceiving short days (<13 h) trigger larval diapause (Nylin et al., 1989). Such trajectories reflect changes in larval/pupal growth rate since "diapausing" larvae develop slower than developing ones and diapausing larvae develop into heavy pupae, diapausing pupae have intermediate weight and "nondiapausing" larvae develop into small pupae (Gotthard & Berger, 2010; Nylin et al., 1989). In this species, diapause trajectory is also linked to the appearance of seasonal butterfly morphs (eyespots on the dorsal side of the wings) (Gotthard & Berger, 2010; Van Dyc & Wiklund, 2002; Wiklund & Friberg, 2011).

In contrast to the slowing of development in some species, diapause is linked to a fast larval growth when larvae have to quickly reach the dormant stage when developing in proximity of an imminent winter (i.e., late in the season of "rapidly changing" environments) (Gotthard, 2008). Such a phenomenon is also reported for two nymphalids, *Lasiommata maera* and *Lopinga achine*, whose larvae hatch in late summer and diapause as half-grown larvae in autumn. Such larvae accelerate their growth to quickly reach diapause when exposed to short days. Once diapause ends, the modulation of developmental time is reversed: larvae resume postdiapause development, but grow at a slower rate under short days (spring—meaning more time available to reach pupation) and a faster one under long days (summer—meaning less time available to develop as adult before autumn) (Gotthard, Nylin, & Wiklund, 1999). Similar regulation also occurs in the nympahalid *Lasiommata petropolitana* whose larvae finely adjust their growth rate in advance of diapause onset. In fact, larvae set for pupal diapause (short days) exhibit faster larval development and a higher larval growth rate, quickly pupating with a smaller size than those set for "long-day" development (about 40 days vs. 70 days to reach pupation) (Gotthard, 1998; Leimar, 1996; Nylin, Gotthard, & Wiklund, 1996). Thus, larvae carefully adjust their growth rate before and after diapause with respect to the predicted time available for growth (Gotthard, 2008).

7. SEASONAL POLYPHENISMS CORRELATE WITH DIAPAUSE

Seasonal polymorphisms (polyphenism) in moths and butterflies often correlate with diapause trajectory and their specification occurs coincidentally to dormancy induction. Developmentally, changes in a hormone-sensitive period or modifications in the strength, or in the timing, of hormone signaling are pivotal in orchestrating these seasonal polyphenisms (Nijhout, 2003; Saunders et al., 2002). Photoperiodic changes also play a fundamental role in such phenomena as reliable cues of the time of the year and, as described above, are used by the larvae to finely adjust their growth rate in advance of diapause onset. As an example, the nymphalid *Anaea andria* exhibits a strong seasonal polyphenism in wing shape that is coincidental with diapause trajectory. The fifth instar larvae ($\sim$4–5 days prior to pupation) grown under long days (>16 h of light) develop into a "summer" form that has blunt forewing apex tails and decreased posterior projection on the hind wing. Conversely, short photoperiods (<12 h of light) induce the appearance of "autumnal/winter" morphs that exhibit forewing apices developed as recurved tails and an increased anal projection on the hind wing. The two forms also differ in the dorsal pigmentation of the wings and in the cryptic color of ventral wing surface, with the "winter" morphs that are darker (mostly brown) than the "summer" ones. Interestingly, the two morphs are also different in their developmental time (at 27 °C) since the "summer" form needs 25–45 days to emerge as a butterfly while the "winter" form requires roughly 5 days more than the time of the summer form (Riley, 1980). Since this species also exhibits adult diapause (reproductive diapause) (Scott, 1979 and references therein), the seasonal polyphenism might be hormonally linked to diapausing functions so that morphs which are adapted to the expected adverse conditions are produced.

In *Pieris napi* (Pieridae), diapausing pupae develop into spring *venosa* morpha that have a fat, dark, and hairy body, and dark scales on the veins of the wings (on all hind wings and on forewing apex). Conversely, nondiapausing pupae produce the summer *castoria* form which has a slender, light, and smooth body with light wings (Oliver, 1970). This pathway decision is linked to modifications of larval growth and it is mostly state-independent since "diapause" larvae develop slower than those set for direct development only after the pathway decision has been made in the fourth larval instar (Friberg, Dahlerus, & Wiklund, 2012; Shapiro, 1975, 1977). Indeed, direct-developing larvae grow slower than "diapausing" larvae during the first two instars, whereas they develop at the same rate during the third and fourth instars. In the fifth (last) instar, when the pathway decision is made, larvae set to be "nondiapausing" grow faster than those set for diapause, revealing once again that many species modify their larval growth rates in relation to their developmental pathway (diapause vs. direct development) (Friberg et al., 2012).

Developmentally, diapause-bound polyphenisms are induced by the differential timing of ECD pulses during a "hormone-sensitive" period. The strength of ECD pulses during such a sensitive period (during either larval or pupal life) determines the induction, or the inhibition, of the alternative developmental pathway. It appears that the timing of the ECD pulse and a threshold in hormonal sensitivity (Hartfelder & Emlen, 2012; Nijhout, 2003; Saunders et al., 2002) are key to determining outcome. A well documented case is that of *Araschnia levana* (Nymphalidae) in which the last larval instar can develop into two butterfly morphs: *levana* and *prorsa* (Koch & Bückmann 1987). Under short days, larvae develop into dark diapausing pupae from which the *levana* morph (orange/dark with black spots) emerges in spring, once dormancy is ended. In contrast, larvae grown under long days develop into light pupae (direct development) from which the *prorsa* form (white and black) ecloses. This phenomenon depends by the timing of the ECD pulse. In "*prorsa*" pupae, it pulses during the "hormone-sensitive" period (between 2 and 10 days after pupation), while it is delayed in "*levana*" pupae because of diapause. Injection of 20-ECD into 3-day-old diapausing pupae (*levana*) induces them to promptly break dormancy and develops into *prorsa* morphs instead of the expected *levana* form. However, injection of 20-ECD into 14-day-old *levana* pupae, while still able to promptly break dormancy, is not able to reverse the morph identity since the "*levana*" developmental pathway has proceeded past a critical point (Koch & Bückmann 1987). Diapause-bound morphs of Araschnia also differ in other aspects of developmental growth. The *prorsa* morph is larger and exhibits a different wing shape, heavier thorax, and smaller abdomen than *levana*, reflecting their differential dispersal rates (Fric & Konvička, 2002). Moreover, diapause trajectory is strictly linked to the duration of fifth larval instar with the *levana* morph having a delayed development (Windig & Lammar, 1999). A simplified model of Araschnia seasonal morphs and diapause is shown in Fig. 8.4.

A diapause-linked diphenism also occurs in *B. mori* in which a moth laying diapausing embryos exhibits a greater number of dark scales on the wings (autumnal morphs) than those laying developing eggs (spring–summer morphs) and the diapausing embryos are darker with a thicker chorion than directly developing ones (Tsurumaki, Ishiguro, & Yamanaka, 1999). "Nondiapause" pupae develop as autumnal or intermediate moths (instead of summer moths) upon implantation, during the early fifth larval instar, of the Brain-SOG complexes of "diapause" larvae. In contrast, no morph shifting is provoked when autumn moths ("diapause") receive the Brain-SOG complexes of fifth instar larvae set to develop into summer morph ("nondiapausing") (Yamanaka, Tsurumaki, & Endo, 2000). Such effects are due to DH since its injections into young "nondiapause" pupae induce them to develop

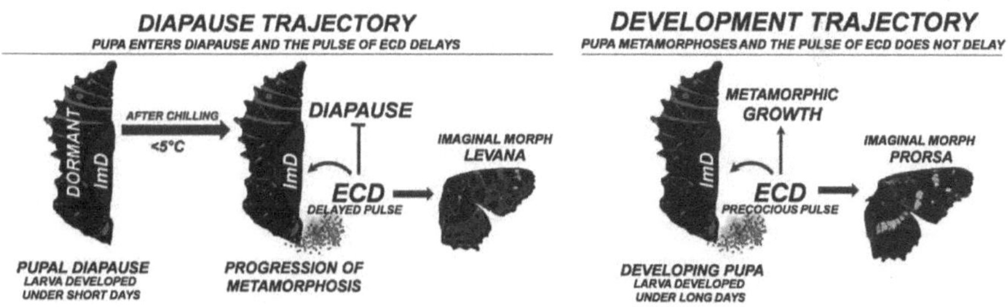

Figure 8.4 Hormonal control of "diapause-bound" morphs in *Araschnia levana*. In short days of autumn, the last larval instar develops into a dark diapausing pupa (*diapause trajectory*) carrying numerous tubercles, upon failure of the pupal ECD signaling. After chilling, the dormant pupa breaks diapause by eliciting a pulse of ECD, which is delayed because of dormancy. This "delayed" pupal pulse of ECD induces pupae to develop into *levana* morphs. In contrast, the light pigmented nondiapausing pupae (*development trajectory*) elicit a precocious pulse of ECD which induces the progression through metamorphosis. Hence, the developing pupa ecloses as *prorsa* butterfly morph. ImD, imaginal wing discs. *Wings modified from Nijhout (2003) with the permission of John Wiley & Sons Ltd.* (For color version of this figure, the reader is referred to the online version of this chapter.)

as autumnal or intermediate morphs instead of the presumptive summer ones, and the effects depend on the timing of the injections (Yamanaka et al., 2000). Thus, DH also seems to control seasonal polyphenism of Bombyx, presumably by signaling to imaginal discs early during pupal development (Fig. 8.1), in addition to its role on diapause.

Like Bombyx, the tussock moth *Orgyia thyellina* (Lymantriidae) diapauses as an early embryo under the control of DH. This species also exhibits a "multistage" polyphenism that is linked to dormancy (Fig. 8.5A). Larvae grown during the long days of summer emerge as winged *Macropterus* morphs, while autumn reared larvae develop into wingless *Brachypterus*. Moreover, pupae are also diphenic: pale pupae develop into *Macropteres*, while dark ones become *Brachypteres*. *Macropteres* lay direct developing eggs,

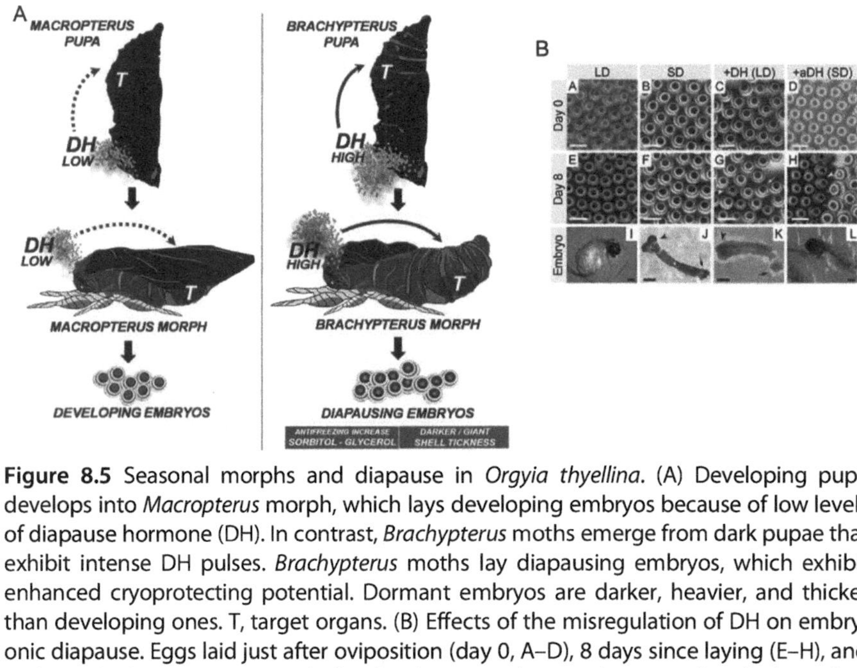

Figure 8.5 Seasonal morphs and diapause in *Orgyia thyellina*. (A) Developing pupa develops into *Macropterus* morph, which lays developing embryos because of low levels of diapause hormone (DH). In contrast, *Brachypterus* moths emerge from dark pupae that exhibit intense DH pulses. *Brachypterus* moths lay diapausing embryos, which exhibit enhanced cryoprotecting potential. Dormant embryos are darker, heavier, and thicker than developing ones. T, target organs. (B) Effects of the misregulation of DH on embryonic diapause. Eggs laid just after oviposition (day 0, A–D), 8 days since laying (E–H), and embryos (I–L). *Macropterus* moths lay developing embryos, when reared under long days (LD) during larval life (A, E, I). *Brachypterus* moths reared under short days (SD) lay diapausing embryos (B, F, J). Injection of DH into *Macropterus* moth induces the production of diapausing embryos (C, G, K), instead of the developing ones. In contrast, injections of anti-DH induce *Brachypterus* moths to lay developing eggs (D, H, L). The white arrow and arrowhead indicate diapause and nondiapause eggs, respectively. The black arrows and arrowheads indicate telson and cephalic lobes, respectively. Scale bar = 5 mm. *Reproduced from Uehara et al. (2011) with the permission of PLoS One.* (For color version of this figure, the reader is referred to the online version of this chapter.)

whereas *Brachypteres* lay diapausing embryos that are heavier and darker than direct developing eggs (Kimura & Masaki, 1977). Although there is a correlation between the polyphenic trait and the type of egg laid, they are not inextricably linked. Manipulation of DH signaling during pupal life, for example, can revert the developmental pathway without switching the morph type. *Macropterus* pupae develop into moths laying diapausing embryos when injected with DH, whereas *Brachypterus* pupae develop into moths laying direct developing embryos when injected with anti-DH (DH antagonist) (Fig. 8.5B) (Uehara et al., 2011).

Interestingly, Bombyx embryonic diphenism is also induced by ECD since injections of 20-ECD into "nondiapause" moths induce them to lay developing, but giant, eggs that are larger and heavier than normal, and are similar in appearance to the diapausing type (Kawaguchi, Banno, Doira, & Fujii, 1989). Hence, diapausing signals might orchestrate the timing, or the levels, of developmental hormones for the induction of seasonal polyphenism although the precise interactions between DH and ECD remain ill defined.

7.1. Adaptive advantages of seasonal polyphenisms

As with diapause, the seasonal polyphenisms are thought to provide adaptive advantages. The swallowtail *Papilio polyxenes* diapauses as freeze-tolerant pupa specified under autumnal short days. Larvae set for pupal diapause (short days) develop into autumnal dark larvae upon exposure to short photoperiods during the first three instars. In this case, the dark form has adaptive advantages since it helps to elevate larval body temperatures to grow rapidly during the autumn so that the diapausing pupal stage can be reached prior to the first freeze (Hazel, 2002). Another example is the noctuid moth, *Mamestra brassicae*, in which the fourth and fifth instar larvae exhibit seasonal color morphs ranging from a pale green to a dark morph. As above, dark-morph larvae are thought to optimize the capture of solar radiation enhancing their body temperature in autumnal (or mountain) environments such that these larvae grow more rapidly than the green morphs and, although they have light weight at the onset of fifth instar, they undergo a "catch-up" growth during the fifth and sixth larval stages so that they reach pupation without delaying development or impairing pupal size (Goulson, 1994).

7.2. Molecular mechanism for polyphenism induction

Interestingly, the hormonal mechanism underlying the production of polyphenism in Manduca moths might offer insight into the molecular basis underlying both control and evolution of larval "diapause-bound" morphs.

Fourth instar larvae of *Manduca quinquemaculata* develop into fifth instar as black morphs which enter pupal diapause under autumnal temperatures (<20 °C) but as cryptic green forms in summer (≈28 °C or more) (Hudson, 1966). In contrast, the related species *Manduca sexta* lacks polyphenism exhibiting only the cryptic green morph which triggers pupal diapause under autumnal short days (see above). Yet, the *Manduca sexta* "black" mutant has black larvae that can develop into the normal cryptic form only when exposed to "summer-like" temperatures (≈28 °C or more) during the "hormone-sensitive" period of the fourth larval instar. Such mutants exhibit lower levels of JH than the normal green larvae, indicating that the developmental output can be sensitized by modifying "normal" transitional hormonal signals to reveal a hidden developmental plasticity (Suzuki & Nijhout, 2006, 2007). This mechanism might also underlie the appearance of diapausing larval morphs in other lepidopteran species, since JH, as a major developmental hormone, can influence two different developmental trajectories (direct development vs. diapause) by modulating ECD signaling output (see above larval diapause).

In sum, diapause is mainly controlled by the strength and the timing of developmental hormones during "hormone-sensitive" periods which also set the appearance of diapause-bound polyphenisms. Similarly to the polyphenisms exhibited in other organisms (Nijhout, 2003; Rajakumar et al., 2012; Suzuki & Nijhout, 2006), lepidopteran diapause might evolve by "genetic accommodation". This is a evolutionary mechanism wherein a novel alternative trait (e.g., diapause or polyphenism) is molded into an adaptive phenotype through quantitative changes in a hormonal regulatory pathway which then modifies the levels, or the timing, of a developmental hormone (developmental sensitizing) in a such a way that transient environmental variations uncover the cryptic genetic variation (genetic variability that is invisible to selection since it does not produce phenotypic variants, until, under certain conditions, it is exposed) to natural selection. Indeed, the environmental variation alters hormonal signaling either above or below a signal intensity that elicits the novel alternative phenotypes (Nijhout, 2003; Suzuki & Nijhout, 2006; West-Eberhard, 2003). Hence, both developmental routes, diapause and direct-development, may be inducible (accommodated) depending on the environmental conditions. By contrast, without the "sensitizing" modifications, the developmental system buffers the environmental perturbing effects and it fails in inducing the novel trait. Besides, depending on the adaptive pressure, direct-development may be fixed, genetically assimilated (lack of diapause) or diapause may become an obligate period (i.e., obligate diapause) in a developmental program. Some evidence supporting diapause evolution by genetic accommodation might be provided by "nondiapausing" strains of some lepidopteran species (e.g. *Lymantria* or *Bombyx* strains) in which diapause can be elicited by mis-regulating a specific hormone (see above).

Mechanisms underlying the evolution of diapause trajectory might also enhance the environmental sensitivity of other developmental modules and promote the evolution of multiple polyphenic, diapause-linked traits (such as those found for *B. mori* and *Araschnia levana*). Indeed, sensitizing changes altering the timing, or the levels, of developmental hormones during diapause evolution might shift, or sensitize, the hormonally sensitive period of a diapause-linked trait, promoting genetic accommodation. These altered hormonal dynamics might facilitate the evolution of different life cycles, as discussed for other issues (Heyland et al., 2005; Pfennig et al., 2010).

8. THE KILLIFISH DIAPAUSING EMBRYO: A DEVELOPMENTAL ANALOGY IN A VERTEBRATE

Diapause is also induced in some vertebrate embryos without causing any developmental aberration, analogous to insect diapause. An example is the annual killifish, *Austrofundulus limanaeus* (Cyprinodontiformes; Apocheiloidei). Although its diapause is not a hibernanal dormancy in response to an upcoming winter but, is instead a "hypoxia-induced" diapause, the developmental mechanism is, nevertheless, reminiscent of what occurs in lepidopteran species. Killifish live in ephemeral ponds that seasonally undergo fast desiccation during the dry season. Hence, survival of the species depends entirely on the buried (dormant) embryos that develop during the rainy period, once the ponds are refilled by seasonal rain (Berois, Arezo, Papa, & Clivio, 2012; Podrabsky, Carpenter, & Hand, 2001, Podrabsky & Hand, 1999; Podrabsky, Lopez, Fan, Higashi, & Somero, 2007). Killifish embryos block developmental growth and metabolism ($\approx$80–90% reduction) by triggering diapause potentially at three different stages: as an early embryo (diapause I, rare), as 38-somite embryos (diapause II, the embryo has developing endoderm-derived organs, a beating heart, circulating red blood cells, optic cups and a maturating nervous system), or as "hatch-ready" embryo (diapause III, obligate, fully developed embryo waiting the appropriate cues to hatch) (Berois et al., 2012; Podrabsky, Garrett, & Kohl, 2010). Diapause trajectory is strictly associated with modifications of the timing of developmental growth (Fig. 8.6) (Podrabsky et al., 2010). Diapausing embryos are

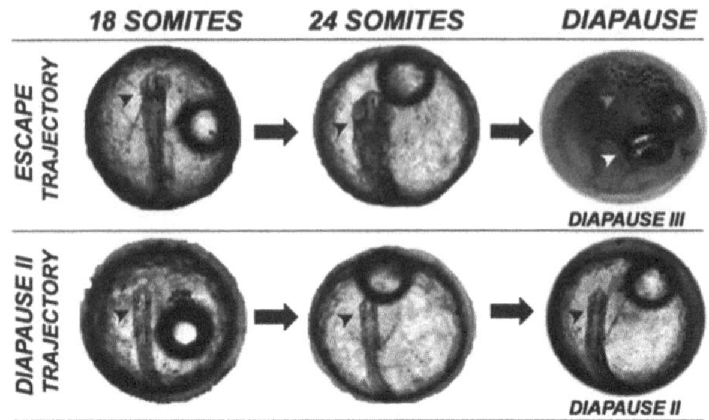

Figure 8.6 Developmental growth of the killifish *Austrofundulus limnaeus* in relation to diapause. Model of the two developmental trajectories in embryos of *A. limnaeus*. Diapause is strictly associated with modifications of development. Early embryogenesis normally occurs until 10-somite stage, without any difference between the two trajectories. By the 18-somite stage, nondiapausing embryos (escape trajectory, diapause III) differ morphologically from diapausing ones (diapause II trajectory). At the end of the "escape" trajectory, the fully developed embryos enter obligate diapause III, which occurs directly prior to hatching. Note the morphological differences between the two developmental trajectories at level of trunk and cephalic development (red arrowheads). Blue arrowhead: melanocytes; white arrowhead: eyes. *Modified from Podrabsky et al. (2010) with the permission of The Journal of Experimental Biology.* (See color plate.)

extremely resistant to both dehydratation and hypoxia still remain alive for months in the complete absence of oxygen (Podrabsky et al., 2010; Podrabsky & Hand, 1999; Podrabsky, Menze, & Hand, 2012). Such tolerance to anoxia, mostly exhibited by diapause II embryos, is maintained for several days of postdiapause II development, providing a condition for which the embryos develop actively but still being able to resist anoxia (Podrabsky et al., 2007).

Austrofundulus diapause seems to be overridden by steroid hormone. In fact, nondiapausing embryos differ from diapausing ones in having higher titers of estrogen (17-β-estradiol, E2) at 5/10-somites stage as well as in the timing of morphological traits which do not develop until several days after diapause (i.e., melanocytes, yolk vasculature, and otholite primordia). Consistent with this observation, most diapausing embryos escape diapause II when incubated with exogenous E2 (Pri-Tal, Blue, Pau, & Podrabsky, 2011). Intriguingly, as in *Austrofundulus*, diapause seems to be maternally primed in embryos of killifish *Nothobranchius guentheri*, in which diapause is induced by "pheromonal-like" signals produced by adult fish since all the embryos trigger diapause I in the presence of adults (Inglima, Perlmutter, & Markofsky, 1981).

Diapause occurs also in *Austrolebias* (diapauses I and II are facultative and diapause III is obligate), in *Pterolebias* and *Rachovia* killifish (similar to *Austrofundulus*, diapause II is obligate) and in African *Nothobranchius* killifish where diapause has evolved independently (Arezo, Pereiro, & Berois, 2005; Inglima et al., 1981; Levels, Van Tits, & Denucé, 1986; Murphy & Collier, 1997; Pri-Tal et al., 2011; Wourms, 1972a, 1972b, 1972c). Annual fish diapauses are an evolutionary innovation in the Cyprinodontiformes (Berois et al., 2012) and they might result by environmentally sensitizing a plastic preexisting developmental system or by modifications in the timing of developmental hormones. Some nonannual Cyprinodontiformes can delay the hatching in intertidal environments, suggesting that diapause III could originate from this feature (Berois et al., 2012; Taylor, 1999). In addition, the embryo of nonannual Cyprinodontiformes *Fundulus heteroclitus* can avoid dehydratation when exposed to air (Tingaud-Sequeira, Zapater, Chauvigne, Otero, & Cerda, 2009). Interestingly, zebrafish embryos (*Danio rerio*) lack diapause but, when exposed to chronic hypoxia, they slow down developmental growth (at about 30 hpf stage) without any developmental aberration. "Hypoxic" embryos inhibit insulin growth factor 1 (IGF1) signaling by enhancing IGF binding protein 1 (IGFBP1) expression. Upon reoxygenation, embryos accelerate development and growth rate ("catch-up" growth) by restoring IGF1 signaling (Kamei et al., 2011). Despite the fact that loss of IGF1 signaling produces developmental delayed dwarf embryos that die at a stage corresponding to the wild-type 18 hpf stage (Schlueter et al., 2006), the transient and timed loss of IGF signaling can enhance developmental plasticity during hypoxia (Kamei et al., 2011), and resembles the outcome observed in killifish diapausing embryos.

8.1. Do insulin/IGFs play an evolutionary conserved function in diapause?

Similar to the hypoxia-induced response of the killifish, insulin/IGF-type factors likely also influence diapause trajectory in Lepidoptera, suggesting that this signaling pathway has been evolutionary recruited for diapause in both vertebrates and invertebrates. In support of this scenario, insect insulin/IGFs like ligands have been identified in several lepidopteran species (Antonova, Arik, Moore, & Riehle, 2012) and are important developmental hormones that play pivotal roles in modulating holometabolous growth. In the case of the monarch butterfly, *Danaus plexippus*, which exhibits seasonal developmental changes in relation to adult diapause (Zhan, Merlin, Boore, & Reppert, 2011), microarray analysis has revealed that insulin/IGF-1 signaling (IIS) is downregulated in diapausing migrants, suggesting that the activation of the "anti-growth" Forkhead Box-O (FOXO) transcription factor, which is antagonized by IIS signaling, might be a key molecule for inducing the diapausing developmental switch (Zhan et al., 2011). Further evidence for an insulin/IGFs role in diapause of some lepidopteran species (e.g., *Samya cynthia* and *Pieris brassicae*) is provided by experiments in which the injections of mammalian insulin into dormant pupae breaks diapause and stimulates a resumption of "ECD-dependent" pupal growth (Arpagaus, 1987; Wang, Zheng, & Guo, 1986). Lastly, insulin signaling has been proposed to play a major role in directing reproductive diapause, in several insects (Tatar and Yin, 2001). Reproductive diapause is a condition in which the energetically expensive process of egg production by mature adults is arrested during winter presumably inducing a more stress resistant physiology. Additional research on the roles of insulin/IGF-type factors in both normal development (e.g., metamorphosis, larval growth, etc.) and in dormancy induction, utilizing both vertebrate and invertebrate models will likely help identify additional genetic and evolutionary mechanisms underlying seasonal adaptations.

9. SUMMARY AND PERSPECTIVE

Diapause is an alternative developmental pathway that blocks developmental growth by modifying the levels, or the timing, of developmental hormones in response to environmental cues. The final result is a dormant entity that is able to resist extremely low temperatures for several months. Once the decision-making has been made, diapause trajectory drastically modifies not only the physiology of diapausing stage, but also the larval and pupal growth by altering either ECD signaling or other growth signals. The developmental hormones that trigger diapause are also responsible for the induction of seasonal morphs (polyphenism). Hence, diapause might accelerate the evolution by genetic modifications of multiple polyphenic traits that can be elicited upon exposure to the same, or temporally different, environmental perturbations. Insights from species exhibiting multiple polyphenic traits linked to diapause (i.e., *B. mori* or *Araschnia levana*) will provide a framework for future research on this hypothesis and the inducibility of alternative developmental pathways (developmental plasticity).

Another important issue is how photoperiodic changes are decoded and transduced to modify hormonal signals and resulting in the activation of alternative genetic programs. The perception of daylight duration and average temperature are potential reliable cues for an organism to perceive seasonal time and to induce diapause and/or modifications of larval/pupal growth. Is the endogenous circadian clock pivotal in mediating the hormonal sensitivity to light changes or is the visual system the primary mediator, or do both play a role? Insights from *B. mori*, which exhibits a strong photoperiodic sensitivity in specifying developmental fate, during early larval life, the DH-based diapause trajectory will provide a good model to study the regulatory mechanism upstream of developmental hormones. What role and by what mechanism could changes in average temperature be perceived with time is also an important issue to explore?

As described, diapause is also triggered in the vertebrate killifish embryo by a mechanism that, presumably, is hormonally based, as is the IGF-1-based block in development of zebrafish embryos under certain conditions. A functional analysis of IGF-1 signaling in both zebrafish and killifish embryos, together with engineering of cryoprotecting potential, may provide a framework for deciphering how hibernal diapause evolved in vertebrates within a preexisting developmental program.

ACKNOWLEDGMENTS

L.S. is supported by a doctoral fellowship from the University of Padova (Italy). L.S. also thanks Fondazione Ing. "A.Gini" of University of Padova (Italy) and Rudi Costa (University of Padova). M.B.O. is supported by NIH grant GM R01 GM0933001. The authors thank Emilie Snell-Rood for helpful comments on the manuscript.

REFERENCES

Agui, N. (1975). Activation of prothoracic glands by brains in vitro. *Journal of Insect Physiology*, *21*, 903–907.

Antonova, Y., Arik, A. J., Moore, W., & Riehle, M. A. (2012). Insulin-like peptides: Structure, signaling, and function. In L. I. Gilbert (Ed.), *Insect endocrinology* (pp. 63–92). San Diego: Academic Press, Chapter 2.

Arezo, M. J., Pereiro, L., & Berois, N. (2005). Early development in the annual fish *Cynolebias viarius*. *Journal of Fish Biology*, *66*, 1357–1370.

Arpagaus, M. (1987). Vertebrate insulin induces diapause termination in *Pieris brassicae* pupae. *Roux's Archives of Developmental Biology*, *196*, 527–530.

Bean, D. W., & Beck, S. D. (1980). The role of juvenile hormone in the larval diapause of the European com borer, *Ostrinia nubilalis*. *Journal of Insect Physiology*, *26*, 579–584.

Bean, D. W., & Beck, S. D. (1983). Haemolymph ecdysteroid titres in diapause and nondiapause larvae of the European corn borer, *Ostrinia nubilalis*. *Journal of Insect Physiology*, *29*, 687–693.

Berois, N., Arezo, M. J., Papa, N. G., & Clivio, G. A. (2012). Annual fish: Developmental adaptations for an extreme environment. *Developmental Biology*, *1*, 595–602.

Bodnaryk, R. P. (1985). Ecdysteroid levels during postdiapause development and 20-hydroxyecdysoneinduced development in male pupae of *Mamestra configurata* Wlk. *Journal of Insect Physiology*, *31*, 53–58.
Bowen, M. F., Bollenbacher, W. E., & Gilbert, L. I. (1984). In vitro studies on the role of the brain and prothoracic glands in the pupal diapause of *Manduca sexta*. *Journal of Experimental Biology*, *108*, 9–24.
Bowen, M. F., Irish, R., Whishenton, L. R., Granger, N. A., Gilbert, L. I., & Bollenbacher, W. E. (1985). Endocrine events during pre-diapause and non-diapause larval–pupal development of the tobacco hornworm, *Manduca sexta*. *Journal of Insect Physiology*, *31*, 83–90.
Calvez, B. (1976). Circulating ecdysone levels in late larval stages and the induction of the nymphal diapause in *Pieris brassicae* [Taux d'ecdysone circulante aux derniers ages larvaires et induction de la diapause nymphales chez *Pieris brassicae*]. *Comptes rendus hebdomadaires des séances de l'Académie des sciences (Paris, series D)*, *282*, 1367–1370.
Chippendale, G. M. (1977). Hormonal regulation of larval diapause. *Annual Review of Entomology*, *22*, 121–138.
Chippendale, G. M. (1984). Environmental signals, the neuroendocrine system, and the regulation of larval diapause in the southwestern com borer, *Diatraea grandiosella*. In *Photoperiodic regulation of insect and molluscan hormones, Ciba Foundation Symposium 104* (pp. 259–274).
Chippendale, G. M., & Yin, C.-M. (1973). Endocrine activity retained in diapause insect larvae. *Nature*, *246*, 511–513.
Denlinger, D. L., Yocum, G. P., & Rinehart, J. P. (2012). Hormonal control of diapause. In L. I. Gilbert (Ed.), *Insect endocrinology* (pp. 430–463). Elsevier, Chapter 10.
Eizaguirre, M., Prats, J., Abellana, M., Carmen, L., Llovera, M., & Canela, R. (1998). Juvenile hormone and diapause in the Mediterranean corn borer, *Sesamia nonagrioides*. *Journal of Insect Physiology*, *44*, 419–425.
Eizaguirre, M., Schafellner, C., Lopez, C., & Sehnal, F. (2005). Relationship between an increase of juvenile hormone titer in early instars and the induction of diapause in fully grown larvae of *Sesamia nonagrioides*. *Journal of Insect Physiology*, *51*, 1127–1134.
Friberg, M., Dahlerus, J., & Wiklund, C. (2012). Strategic larval decision-making in a bivoltine butterfly. *Oecologia*, *169*, 623–635.
Friberg, M., Haugen, I. M. A., Dahlerus, J., Gotthard, K., & Wiklund, C. (2011). Asymmetric life-history decision-making in butterfly larvae. *Oecologia*, *165*, 301–310.
Fric, Z., & Konvička, M. (2002). Generations of the polyphenic butterfly *Araschnia levana* differ in body design. *Evolutionary Ecology Research*, *4*, 1017–1032.
Fujiwara, Y., Shindome, C., Takeda, M., & Shiomi, K. (2006). The roles of ERK and P38 MAPK signaling cascades on embryonic diapause initiation and termination of the silkworm, *Bombyx mori*. *Insect Biochemistry and Molecular Biology*, *36*, 47–53.
Fujiwara, Y., & Shiomi, K. (2006). Distinct effects of different temperatures on diapause termination, yolk morphology and MAPK phosphorylation in the silkworm, *Bombyx mori*. *Journal of Insect Physiology*, *54*, 1194–1201.
Fujiwara, Y., Tanaka, Y., Iwata, K., Rubio, R., Yaginuma, T., Yamashita, O., et al. (2006). ERK/MAPK regulates ecdysteroid and sorbitol metabolism for embryionic diapause termination in the silkworm, *Bombyx mori*. *Journal of Insect Physiology*, *52*, 569–575.
Fukuda, S. (1951). Production of the diapause eggs by transplanting the suboesophageal ganglion in the silkworm. *Proceedings of the Imperial Academy of Japan*, *27*, 672.
Gadenne, C., Varjas, L., & Mauchamp, B. (1990). Effects of the non-steroidal ecdysone mimic, RH-5849, on diapause and non-diapause larvae of the European corn borer, *Ostrinia nubilalis*. *Journal of Insect Physiology*, *36*, 555–559.
Gelman, D. B., & Brents, L. A. (1984). Haemolymph ecdysteroid levels in diapause- and nondiapause-bound fourth and fifth instars and in pupae of the European corn borer, *Ostrinia nubilalis* (Hubner). *Comparative Biochemistry and Physiology: Part A*, *78*, 319–323.

Gelman, D. B., Thyagaraja, B. S., Kelly, T. J., Masler, E. P., Bell, R. A., & Borkovec, A. B. (1992). Prothoracicotropic hormone levels in brains of the European corn borer, *Ostrinia nubilalis*: diapause vs. the non-diapause state. *Journal of Insect Physiology, 38*, 383–395.

Gelman, D. B., & Woods, C. W. (1983). Haemolymph ecdysteroid titers of diapause- and nondiapause-bound fifth instars and pupae of the European corn borer, *Ostrinia nubilalis* (Hubner). *Comparative Biochemistry and Physiology: Part A, 76*, 367–376.

Gotthard, K. (1998). Life-history plasticity in the satyrine butterfly *Lasiommata petropolitana*: Investigating an adaptive reaction norm. *Journal of Evolutionary Biology, 11*, 21–39.

Gotthard, K. (2008). Adaptive growth decisions in Butterflies. *Bioscience, 58*, 222–230.

Gotthard, K., & Berger, D. (2010). The diapause decision as a cascade switch for adaptive developmental plasticity in body mass in a butterfly. *Journal of Evolutionary Biology, 23*, 1129–1137.

Gotthard, K., Nylin, S., & Wiklund, C. (1999). Seasonal plasticity in two satyrine butterflies: State-dependent decision making in relation to day length. *Oikos, 84*, 453–462.

Goulson, D. (1994). Determination of larval melanization in the moth, *Mamestra brassicae*, and the role of melanin in thermoregulation. *Heredity, 73*, 471–479.

Hahn, D. A., & Denlinger, D. L. (2010). Energetics of insect diapause. *Annual Review of Entomology, 56*, 103–121.

Hartfelder, K., & Emlen, D. J. (2012). Endocrine control of insect polyphenism. In L. I. Gilbert (Ed.), *Insect endocrinology* (pp. 465–522). San Diego: Academic Press, Chapter 10.

Hasegawa, K. (1951). Studies in voltinism in the silkworm, *Bombyx mori*, with special reference to the organs concerning voltinism (a preliminary note). *Proceedings of the Imperial Academy of Japan, 27*, 667–671.

Hazel, W. N. (2002). The environmental and genetic control of seasonal polyphenism in larval color and its adaptive significance in a swallowtail butterfly. *Evolution, 56*, 342–348.

Heyland, A., Hodin, J., & Reitzel, A. M. (2005). Hormone signaling in evolution and development: A non-model system approach. *Bioessays, 27*, 64–75.

Highnam, K. C. (1958). Activity of the brain/corpora cardiaca system during pupal diapause 'break' in *Mimas tiliae* (Lepidoptera). *Quarterly Journal of Microscopical Science, 99*, 73–88.

Homma, T., Watanabe, K., Tsurumaru, S., Kataoka, H., Imai, K., Kamba, M., et al. (2006). G protein-coupled receptor for diapause hormone, an inducer of *Bombyx* embryonic diapause. *Biochemical and Biophysical Research Communications, 344*, 386–393.

Horie, Y., Kanda, T., & Mochida, Y. (2000). Sorbitol as an arrester of embryonic development in diapausing eggs of the silkworm, *Bombyx mori*. *Journal of Insect Physiology, 46*, 1009–1016.

Hudson, A. (1966). Proteins in the haemolymph and other tissues of the developing tomato hornworm, *Protoparce quinquemaculata* Haworth. *Canadian Journal of Zoology, 44*, 541–555.

Inglima, K., Perlmutter, A., & Markofsky, J. (1981). Reversible stage-specific embryonic inhibition mediated by the presence of adults in the annual fish *Nothobranchius guentheri*. *The Journal of Experimental Zoology, 215*, 23–33.

Isabel, G., Gourdoux, L., & Moreau, R. (2001). Changes of biogenic amine levels in haemolymph during diapausing and non-diapausing status in *Pieris brassicae* L. *Comparative Biochemistry and Physiology: Part A, 28*, 117–123.

Islam, A. T., Nankaku, N., Marui, Y., Shahjahan, R., Kong, W., Yamanaka, A., et al. (2005). Neuroendocrine roles of the brain in the regulation of 20-hydroxyecdysone responsiveness in two types of diapause pupae of the cabbage armyworm, *Mamestra brassicae*. *Zoological Science, 22*, 775–784.

Iwata, K., Shindome, C., Kobayashi, Y., Takeda, M., Yamashita, O., Shiomi, K., et al. (2005). Temperature-dependent activation of ERK/MAPK in yolk cells and its role in embryonic diapause termination in the silkworm *Bombyx mori*. *Journal of Insect Physiology, 51*, 1306–1312.

Judy, K. J. (1972). Diapause termination and metamorphosis in brainless tobacco hornworms (Lepidoptera). *Life Sciences, 11*, 605–611.

Kamei, H., Ding, Y., Kajimura, S., Wells, M., Chiang, P., & Duan, C. (2011). Role of IGF signaling in catch-up growth and accelerated temporal development in zebrafish embryos in response to oxygen availability. *Development, 138*, 777–786.

Kawaguchi, Y., Banno, Y., Doira, H., & Fujii, H. (1989). Manifestation of characteristics in the "Giant Egg" mutant of *Bombyx mori* (Lepidoptera: Bombycidae). 2. Artificial induction of large eggs by application of large amounts of 20-hydroxyecdysone. *Odokon, 33*, 63–68.

Kevan, D. K. M. (1944). The bionomics of the neotropical cornstalk borer, *Diatraea lineolata* Wlk in Trinidad, B.W.I.. *Bulletin of Entomological Research, 35*, 23–30.

Kimura, T., & Masaki, S. (1977). Brachypterism and seasonal adaptation in *Orgyia thyellina* Butler (Lepidoptera, Lymantriidae). *Kontyu, 45*, 97–106.

Kind, T. V. (1978). A study of the reactivation of diapausing pupae of *Acronycta rumicis* L. (Lepidoptera, Noctuidae). II. Reactivation of intact and brainless pupae at various temperatures. *Entomological Review, 57*, 163–167.

Kivelä, S. M., Välimäki, P., & Mäenpää, M. I. (2012). Genetic and phenotypic variation in juvenile development in relation to temperature and developmental pathway in a geometrid moth. *Journal of Evolutionary Biology, 25*, 881–891.

Koch, P. B., & Bückmann, D. (1987). Hormonal control of seasonal morphs by the timing of ecdysteroid release in *Araschnia levana* L. (Nymphalidae: Lepidoptera). *Journal of Insect Physiology, 33*, 823–829.

Kukal, O., Serianni, A. S., & Duman, G. J. (1988). Glycerol metabolism in a freeze-tolerant arctic insect: An in vivo 13 C NMR study. *Journal of Comparative Physiology, 158*, 175–183.

Lee, E., Jr., & Denlinger, D. L. (1991). *Insects at low temperature* (513 pp.). New York: Chapman and Hall.

Lee, K.-Y. , & Denlinger, D. L. (1996). Diapause-regulated proteins in the gut of pharate first instar larvae of the gypsy moth, *Lymantria dispar*, and the effects of KK-42 and neck ligation on expression. *Journal of Insect Physiology, 42*, 423–431.

Lee, K.-Y. , & Denlinger, D. L. (1997). A role for ecdysteroids in the induction and maintenance of the pharate first instar diapause of the gypsy moth, *Lymantria dispar*. *Journal of Insect Physiology, 43*, 289–296.

Lee, K.-Y. , Horodyski, F. M., Valaitis, A. P., & Denlinger, D. L. (2002). Molecular characterization of the insect immune protein hemolin and its high induction during embryonic diapause in the gypsy moth, *Lymantria dispar*. *Insect Biochemistry and Molecular Biology, 32*, 1457–1467.

Lee, K.-Y. , Valaitis, A. P., & Denlinger, D. L. (1997). Further evidence that diapause in the gypsy moth, *Lymantria dispar*, is regulated by ecdysteroids: A comparison of diapause and non-diapause strains. *Journal of Insect Physiology, 43*, 897–903.

Leimar, O. (1996). Life history plasticity: Influence of photoperiod on growth and development in the common blue butterfly. *Oikos, 76*, 228–234.

Levels, P. J., Van Tits, L. J. H., & Denucé, J. M. (1986). The effect of the presence of adult fishes, gonad homogenates, and embryo homogenates on the dispersionreaggregation phase during early embryonic development of the annual fish *Nothobranchius korthausae*. *The Journal of Experimental Zoology, 240*, 259–264.

Loeb, M. J. (1982). Diapause and development in the tobacco budworm, *Heliothis virescens*: A comparison of haemolymph ecdysteroid titres. *Journal of Insect Physiology, 28*, 667–673.

Maslennikova, V. A. (1970). Hormonal regulation of diapause in *Pieris brassicae* L. *Doklady Akademii Nauk SSSR, 192*, 942–945.

Matsumoto, M., & Takeda, M. (2002). Changes in brain monoamine contents in diapause pupae of *Antheraea pernyi* when activated under long-day and by chilling. *Journal of Insect Physiology, 48*, 765–771.

McDaniel, C. N., & Berry, S. J. (1967). Activation of the prothoracic glands of *Antheraea polyphemus*. *Nature*, *214*, 1032–1033.

McNeil, J. N., & Fields, P. G. (1985). Seasonal diapause development and diapause termination in the European skipper, *Thymelicus lineola* (Ochs.). *Journal of Insect Physiology*, *31*, 467–470.

Meola, R. W., & Adkisson, P. L. (1977). Release of prothoracicotropic hormone and potentiation of developmental ability during diapause in the bollworm, *Heliothis zea*. *Journal of Insect Physiology*, *23*, 683–688.

Mishra, P. K., Sharan, S. K., Kumar, D., Singh, B. M. K., Subrahmanyam, B., & Suryanarayana, N. (2008). Effect of ecdysone on termination of pupal diapause and egg production in *Antheraea mylita* Drury. *Journal of Advanced Zoology*, *29*, 128.

Mizoguchi, A., & Ishizaki, H. (1982). Prothoracic glands of the saturniid moth *Samia cynthia ricini* possess a circadian clock controlling gut purge timing. *Proceedings of the National Academy of Sciences of the United States of America*, *79*, 2726–2730.

Morita, A., Niimi, T., & Yamashita, O. (2004). Physiological differentiation of DH-PBAN-producing neurosecretory cells in the silkworm embryo. *Journal of Insect Physiology*, *49*, 1093.

Munyiri, F. N., & Ishikawa, Y. (2004). Endocrine changes associated with metamorphosis and diapause induction in the yellow-spotted longicorn beetle, *Psacothea hilaris*. *Journal of Insect Physiology*, *50*, 1075–1081.

Murphy, W. J., & Collier, G. E. (1997). A molecular phylogeny for aplocheiloid fishes (Atherinomorpha, Cyprinodontiformes): The role of vicariance and the origins of annualism. *Molecular Biology and Evolution*, *14*, 790–799.

Nakagaki, M., Takei, R., Nagashima, E., & Yaginuma, T. (1991). Cell-cycles in embryos of the silkworm, *Bombyx mori*: G2 arrest at diapause stage. *Roux's Archives of Developmental Biology*, *200*, 223–229.

Nakamura, A., Miyado, K., Takezawa, Y., Ohnami, N., Sato, M., Ono, C., et al. (2011). Innate immune system still works at diapause, a physiological state of dormancy in insects. *Biochemical and Biophysical Research Communications*, *410*, 351–357.

Nijhout, H. F. (2003). Development and of evolution of adaptive polyphenisms. *Evolution and Development*, *5*, 9–18.

Noguchi, H., & Hayakawa, Y. (1997). Role of dopamine at the onset of pupal diapause in the cabbage armyworm *Mamestra brassicae*. *FEBS Letters*, *413*, 157–161.

Noguchi, H., & Hayakawa, Y. (2001). Dopamine is a key factor for the induction of egg diapause of the silkworm, *Bombyx mori*. *European Journal of Biochemistry*, *268*, 774–780.

Nylin, S., Gotthard, K., & Wiklund, C. (1996). Reaction norms for age and size at maturity in *Lasiommata* butterflies: Predictions and tests. *Evolution*, *50*, 1351–1358.

Nylin, S., Wickman, P., & Wiklund, C. (1989). Seasonal plasticity in growth and development of the speckled wood butterfly, *Pararge aegeria* (Satyrinae). *Biological Journal of the Linnean Society*, *38*, 155–171.

Oliver, C. G. (1970). The environmental regulation of seasonal dimorphism in *Pieris napi oleracea* (Pieridae). *Journal of the Lepidopterists' Society*, *24*, 77–81.

Ozeki, K. (1954). Experiments on the formation of imaginal structures in the pupae of the swallowtail, *Papilio xuthus* Linnaeus. *Scientific Papers of the College of General Education (Tokyo)*, *4*, 47.

Peypelut, L., Beydon, P., & Lavenseau, L. (1990). 20-Hydroxyecdysone triggers the resumption of imaginal wing disc development after diapause in the European corn borer, *Ostrinia nubilalis*. *Archives of Insect Biochemistry and Physiology*, *15*, 1–19.

Pfennig, D. W., Wund, M. A., Snell-Rood, E. C., Cruickshank, T., Schlichting, C. D., & Moczek, A. P. (2010). Phenotypic plasticity's impact on diversification and speciation. *Trends in Ecology & Evolution*, *25*, 459–467.

Podrabsky, J. E., Carpenter, J. F., & Hand, S. C. (2001). Survival of water stress by annual killifish embryos: dehydration avoidance and amyloid fibrils in the egg envelope. *American Journal of Physiology. Regulatory, Integrative and Comparative Physiology, 280,* R123–R131.

Podrabsky, J. E., Garrett, I. D. F., & Kohl, Z. F. (2010). Alternative developmental pathways associated with diapause regulated by temperature and maternal influences in embryos of the annual killifish *Austrofundulus limnaeus. Journal of Experimental Biology, 213,* 3280–3288.

Podrabsky, J. E., & Hand, S. C. (1999). The bioenergetics of embryonic diapause in an annual killifish, *Austrofundulus limnaeus. Journal of Experimental Biology, 202,* 2567–2580.

Podrabsky, J. E., Lopez, J. P., Fan, T. W. M., Higashi, R., & Somero, G. N. (2007). Extreme anoxia tolerance in embryos of the annual killifish *Austrofundulus limnaeus*: Insights from a metabolomics analysis. *Journal of Experimental Biology, 210,* 2253–2266.

Podrabsky, J. E., Menze, M. A., & Hand, S. C. (2012). Long-term survival of anoxia despite rapid ATP decline in embryos of the annual killifish *Austrofundulus limnaeus. Journal of Experimental Zoology. Part A, Ecological Genetics and Physiology, 317,* 524–532.

Pri-Tal, B. M., Blue, S., Pau, F. K., & Podrabsky, J. E. (2011). Hormonal components of altered developmental pathways in the annual killifish, *Austrofundulus limnaeus. General and Comparative Endocrinology, 174,* 166–174.

Puiroux, J., Moreau, R., & Gourdoux, L. (1990). Variations of biogenic amine levels in the brain of *Pieris brassicae* pupae during non-diapausing and diapausing development. *Archives of Insect Biochemistry and Physiology, 14,* 57–69.

Pullin, A. S., & Bale, J. S. (1989). Effects of ecdysone, juvenile hormone and haemolymph transfer on cryoprotectant metabolism in diapausing and non-diapausing pupae of *Pieris brassicae. Journal of Insect Physiology, 35,* 911–918.

Rajakumar, R., San Mauro, D., Dijkstra, M. B., Huang, M. H., Wheeler, D. E., Hiou-Tim, F., et al. (2012). Ancestral developmental potential facilitates parallel evolution in ants. *Science, 335,* 79–82.

Riley, T. J. (1980). Effects of long and short day photoperiods on the seasonal dimorphism of *Anaea andria* (Nymphalidae) from central Missouri. *Journal of the Lepidopterists' Society, 34,* 330–337.

Roxström-Lindquist, K., Assefaw-Redda, Y., Rosinska, K., & Faye, I. (2005). 20-Hydroxyecdysone indirectly regulates Hemolin gene expression in *Hyalophora cecropia. Insect Molecular Biology, 14,* 645–652.

Sato, Y., Ikeda, M., & Yamashita, O. (1994). Neurosecretory cells expressing the gene for common precursor for diapause hormone and pheromone biosynthesis-activating neuropeptide in the suboesophageal ganglion of the silkworm, *Bombyx mori. General and Comparative Endocrinology, 96,* 27–36.

Sato, Y., Oguchi, M., Menjo, N., Imai, K., Saito, H., Ikeda, M., et al. (1993). Precursor polyprotein for multiple neuropeptides secreted from the suboesophageal ganglion of the silkworm *Bombyx mori*: Characterization of the cDNA encoding the diapause hormone precursor and identification of additional peptides. *Proceedings of the National Academy of Sciences of the United States of America, 90,* 3251–3255.

Sato, Y., Shiomi, K., Saito, H., Imai, K., & Yamashita, O. (1998). Phe-X-Pro-Arg-Leu-NH2 peptide producing cells in the central nervous system of the silkworm, *Bombyx mori. Journal of Insect Physiology, 44,* 333–342.

Sauman, I., & Reppert, S. M. (1996). Molecular characterization of prothoracicotropic hormone (PTTH) from the giant silkmoth *Antheraea pernyi*: Developmental appearance of PTTH-expressing cells and relationship to circadian clock cells in central brain. *Developmental Biology, 178,* 418–429.

Saunders, D. S., Steel, C. G. H., Vafopoulou, X., & Lewis, R. D. (2002). Chapter 9: Photoperiodism and seasonal cycles of, development. In *Insect Clocks* (pp. 271–298). Amsterdam: Elsevier.

Schlueter, P. J., Royer, T., Farah, M. H., Laser, B., Chan, S. J., Steiner, D. F., et al. (2006). Gene duplication and functional divergence of the zebrafish insulin-like growth factor 1 receptors. *The FASEB Journal, 20*, 1230–1232.

Scott, J. A. (1979). Hibernal diapause of North American Papillionidae and Hesperioidea. *Journal of Research on the Lepidoptera, 18*, 171–200.

Shapiro, A. M. (1975). Developmental and phenotypic responses to photoperiod in uni- and bivoltine *Pieris napi* in California. *Transactions of the Entomological Society of London, 127*, 65–71.

Shapiro, A. M. (1977). Phenotypic induction in *Peris napi* L.: Role of temperature and photoperiod in a coastal California population. *Ecological Entomology, 2*, 217–224.

Shimada, K. (1980). Some physiological properties associated with freeze-tolerance in diapausing pupae of *Papilio machaon*. *Low Temperature Science, 38*, 53–60.

Shiomi, K., Fujiwara, Y., Yasukochi, Y., Kajiura, Z., Nagagaki, M., & Yaginuma, T. (2007). The Pitx homeobox gene in *Bombyx mori*: Regulation of DH-PBAN neuropeptide hormone gene expression. *Molecular and Cellular Neuroscience, 34*, 209–218.

Shionoya, M., Matsubayashi, H., Asahina, M., Kuniyoshi, H., Nagata, S., Riddiford, L. M., et al. (2003). Molecular cloning of the prothoracicotropic hormone from the tobacco hornworm, *Manduca sexta*. *Insect Biochemistry and Molecular Biology, 33*, 795–801.

Sieber, R., & Benz, G. (1977). Juvenile hormone in larval diapause of the codling moth, *Laspeyresia pomonella* L. (Lepidoptera, Tortricidae). *Experientia, 33*, 1598–1599.

Sieber, R., & Benz, G. (1980). The hormonal regulation of the larval diapause in the codling moth, *Laspeyresia pomonella* (Lep. Tortricidae). *Journal of Insect Physiology, 26*, 213–218.

Sielezniew, M., & Cymborowski, B. (1997). Effects of ecdysteroid agonist RH-5849 on pupal diapause of the tobacco hornworm (*Manduca sexta*). *Archives of Insect Biochemistry and Physiology, 35*, 191–197.

Smith, W. A., Bowen, M. F., Bollenbacher, W. E., & Gilbert, L. I. (1986). Cellular changes in the prothoracic glands of diapausing pupae of *Manduca sexta*. *Journal of Experimental Biology, 120*, 131–142.

Smith, W., & Rybczynski, R. (2012). Prothoracicotropic hormone. In L. I. Gilbert (Ed.), *Insect endocrinology* (pp. 1–62). San Diego: Academic Press, Chapter 1.

Sugiki, T., & Masaki, S. (1972). Photoperiodic control of larval and pupal development in *Spilarctia imparilis* Butler (Lepidoptera: Arctiidae). *Kontyu, 40*, 269–278.

Sun, J. S., Zhang, T. Y., Zhang, Q. R., & Xu, W. H. (2003). Effect of the brain and suboesophageal ganglion on pupal development in *Helicoverpa armigera* through regulation of FXPRLamide neuropeptides. *Regulatory Peptides, 116*, 163–171.

Sun, J. S., Zhang, Q. R., Zhang, T. Y., Zhu, Z. L., Zhang, H. M., Teng, M. K., et al. (2005). Developmental expression of FXPRLamide neuropeptides in peptidergic neurosecretory cells of diapause- and nondiapause-destined individuals of the cotton bollworm, *Helicoverpa armigera*. *General and Comparative Endocrinology, 141*, 48–57.

Suwan, S., Isobe, M., Yamashita, O., Minakata, H., & Imai, K. (1994). Silkworm diapause hormone, structure activity relationships, indispensable role of C-terminus amide. *Insect Biochemistry and Molecular Biology, 24*, 1001–1007.

Suzuki, K., Fujita, M., & Miya, K. (1983). Changes in supercooling point of silworm eggs. *Journal of Sericultural Science of Japan, 52*, 185–190.

Suzuki, K., Minagawa, T., Kumagai, T., Naya, S.-I. , Endo, Y., Osanai, M., et al. (1990). Control mechanism of diapause of the pharate first-instar larvae of the silkmoth *Antheraea yamamai*. *Journal of Insect Physiology, 36*, 855–860.

Suzuki, K., Nakamura, T., Yanbe, T., Kurihara, M., & Kuwano, E. (1993). Termination of diapause in pharate first instar larvae of the gypsy moth *Lymantria dispar japonica* by an imidazole derivative KK-42. *Journal of Insect Physiology, 39*, 107–110.

Suzuki, Y., & Nijhout, H. F. (2006). Evolution of a polyphenism by genetic accommodation. *Science, 311*, 650–652.

Suzuki, Y., & Nijhout, H. F. (2007). Genetic basis of adaptive evolution of a polyphenism by genetic accommodation. *Journal of Evolutionary Biology, 21*, 57–66.
Tatar, M., & Yin, C. (2001). Slow ageing during insect reproductive diapause: Why butterflies, grasshoppers and flies are like worms. *Experimental Gerontology, 36*, 723–738.
Taylor, M. H. (1999). A suite of adaptations for intertidal spawning. *American Zoologist, 39*, 313–320.
Tingaud-Sequeira, A., Zapater, C., Chauvigne, F., Otero, D., & Cerda, J. (2009). Adaptive plasticity of killifish (*Fundulus heteroclitus*) embryos: Dehydration-stimulated development and differential aquaporin-3 expression. *American Journal of Physiology. Regulatory, Integrative and Comparative Physiology, 296*, R1041–R1052.
Tsurumaki, J., Ishiguro, J., & Yamanaka, A. (1999). Effects of photoperiod and temperature on seasonal morph development and diapause egg oviposition in a bivoltine race (Daizo) of the silkmoth, *Bombyx mori*. *Journal of Insect Physiology, 45*, 101–106.
Uehara, H., Senoh, Y., Yoneda, K., Kato, Y., & Shiomi, K. (2011). An FXPRLamide neuropeptide induces seasonal reproductive polyphenism underlying a life-history tradeoff in the tussock moth. *PLoS One, 6*, e24213.
Usua, E. J. (1973). Induction of diapause in the maize stemborer, *Busseola fusca*. *Entomologia Experimentalis et Applicata, 16*, 322–328.
Välimäki, P., Kivelä, S. M., Mäenpää, M. I., & Tammaru, T. (2013). Latitudinal clines in alternative life histories in a geometrid moth. *Journal of Evolutionary Biology, 26*, 118–129.
Van Dyc, H., & Wiklund, C. (2002). Seasonal butterfly design: Morphological plasticity among three developmental pathways relative to sex, flight and thermoregulation. *Journal of Evolutionary Biology, 15*, 218–225.
Wang, X., Yang, Q., Zhou, X., Zhao, F., & Lei, C. (2007). Effect of photoperiod associated with diapause induction on the accumulation of metabolites in *Sericinus montelus* (Lepidoptera: Papilionidae). *Applied Entomology and Zoology, 42*, 419–424.
Wang, Z. S., Zheng, W. H., & Guo, F. (1986). Effect of Bombyx 4K-PTTH and bovine insulin on testis development of Indian silkworm, *Philosamia cynthia ricini*. In A. B. Borkovec & D. B. Gelman (Eds.), *Insect Neurochemistry and Neurophysiology* (pp. 335–338). New York: Humana Press.
Wei, Z. J., Zhang, Q. R., Kang, L., Xu, W. H., & Denlinger, D. L. (2005). Molecular characterization and expression of prothoracicotropic hormone during development and pupal diapause in the cotton bollworm, *Helicoverpa armigera*. *Journal of Insect Physiology, 51*, 691–700.
West-Eberhard, M. J. (2003). *Developmental plasticity and evolution*. New York: Oxford University Press.
Wiklund, C., & Friberg, M. (2011). Seasonal development and variation in abundance among four annual flight periods in a butterfly – A 20 year study of the speckled wood (*Pararge aegeria*). *Biological Journal of the Linnean Society, 102*, 635–649.
Williams, C. M. (1946). Physiology of insect diapause. I. The role of the brain in the production and termination of pupal dormancy in the giant silkworm *Platysamia cecropia*. *The Biological Bulletin: Marine Biological Laboratory (Woods Hole), 90*, 234–243.
Williams, C. M. (1947). Physiology of insect diapause. II. Interaction between the pupal brain and prothoracic glands in the metamorphosis of the giant silkworm, *Platysamia cecropia*. *The Biological Bulletin, 93*, 89–98.
Williams, C. M. (1952). Physiology of insect diapause. IV. The brain and prothoracic glands as an endocrine system in the cecropia silkworm. *The Biological Bulletin: Marine Biological Laboratory (Woods Hole), 103*, 120–138.
Williams, C. M. (1968). Ecdysone and ecdysone-analogues: Their assay and action on diapausing pupae of the Cynthia silkworm. *The Biological Bulletin, 134*, 344–355.
Wilson, G. R., & Larsen, J. R. (1974). Debraining and diapause development in *Manduca sexta* pupae. *Journal of Insect Physiology, 20*, 2459–2473.

Windig, J., & Lammar, P. (1999). Evolutionary genetics of seasonal polyphenism in the map butterfly *Araschnia levana* (Nymphalidae: Lepidoptera). *Evolutionary Ecology Research, 1*, 875–894.

Wourms, J. P. (1972a). The developmental biology of annual fishes. I. Stages in the normal development of *Austrofundulus myersi* Dahl. *The Journal of Experimental Zoology, 182*, 143–168.

Wourms, J. P. (1972b). The developmental biology of annual fishes. II. Naturally occurring dispersion and reaggregation of blastomeres during the development of annual fish eggs. *The Journal of Experimental Zoology, 182*, 169–200.

Wourms, J. P. (1972c). The developmental biology of fishes. III. Preembryonic diapause of variable duration in the eggs of annual fishes. *The Journal of Experimental Zoology, 182*, 389–414.

Xu, W.-H. , & Denlinger, D. L. (2003). Molecular characterization of prothoracicotropic hormone and diapause hormone in *Heliothis virescens* during diapause, and a new role for diapause hormone. *Insect Molecular Biology, 12*, 509–516.

Xu, W.-H. , & Denlinger, D. L. (2004). Identification of a cDNA encoding DH, PBAN and other FXPRL neuropeptides from the tobacco hornworm, *Manduca sexta*, and expression associated with pupal diapause. *Peptides, 25*, 1099–1106.

Xu, W.-H. , Sato, Y., Ikeda, M., & Yamashita, O. (1995a). Stage-dependent and temperature controlled expression of the gene encoding the precursor protein of diapause hormone and pheromone biosynthesis activating neuro peptide in the silkworm, *Bombyx mori*. *Journal of Biological Chemistry, 270*, 3804–3808.

Xu, W.-H. , Sato, Y., Ikeda, M., & Yamashita, O. (1995b). Molecular characterization of the gene encoding the precursor protein of diapause hormone and pheromone biosynthesis activating neuropeptide (DH-PBAN) of the silkworm, *Bombyx mori* and its distribution in some insects. *Biochimica et Biophysica Acta, 1261*, 83–89.

Yagi, S., & Fukaya, M. (1974). Juvenile hormone as a key factor regulating larval diapause of the rice stem borer, *Chilo suppressalis* (Lepidoptera: Pyralidae). *Applied Entomology and Zoology, 9*, 247–255.

Yamanaka, A., Tsurumaki, J., & Endo, K. (2000). Neuroendocrine regulation of seasonal morph development in a bivoltine race (Daizo) of the silkmoth, *Bombyx mori* L. *Journal of Insect Physiology, 46*, 803–808.

Yamashita, O. (1996). Diapause hormone of the silkworm, *Bombyx mori*: Structure, gene expression and function. *Journal of Insect Physiology, 42*, 669–679.

Yamashita, O., Shiomi, K., Ishida, Y., Katagiri, N., & Niimi, T. (2001). Insights for future studies on embryonic diapause promoted by molecular analyses of diapause hormone and its action in *Bombyx mori*. In D. L. Denlinger, J. Giebultowicz, & D. S. Saunders (Eds.), *Insect timing: Circadian rhythmicity to seasonality* (pp. 145–153). Amsterdam: Elsevier Science B. V.

Yin, C. M., & Chippendale, G. M. (1973). Juvenile hormone regulation of the larval diapause of the southwestern com borer, *Diatraea grandiosella*. *Journal of Insect Physiology, 19*, 2403–2420.

Yin, C. M., & Chippendale, G. M. (1974). Juvenile hormone and the inducion of larval polymorphism and diapause of the southwestern corn borer, *Diatrea grandiosella*. *Journal of Insect Physiology, 20*, 1833–1847.

Yin, C. M., & Chippendale, G. M. (1975). Insect prothoracic glands: Function and ultrastructure in diapause and non-diapause larvae of *Diatrea grandiosella*. *Canadian Journal of Zoology, 53*, 124–131.

Yin, C. M., & Chippendale, G. M. (1976a). Hormonal control of larval diapause and metamorphosis of the southwestern corn borer *Diatrea grandiosella*. *Journal of Experimental Biology, 64*, 303–310.

Yin, C. M., & Chippendale, G. M. (1976b). Diapause of the southwestern corn borer, *Diatrea grandiosella*: Further evidence showing juvenile hormone to be the regulator. *Journal of Insect Physiology*, *25*, 513–524.

Zhan, S., Merlin, C., Boore, J. L., & Reppert, S. M. (2011). The monarch butterfly genome yields into long-distance migration. *Cell*, *147*, 1171–1185.

Zhang, Q., & Denlinger, D. L. (2012). Dynamics of diapause hormone and prothoracicotropic hormone transcript expression at diapause termination in pupae of the corn earworm, *Helicoverpa zea*. *Peptides*, *34*, 120–126.

Zhang, T. Y., Kang, L., Zhang, Z. F., & Xu, W. H. (2004). Identification of a POU factor involved in regulating the neuron-specific expression of the gene encoding diapause hormone and pheromone biosynthesis-activating neuropeptide in *Bombyx mori*. *Biochemical Journal*, *380*, 255–263.

Zhang, Q., Nachman, R. J., Kaczmarek, K., Zabrocki, J., & Denlinger, D. L. (2011). Disruption of insect diapause using agonists and an antagonist of diapause hormone. *Proceedings of the National Academy of Sciences of the United States of America*, *108*, 16922–16926.

Zhang, T. Y., Sun, J. S., Zhang, L. B., Shen, J. L., & Xu, W. H. (2004). Cloning and expression of the cDNA encoding the FXPRL family of peptides and a functional analysis of their effect on breaking pupal diapause in *Helicoverpa armigera*. *Journal of Insect Physiology*, *50*, 25–33.

Zhang, T. Y., Sun, J. S., Zhang, Q. R., Xu, J., Jiang, R. J., & Xu, W. H. (2004). The diapause hormone–pheromone biosynthesis activating neuropeptide gene of *Helicoverpa armigera* encodes multiple peptides that break, rather than induce, diapause. *Journal of Insect Physiology*, *50*, 547–554.

Zhang, Q., Zdarek, J., Nachman, R. J., & Denlinger, D. L. (2008). Diapause hormone in the corn earworm, *Helicoverpa zea*: Optimum temperature for activity, structureactivity relationships, and efficacy in accelerating flesh fly pupariation. *Peptides*, *29*, 196–205.

CHAPTER NINE

Circannual Transitions in Gene Expression: Lessons from Seasonal Adaptations

Christine Schwartz, Matthew T. Andrews[1]
Department of Biology, University of Minnesota Duluth, Duluth, Minnesota, USA
[1]Corresponding author: e-mail address: mandrews@d.umn.edu

Contents

1. Introduction 248
2. Seasonal Courtship and Territorial Behavior 249
3. Migration 251
4. Hibernation 253
   4.1 Satiety and feeding behavior 256
   4.2 Fuel switch 257
   4.3 Central nervous system in hibernation 258
5. Potential Pathways Involved in Circannual Timing 259
   5.1 Thyroid hormone signaling 259
   5.2 VGF signaling 260
   5.3 Modulation of neuronal cilia 261
   5.4 Retinoic acid signaling 263
   5.5 Melatonin and clock genes 264
6. Recent Advances in Circannual Timing Research 266
7. Conclusions: Future Directions and Potential Applications 267
Acknowledgments 267
References 268

Abstract

Circannual timing is important for the coordination of seasonal activities, particularly promoting the survival of individuals in adverse conditions through adaptive physiological and behavioral changes. This includes optimizing the survival of offspring by coordinating reproductive efforts at appropriate times. Thus, timing is very important for overall fitness. In this chapter, we provide several examples of circannually timed events, including mammalian hibernation, discussing the physiological changes that accompany these events, and some of the known genes and pathways underlying these changes. We then describe five candidate systems that are potentially involved in circannual timing. Finally, we discuss several recent advances in molecular biology

Current Topics in Developmental Biology, Volume 105
ISSN 0070-2153
http://dx.doi.org/10.1016/B978-0-12-396968-2.00009-9

and animal husbandry that have made the use of nonmodel organisms for research more feasible, which will hopefully promote and encourage further advancement in the knowledge of circannual timing.

1. INTRODUCTION

Circannual behavior is an adaptation to the physical world. Animals have no control over the physical environment so they have developed many different strategies to anticipate seasonal changes and prepare accordingly for survival and/or maximum fitness. One example of an adaptation to environmental stresses is diapause, a developmental arrest mechanism implemented by some invertebrate species, including the nematode *Caenorhabditis elegans* (reviewed in Carey, Andrews, & Martin, 2003) and some insects (discussed in chapter 8). Here, we focus on circannual events utilized primarily by vertebrates. Examples of some of these events are provided in Fig. 9.1. The molecular pathways underlying the timing of these circannual events are not well characterized, but active research on this subject has provided many advances in the field. Here, we will describe some well-known examples of

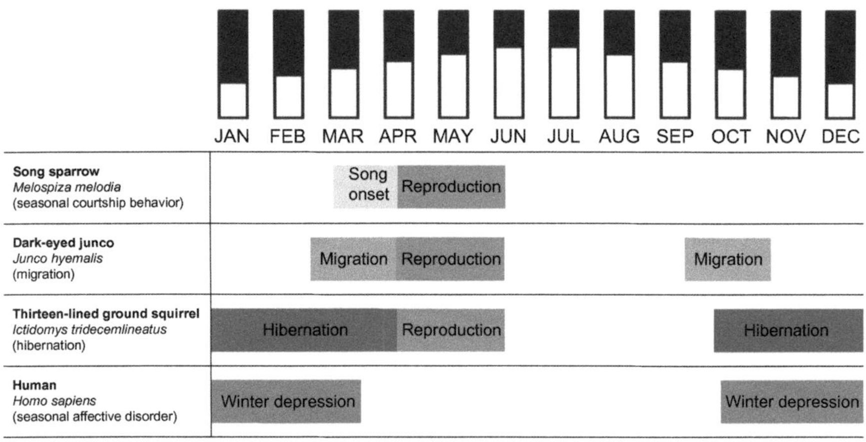

Figure 9.1 Circannually timed events. Representative examples and approximate annual timing of different seasonal events are provided along with an example species. Light/dark cycles illustrated above the months are representative of the annual changes in photoperiod in the northern United States of America. Reproduction represents both mating behavior and gestation. (For color version of this figure, the reader is referred to the online version of this chapter.)

circannual events, and provide some ideas about how they could be coordinated and controlled on a seasonal basis.

2. SEASONAL COURTSHIP AND TERRITORIAL BEHAVIOR

Seasonally reproducing animals often display courtship or territorial behavior associated with their mating season. For instance, male songbirds sing complex courtship songs robustly in their breeding season both to attract mates and deter other males. Consequently, singing is a circannual behavior in various species, including canaries (*Serinus canaria*), song sparrows (*Melospiza melodia*), and many others. Songbirds have a system of nuclei (clusters of neurons in the central nervous system) that control song production and learning (Fig. 9.2). The motor pathway controlling song production consists of projections from the interfacial nucleus of the nidopallium (Nif) to the nidopallial vocal nucleus HVC (used as a proper name). HVC projects to the robust nucleus of the archipallium (RA), which then projects to the midbrain and brainstem syringeal and respiratory motoneuronal pools that activate the muscles used for vocalizing (Brenowitz, Margoliash, & Nordeen,

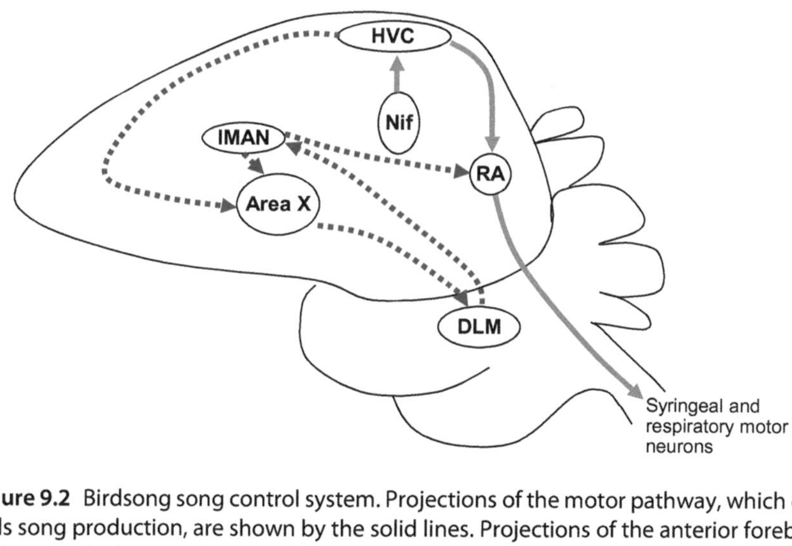

Figure 9.2 Birdsong song control system. Projections of the motor pathway, which controls song production, are shown by the solid lines. Projections of the anterior forebrain pathway, which controls song learning and development, are shown by the dotted lines. DLM, dorsal lateral nucleus; HVC, used as proper name; lMAN, lateral magnocellular nucleus of the anterior nidopallium; Nif, interfacial nucleus of the nidopallium; RA, robust nucleus of the arcopallium. *Model is based on Nottebohm (2005) and Tramontin and Brenowitz (2000).* (For color version of this figure, the reader is referred to the online version of this chapter.)

1997). In the anterior forebrain pathway, which plays an important role in song learning and development, HVC projects to striatal Area X, which then projects to the medial portion of the dorsal lateral nucleus (DLM) of the thalamus. DLM projects to the lateral magnocellular nucleus of the anterior nidopallium (lMAN), which then projects both back to Area X and also to RA to influence the song production pathway. Extensive seasonal plasticity has been reported in these song nuclei, with a significant increase in nuclei size during the breeding season when the singing behavior occurs (Nottebohm, 1981; Smith, 1996; Smith, Brenowitz, Beecher, & Wingfield, 1997; Smith, Brenowitz, Wingfield, & Baptista, 2004; Fig. 9.3A). This increase in size is

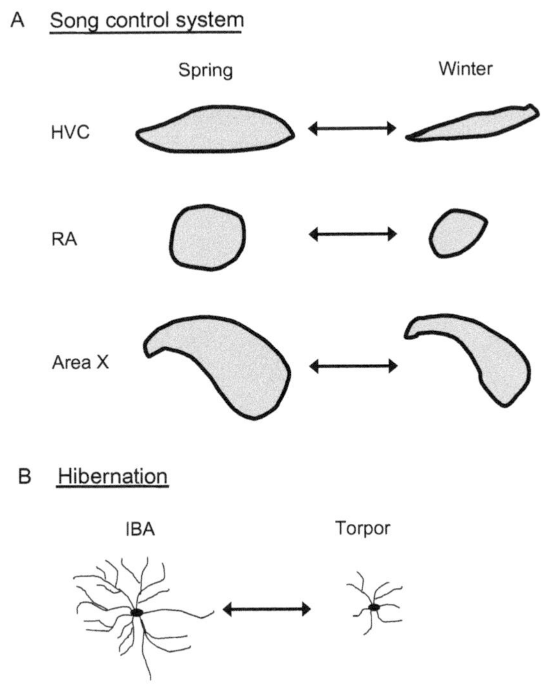

Figure 9.3 Seasonal plasticity in the brain. (A) Illustration of the representative changes in song nuclei volume in the brains of songbirds from spring, when courtship singing occurs, to winter, when singing behavior ceases. Model is based on data from Smith (1996). (B) Illustration of the plasticity and remodeling that occurs during hibernation in the ground squirrel. Dendrites regress and synapses dissociate during hypothermic torpor, but reconnect upon return to normothermic interbout arousals (IBAs). Model is based on data from several sources (Magariños, McEwen, Saboureau, & Pevet, 2006; Popov, Bocharova, & Bragin, 1992; von der Ohe, Darian-Smith, Garner, & Heller, 2006; von der Ohe, Garner, Darian-Smith, & Heller, 2007). HVC, used as proper name; RA, robust nucleus of the arcopallium; IBA, interbout arousal.

due to neurite growth, increased synapse formation, and neurogenesis (Nottebohm, 2005; Tramontin & Brenowitz, 2000).

Several parameters have an effect both on song behavior and the underlying neural mechanisms. Testosterone triggers singing behavior and an increase in song nuclei volume in female canaries, which do not usually sing or exhibit seasonal plasticity (Nottebohm, 1980). House sparrows (*Passer domesticus*) housed in constant conditions and administered melatonin mimicking long days comparable to their natural breeding season exhibited large song control nuclei. Conversely, sparrows administered short day melatonin durations exhibited small song nuclei (Cassone, Bartell, Earnest, & Kumar, 2008). Melatonin receptors are found in the song control system of house sparrows, including HVC, RA, and Area X (Whitfield-Rucker & Cassone, 1996), suggesting that melatonin signaling could influence seasonal song production.

On the molecular level, a recent study examined seasonal gene expression changes in HVC and RA in Gambel's white-crowned sparrows (*Zonotrichia leucophrys gambelii*) (Thompson et al., 2012). They identified over 300 differentially expressed genes, including cerebellin precursor protein 2 (CBLN2), which is important for synaptic plasticity and is upregulated during long days when the song nuclei are increased in volume, but downregulated during short days. Similarly, brain-specific angiogenesis inhibitor 1-associated protein 2 (BAIAP2), important for positive neurite extension and axonogenesis, is upregulated at the beginning of exposure to long days, but is downregulated after several weeks in long days and in short days. In addition to plasticity-related genes, there was also differential expression of genes associated with apoptosis, angiogenesis, and proliferation, along with transcripts for growth factors and hormones. Mukai et al. (2009) used the same microarray methods in song sparrows to examine seasonal changes in gene expression in the hypothalamus in relation to territorial aggression behavior, which included both singing and aggressive displays. They found differential expression of genes associated with thyroid hormone signaling, including glycoprotein hormones alpha chain precursor (*CGA*), mu-crystallin homolog (*CRYM*), and transthyretin precursor (*TTR*). *CGA* and *CRYM* were increased in spring, while *TTR* was increased in autumn. These changes occurred even in control animals and were therefore associated with season and not territorial behavior.

3. MIGRATION

While seasonal courtship behavior specifically promotes successful reproductive behavior by attracting mates and defending resources, there are also several examples of circannually timed events that are important for individual survival in order to get to the next reproductive season. One example is migration. Migration refers to general animal movement, and is widely found in Kingdom Animalia, from mammals to arthropods (Dingle & Drake, 2007; Drickamer, Vessey, & Jakob, 2002). Seasonal migration is an adaptive behavior where animals relocate prior to the onset of unfavorable conditions, such as cold temperatures and low food availability. In this section, we will focus on bird migration, one of the most well-known and well-studied examples. The arctic tern (*Sterna paradisaea*) is the most extreme case, with some individuals annually flying over 80,000 km roundtrip between the Arctic and the Antarctic (Egevang et al., 2010).

Migration consists of two phases, a preparation phase and the actual migration phase. Both phases are seasonally timed according to photoperiod, a consistent and reliable seasonal cue (Drickamer et al., 2002). In the preparation phase, there are substantial behavioral, morphological, and physiological changes that occur, including hyperphagia (Odum, 1960), accumulation of fat reserves (Ramenofsky, Savard, & Greenwood, 1999), muscle hypertrophy (Piersma, 1998), and changes in energy metabolism that specifically increase capacity for fatty acid utilization (Guglielmo, Haunerland, Hochachka, & Williams, 2002; Guglielmo, Williams, Zwingelstein, Brichon, & Weber, 2002). Prior to migration, birds exhibit increased activity levels at the time of day they would normally migrate, termed migratory restlessness, which occurs even in captive birds kept in constant conditions (Drickamer et al., 2002).

During the migration phase, there is considerable species-specific variation, particularly in trip length, direction of travel, final destination, and frequency/duration of stopovers for feeding. Stopovers serve an important purpose in allowing the bird to replenish energy reserves lost during flight, and the frequency and duration of these stopovers has a great effect on total migration time (Schaub, Jenni, & Bairlein, 2008). Selection of stopover sites is important, because stopping at ideal areas with greater food availability can decrease the overall amount of time spent at the site. However, other factors can affect site selection as well, including the migrant's energetic condition, wind, and weather conditions (Moore, Woodrey, Buler, Woltmann, & Simons, 2005). Some species fly over large bodies of water making it impossible to stop until reaching the other side (Larkin, Griffin, Torre-Bueno, & Teal, 1979; Lowery, 1946). Therefore, stopovers must be precisely timed and coordinated to maximize fitness, both to accumulate enough energy for flight and to arrive at the final destination at the appropriate time.

In spite of the wealth of information available about migration physiology, there is little known about how migration is seasonally controlled at the molecular level, presumably due to the difficulty in studying such a dynamic phenotype in a laboratory setting. Variants in the circadian gene *Clock* have been implicated in the seasonal timing of breeding in barn swallows (*Hirundo rustica*) and are also predicted to be involved in migration (Caprioli et al., 2012). A microsatellite approach in European blackcaps (*Sylvia atricapilla*) found an association between migratory behavior and a microsatellite polymorphism within the 3′UTR of the *ADCYAP1* gene locus (Mueller, Pulido, & Kempenaers, 2011). *ADCYAP1* encodes pituitary adenylate cyclase activating polypeptide 1 (PACAP) which stimulates adenylate cyclase and increases the levels of cyclic adenosine monophosphate (cAMP). PACAP is involved in conveying photoperiod information via the retinohypothalamic tract (RHT), which directly connects the photosensitive ganglion cells of the retina to the suprachiasmatic nucleus (SCN) of the hypothalamus (location provided in Fig. 9.4) (Hannibal et al., 1997). The SCN is considered the master circadian clock in mammals and birds, and is entrained to the external environment by light input from the RHT. Migration occurs on a circannual basis, but it is likely that the circadian system plays a role in this timing as well. Photoperiod varies according to the time of year, therefore the light signal received in the SCN would also vary seasonally.

The SCN also regulates the hormone melatonin, which is produced and released into the bloodstream by the pineal gland at night. Melatonin acts as both a clock and a calendar in the body, because melatonin is only present at night and the duration of melatonin is directly related to the length of the night, which varies seasonally. PACAP stimulated melatonin synthesis in pinealocytes (Simonneaux & Ouichou, 1993) and also modulated the expression of clock genes in the pineal (Nagy & Csernus, 2007), and thus clearly plays a role in circadian and/or circannual timing. PACAP could be important for tracking photoperiod changes over seasons for the coordination of seasonal activities in migrating birds and other animals with seasonal activity. Additionally, intracerebroventricular (ICV) injection of PACAP into chicks (*Gallus gallus*) resulted in changes in energy metabolism, including increased lipid utilization (Tachibana, Oikawa, Adachi, Boswell, & Furuse, 2007), which indicates that PACAP could also be playing a role in migration preparation, helping to facilitate a fuel switch to lipid metabolism.

4. HIBERNATION

Hibernation, like migration, is another survival adaptation to periods of harsh conditions and low food availability. Obligate hibernation, the main focus of this section, is an annual period of heterothermy where mammals

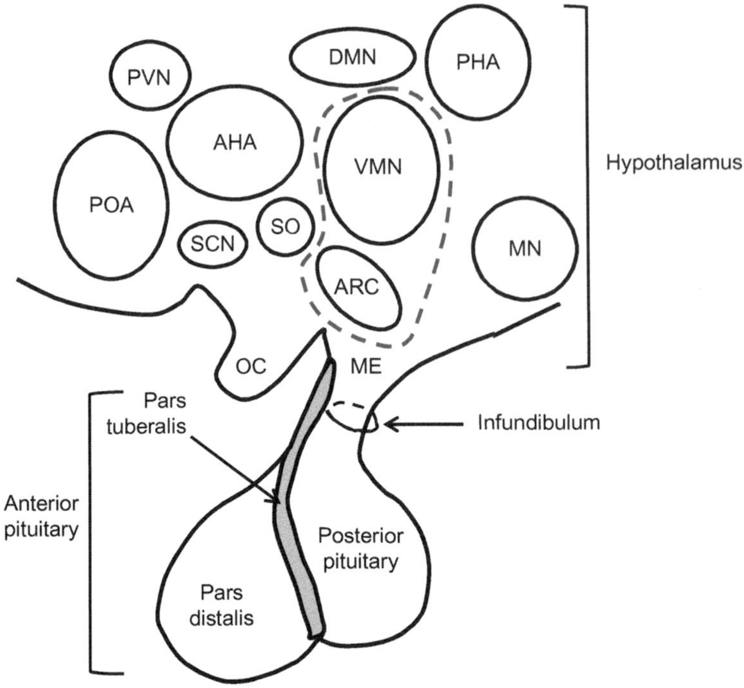

Figure 9.4 Hypothalamic and pituitary anatomy. Sagittal view of mammalian hypothalamic and pituitary anatomy. The mediobasal hypothalamus is encompassed by red dashed lines. The pars tuberalis, part of the anterior pituitary, is shaded in gray. AHA, anterior hypothalamic area; ARC, arcuate nucleus; DMN, dorsomedial nucleus; ME, median eminence; MN, mammillary nuclei; OC, optic chiasm; POA, preoptic area; PHA, posterior hypothalamic area; PVN, paraventricular nucleus; SCN, suprachiasmatic nucleus; SO, supraoptic nucleus; VMN, ventromedial nucleus. (For interpretation of the references to color in this figure legend, the reader is referred to the online version of this chapter.)

undergo a controlled reduction in body temperature, heart rate, and metabolic rate to conserve energy (Fig. 9.5A; Carey et al., 2003; Geiser, 2004). Hibernation is widespread in Class Mammalia suggesting that the ability to hibernate is an ancestral mammalian trait (Carey et al., 2003). The hibernation phenotype is variable across species, with the Arctic ground squirrel (*Urocitellus parryii*) on the extreme end of the spectrum with body temperatures recorded below 0 °C (Barnes, 1989).

During obligate hibernation, mammals go through periods of torpor, ranging from days to weeks, where body temperature is only a few degrees above the ambient temperature and metabolic rate is 2–4% of normal

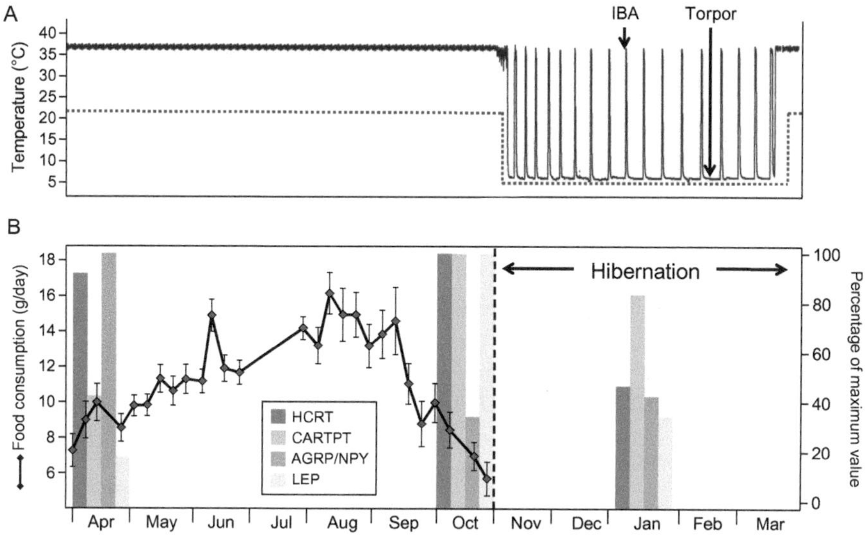

Figure 9.5 Food consumption and satiety signaling in hibernation. (A) Body temperature of a seasonal hibernator. The solid black line displays a representative trace of annual core body temperature taken from a transmitter implanted in a thirteen-lined ground squirrel. The red dashed line represents ambient temperature in the environmental chamber. During hibernation (November 1 to March 31), the ambient temperature is lowered to 5 °C, food is removed, and the animals are kept in constant darkness. During this time squirrels cycle through bouts of torpor and interbout arousals (IBAs). IBAs represent brief returns to normothermia despite low ambient temperature. (B) Food consumption and differential expression of orexigenic (HCRT, AGRP/NPY) and anorexigenic (CARTPT, LEP) transcripts over the hibernation season. Food consumption (g/day) is represented by the black line. Each point represents average daily consumption over a week ($n=11$). During the hibernation period, food is removed, so no consumption is recorded. Error bars are standard error of the mean. Expression values of orexigenic and anorexigenic genes are shown as the percentage of the maximum value. HCRT (blue), AGRP/NPY (green), and CARTPT (orange), are taken from Illumina HiSeq 2000 transcriptome sequencing of the thirteen-lined ground squirrel hypothalamus during October, Hibernation, and April ($n=3$ for each point) (Schwartz, Hampton, & Andrews, 2013). LEP (yellow) is from the same method of sequencing of the white adipose tissue ($n=2$ for each point). Torpor and IBA were analyzed separately, but there was no difference between them for all four genes, so they were combined into one point (Hibernation). AGRP, agouti-related protein; CARTPT, cocaine and amphetamine regulated transcript prepropeptide; HCRT, hypocretin (orexin) neuropeptide precursor; IBA, interbout arousal; LEP, leptin; NPY, neuropeptide Y. (For interpretation of the references to color in this figure legend, the reader is referred to the online version of this chapter.)

(Fig. 9.5A; Carey et al., 2003). These bouts of torpor are interspersed with brief returns to normal body temperature and metabolic rate, called inter-bout arousals (IBAs) that typically last less than 1 day. The reason for these energy costly returns to normothermia is not known, but the IBAs occur regularly throughout the hibernation season, suggesting another specific timing mechanism is in place to coordinate them.

Hibernation of small mammals is more feasible to study in a lab environment than migration, so more is known about the molecular changes underlying the physiology of this phenotype. Several examples are described below.

4.1. Satiety and feeding behavior

Obligate hibernation is circannually timed, with a period of preparation preceding the onset. During this preparation, mammals are hyperphagic (Barnes & Mrosovsky, 1974) and body weight increases dramatically due to the accumulation of white adipose stores (Jameson & Mead, 1964). Roche 454 sequencing of the white adipose tissue transcriptome in thirteen-lined ground squirrels revealed several genes in various tissues important in this preparation period. These include Acyl CoA desaturase (*ACOD*) and adipocyte fatty acid binding protein 4 (*FABP4*), which are involved in lipogenesis and are highly expressed in August when the squirrels are increasing their fuel stores in preparation for hibernation (Hampton et al., 2011). In October, there is an increase in leptin (*LEP*) in white adipose tissue, which encodes a satiety signal received by receptors in the hypothalamus (Fig. 9.5B).

In the hypothalamic transcriptome of the same species, orexigenic hypocretin (*HCRT*), agouti-related protein (*AGRP*), and neuropeptide Y (*NPY*) are highly expressed during April when the ground squirrels have emerged from hibernation and are starting to increase their food consumption (Fig. 9.5B; Schwartz et al., 2013). Anorexigenic cocaine- and amphetamine-regulated protein prepropeptide (*CARTPT*) is low during April, but is highly expressed during October, when food consumption is declining (Fig. 9.5B). *HCRT* remains elevated during October, while *AGRP* and *NPY* levels drop at this time (Fig. 9.5B). This preparatory period is extremely important for survival during hibernation, because the accumulated white adipose stores are used as the primary fuel source for the entire hibernation season.

Food availability is a very strong entrainment stimulus during the active season in brown bears (*Ursus arctos*), a hibernating species that is metabolically suppressed, but only exhibits decreases in body temperature to 32–34 °C, and thus is not considered a deep hibernator (Ware, Nelson, Robbins, & Jansen,

2012). Bears shifted their activity patterns according to feeding times in spite of a natural photoperiod, exhibiting a behavioral flexibility that could be necessary to take advantage of unpredictable periods of food availability in order to gain energy reserves for hibernation (Ware et al., 2012). Additionally, similar to the previously mentioned findings in ground squirrel, leptin levels increased in brown bears prior to hibernation onset (Hissa, Hohtola, Tuomala-Saramaki, Laine, & Kallio, 1998).

4.2. Fuel switch

Generally, hibernating mammals do not use food consumption as a source of energy, although there are some instances of food caching (Gillis, Morrison, Zazula, & Hik, 2005). Some ground squirrel species will not eat any food during hibernation, even if it is offered artificially in the lab (Torke & Twente, 1977). Instead, hibernators rely on a metabolic switch from carbohydrates to lipids by catabolizing their accumulated fat stores for energy.

The fuel switch is particularly important in the heart. In hibernation, the heart beats at a much slower rate during torpor bouts: about 5 beats per minute in the thirteen-lined ground squirrel (*Ictidomys tridecemlineatus*) (Hampton, Nelson, & Andrews, 2010). Northern blot analysis showed that two key genes involved in regulating metabolism were upregulated in the heart during hibernation, pancreatic triacylglycerol lipase (*PTL*), which frees fatty acids from triglycerides at low temperatures, and pyruvate dehydrogenase kinase isozyme 4 (*PDK4*), which prevents the conversion of pyruvate to acetyl CoA, thus blocking glucose oxidation (Andrews, Squire, Bowen, & Rollins, 1998). Protein analysis found that PDK4 and PTL protein were also upregulated in the heart during hibernation (Buck, Squire, & Andrews, 2002; Squire, Lowe, Bauer, & Andrews, 2003), along with succinyl CoA-transferase, the rate limiting enzyme in the catabolism of ketone bodies (Russeth, Higgins, & Andrews, 2006). This fuel switch is evident in other tissues as well, including skeletal muscle, which is inactive during torpor but active during IBAs, and white adipose tissue, the primary fuel source for the body (Hampton et al., 2011).

Further evidence of the fuel switch during hibernation comes from the blood. Analysis of hibernator serum indicated that levels of a fat-derived ketone, D-β-hydroxybutyrate (BHB), are elevated during torpor and is the preferential fuel source of the heart and brain during torpor and arousal, even in the presence of glucose (Andrews, Russeth, Drewes, & Henry, 2009). BHB is also elevated during hibernation in the liver (Serkova,

Rose, Epperson, Carey, & Martin, 2007), where it is produced. A key enzyme in the synthesis pathway of BHB, hydroxymethylglutaryl-CoA synthase 2 (HMGCS2), is also upregulated in the liver during hibernation (Epperson, Dahl, & Martin, 2004).

Seasonal molecular changes also occur in the brain to facilitate this metabolic fuel switch. Monocarboxylic acid transporter 1 (MCT1), also known as solute carrier family 16, member 1 (SLC16A1), is a ketone transporter that facilitates transport of BHB into the brain. This transporter is upregulated at the blood–brain barrier during torpor in ground squirrels (Andrews et al., 2009). *SLC16A1* mRNA levels are also upregulated in the cerebral cortex and hypothalamus during hibernation (Schwartz et al., 2013).

4.3. Central nervous system in hibernation

Like the heart, one area of the brain, the hypothalamus, remains active throughout the hibernation season (Bratincsak et al., 2007; Kilduff, Miller, Radeke, Sharp, & Heller, 1990; Kilduff, Sharp, & Heller, 1982). This region of the brain is involved in many aspects important for hibernation and hibernation preparation, including food intake, circadian rhythms, sleep, and thermoregulation. The SCN of the hypothalamus, containing the circadian clock, is a likely candidate for the control of circannual hibernation timing. *c-fos* mRNA, used as a marker for neuronal activation, increases in the SCN during torpor and peaks during arousal (Bitting et al., 1994; Bratincsak et al., 2007), suggesting that it is involved in torpor bout timing. Broad hypothalamic lesions prevented successful hibernation (Satinoff, 1967) and focused lesions specific to the SCN altered hibernation timing in ground squirrels (Ruby, Dark, Heller, & Zucker, 1996), which suggests that the hypothalamus is important for both hibernation induction and maintenance. However, clock gene rhythms (*per1*, *per2*, *bmal1*) are suppressed during hibernation (Revel et al., 2007), so it is unclear how or if the circadian clock is specifically involved in hibernation timing.

In contrast to the hypothalamus, the rest of the brain is relatively quiescent during hibernation. The cerebral cortex is the first area of the brain to lose activity as the animal goes into a torpor bout, followed by the brainstem and reticular formation, thalamus, and limbic system (Heller, 1979). They regain activity upon arousal in the opposite order. The cerebral cortex, hippocampus, and thalamus all exhibit extensive neurite and synaptic plasticity during hibernation, with synapses dissociating during torpor but reconnecting appropriately upon arousal to an IBA (Magariños et al., 2006; Popov et al.,

1992; von der Ohe et al., 2006, 2007; Fig. 9.3B). Synaptic dissociation could help conserve energy (Drew, Rice, Kuhn, & Smith, 2001) and could also serve a neuroprotective purpose (Schwartz et al., 2013). Recent Illumina HiSeq sequencing of the cerebral cortex transcriptome indicates that the brain upregulates transcripts important for plasticity and remodeling prior to hibernation onset (Schwartz et al., 2013), suggesting that this synaptic plasticity is circannually timed as well.

5. POTENTIAL PATHWAYS INVOLVED IN CIRCANNUAL TIMING

As mentioned previously, the hypothalamus controls many parameters important for seasonal events, including feeding/satiety, circadian rhythms, reproduction, sleep, hormone release, and thermoregulation. Thus, it is likely that this brain region is involved in circannual timing, and there are several candidate systems potentially involved in coordinating this timing.

5.1. Thyroid hormone signaling

Thyroid hormone signaling is affected by photoperiod and is involved in the coordination of seasonal reproduction (Nakao, Ono, & Yoshimura, 2008). Light associated with long days induces conversion of thyroxine (T4) to triiodothyronine (T3) in the mediobasal hypothalamus of Japanese quail (*Coturnix japonica*) by upregulation of *DIO2*, the type 2 iodothyronine deiodinase (Yoshimura et al., 2003). This increased T3 stimulates luteinizing hormone, which initiates the expression of thyroid stimulating hormone *(TSH)* and *DIO2* in the pars tuberalis of the anterior pituitary gland (Fig. 9.4; Nakao, Ono, Yamamura, et al., 2008). ICV injection of TSH into quail kept in short day conditions stimulated a long day-like growth of gonads and *DIO2* expression, triggering seasonal breeding (Nakao, Ono, Yamamura, et al., 2008). Similarly in mammals, high *DIO2* expression is found in the blood vessels surrounding the arcuate nucleus of the hypothalamus and in the ependymal cells of the third ventricle of Siberian hamsters (*Phodopus sungorus*) maintained in long day conditions. This expression decreased after melatonin injections (Watanabe et al., 2004). Ependymal cells separate the brain from the cerebrospinal fluid filled ventricles, acting as a barrier and serving a role in transport (Bruni, 1998). *TSH* and *DIO2* were also shown to be upregulated in mouse pars tuberalis and ependymal cells respectively in response to increasing photoperiod (Ono et al., 2008). Therefore, it appears that thyroid hormone signaling plays an important role in seasonal timing in both mammals and birds. Similarly, thyroid hormone signaling also plays a role in the timing of amphibian metamorphosis, where individuals progress through specific developmental stages, discussed in chapter 10.

Thyroid hormone signaling also appears to play a role in hibernation. Levels of thyroxine-binding globulin (*TBH*) are increased during hibernation in the liver of the golden-mantled ground squirrel (*Callospermophilus lateralis*) (Epperson & Martin, 2002). *TBH* binds both T3 and T4. In the hypothalamus of the thirteen-lined ground squirrel, thyrotrophin releasing hormone (*TRH*) expression peaks in October during the prehibernation period (Schwartz et al., 2013). Additionally, *DIO2* expression peaks in April after completion of hibernation, a time when the squirrels are becoming reproductively active. *DIO2* expression is very low in every other time period observed. *CGA*, the alpha subunit of TSH, increases greater than fourfold in April compared to any other collection point. *CRYM*, a thyroid hormone binding protein, also shows seasonal differential expression with decreased expression in October compared to all other points. *TTR* shows minimal expression in the hypothalamus, but is differentially expressed in cerebral cortex, with increased expression in the hibernation collection points compared to outside hibernation. As mentioned previously, *CGA*, *CRYM*, and *TTR* are all differentially expressed in song sparrow hypothalamus according to season as well (Mukai et al., 2009).

5.2. VGF signaling

VGF (nonacronymic) is a protein involved in energy balance and reproduction (Jethwa & Ebling, 2008). It is processed into several small peptides by prohormone convertases and is expressed throughout the brain, although it is most abundant in the hypothalamus. *VGF* mRNA is induced in the arcuate nucleus of the hypothalamus (Fig. 9.4) in Siberian and Syrian hamsters (*Mesocricetus auratus*) switched from a long day photoperiod to short days (Barrett et al., 2005). *VGF* knockout mice are lean and hypermetabolic, and resisted obesity even with genetic predisposition (Hahm et al., 2002). These mice exhibited cold intolerance, decreased fat storage in both white and brown adipose tissue, upregulation of brown adipose uncoupling proteins 1 and 2 (*UCP1-2*), and an increase in the overall amount of mitochondria and in the density of mitochondrial cristae (Watson et al., 2009).

VGF also could play a role in synaptic plasticity. Long-term depression (LTD) could not be induced in hippocampal slices taken from *VGF* knockout mice, although there were no deficits in long-term potentiation (LTP)

(Bozdagi et al., 2008). LTD and LTP are activity-dependent changes in synaptic efficacy resulting from synaptic plasticity—actual changes in the synaptic structure. LTD weakens synaptic strength, while LTP enhances it. The absence of *VGF* resulted in the inability to weaken synaptic strength, suggesting that *VGF* could be important for the reduction of synaptic connectivity seen in the song control system after the breeding season in songbirds, and also in torpor bouts of hibernating mammals.

Hibernation is characterized by changes in energy metabolism and thermoregulation and also involves coordinated synaptic plasticity, making *VGF* a candidate for involvement in the control of this phenotype. In the thirteen-lined ground squirrel brain, *VGF* expression is increased in October in both the cerebral cortex and the hypothalamus (Schwartz et al., 2013). This time point is characterized by decreased food consumption, increased white adipose stores, and onset of shallow torpor bouts. *VGF* in the hypothalamus could be playing a role in the seasonal switch from the hyperphagic behavior of summer and early fall to the hypophagic behavior seen in hibernation. The fact that the same expression pattern is seen in the cerebral cortex suggests that it could also be playing a role in synaptic plasticity.

5.3. Modulation of neuronal cilia

The primary cilium is a cellular organelle found in almost all eukaryotic cells, including neurons (Louvi & Grove, 2011). Each cell has only one primary cilium, a protrusion of the cell membrane covering a cytoskeletal axoneme core consisting of nine microtubule pairs, which provides structural support to the cilium (Fig. 9.6). The axoneme is anchored by a ciliary basal body, which is a modified centriole. There is a transition zone at the base of the primary cilium for selected protein entry via intraflagellar transport. The exact function of neuronal primary cilia is unclear, but these organelles sense and receive signals from the environment and relay that information to the cell (Adams, 2010).

Deficits in primary cilia are the cause of several disorders, deemed ciliopathies, including Bardet–Biedl syndrome (BBS) and Alstrom syndrome, both characterized by obesity (Badano, Mitsuma, Beales, & Katsanis, 2006). BBS is associated with mutations in at least 11 genes (*BBS1–11*) (Mutch & Clément, 2006). *BBS1* has been implicated in fat storage in *C. elegans* (Mak, Nelson, Basson, Johnson, & Ruvkun, 2006). Loss of primary cilia in the brain using conditional alleles of two ciliogenic genes (*Tg737* and *Kif3a*) induced hyperphagia and obesity in adult mice (Davenport et al., 2007). This

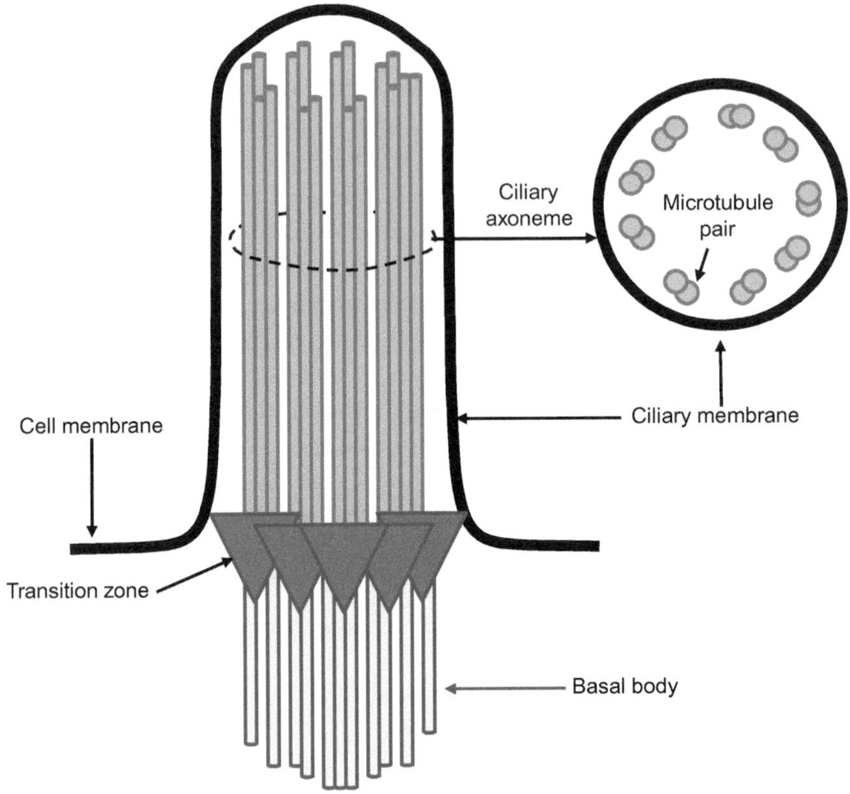

Figure 9.6 Primary cilium structure. *Model is based on Adams (2010) and Louvi and Grove (2011).*

phenotype was induced with the targeted deletion of primary cilia exclusively on pro-opiomelanocortin (POMC)-expressing neurons in the hypothalamus, suggesting that satiety circuits are disrupted. Further, knockout of *BBS2*, *BBS4*, or *BBS6* in mice induces leptin resistance, indicating that BBS proteins are necessary for the reception of this satiety signal (Seo et al., 2009).

Involvement of the neuronal primary cilia in satiety and feeding behavior indicates that they could be very important in regulating several circannually timed events. Both migration and hibernation require periods of hyperphagia and weight gain, along with long periods of hypophagia. In the hypothalamus of the thirteen-lined ground squirrel, several BBS transcripts are differentially expressed across the hibernation season, including *BBS1*, which is elevated during April and October, and *BBS2*, *BBS4*, *BBS5*, and *BBS9*, which are all elevated during hibernation (both torpor and IBA) (Schwartz et al., 2013). Additionally, intraflagellar transcripts *IFT80, IFT81*, and *IFT88* are also upregulated during torpor in the ground squirrel hypothalamus. Ground squirrels will not eat during IBAs even if food is provided artificially in the lab (Torke & Twente, 1977), most likely due to strong circannual satiety signals that are in place to prevent foraging during periods of time when food is scarce and conditions are unfavorable. The expression of BBS and IFT transcripts during hibernation could be playing a role in orchestrating this satiety through modulation of neuronal cilia.

5.4. Retinoic acid signaling

Retinoic acid is a derivative of retinol (vitamin A) which has a wide variety of effects on development, differentiation, and homeostasis (Chambon, 1996). Retinoic acid has two receptor families, RXR and RAR, each with three subtypes (α, β, γ). Retinoic acid signaling is affected by photoperiod and could play an important role in circannual timing. Lengthening photoperiod was shown to enhance retinoic acid signaling in mouse hypothalamus (Shearer et al., 2010). Cellular retinoic acid binding protein 1 (*CRBP1*) is downregulated during short photoperiods (Barrett et al., 2006). Retinoic acid receptors (*RXR, RAR*), *CRBP1*, and *CRBP2* are upregulated in the arcuate nucleus of the hypothalamus in Siberian hamsters when they are switched from short to long photoperiods (Ross et al., 2005). *RXR* appears to be specifically associated with the changes in body weight (Ross et al., 2004).

Retinoic acid signaling is also important for synaptic plasticity. Retinol deficiency resulted in a loss of LTP and LTD in mouse hippocampus, which was reversed by direct application of retinol or retinoic acid (Misner et al., 2001). Deficiency of *RARβ* eliminates LTP and LTD in mouse hippocampus, while *RXRγ* deficiency results in a loss of LTD but not LTP (Chiang et al., 1998), similar to the *VGF* knockout mice discussed above. Retinoic acid and RARα have also been shown to play a role in enhancing synaptic scaling, a method of homeostatic plasticity employed by individual neurons to stabilize their firing rates (Aoto, Nam, Poon, Ting, & Chen, 2008). Synaptic plasticity is a characteristic of several circannual rhythms discussed here, including hibernation and seasonal birdsong.

In the thirteen-lined ground squirrel hypothalamus, several retinoic acid signaling-related genes are differentially expressed across the hibernation season (Schwartz et al., 2013). *RXRα* and *RXRγ* are both elevated during the prehibernation October collection point. *CRBP1* is elevated in April compared to all other collection points. The gene stimulated by retinoic acid gene 6 homolog (*STRA6*) is also elevated in April, while the gene encoding for retinoic acid induced 1 (*RAI1*) is upregulated in both April and October. There are also several transcripts associated with retinol, the precursor to retinoic acid, including retinol binding protein 1 (*RBP1*), and retinol dehydrogenase 11 (*RDH11*). *RBP1* is involved in the transport of retinol and is elevated in April. *RDH11* is elevated during hibernation, and interestingly is involved in adaptation to dark in the retina (Kasus-Jacobi et al., 2005), suggesting a potential role in transducing photoperiod.

5.5. Melatonin and clock genes

Melatonin is produced and released by the pineal gland at night and is inhibited during the day by light via the SCN, which receives a direct photic input from the retina by the RHT. This photoperiodic input entrains the circadian clock to the surrounding environment. A suite of clock genes in the SCN (*Per, Cry, Bmal1, Clk, Rev-Erbα, RORA, DEC1*) form autoregulatory feedback loops that generate the 24-h expression patterns that drive circadian rhythms in all cells (Fig. 9.7; Kawamoto et al., 2004; Ko & Takahashi, 2006). Because the nightly duration of melatonin is representative of the length of the night, melatonin is an excellent molecular indicator for time of year. Additionally, light input to the SCN also varies according to seasonal time, so the circadian clock could also be playing a role in interpreting circannual time.

Melatonin receptors are found throughout the brain, but the most concentrated areas are the pars tuberalis and parts of the hypothalamus (Morgan, Barrett, Howell, & Helliwell, 1994), supporting the role of melatonin in seasonal behavior. Melatonin receptors are also found in the song control system of songbirds (Whitfield-Rucker & Cassone, 1996), and as mentioned previously, exogenous melatonin can stimulate or inhibit singing behavior and the associated neural changes (Cassone et al., 2008). Untreated pinealectomized pied flycatchers (*Ficedula hypoleuca pallas*) failed to orient in the correct direction for migration, but pinealectomized flycatchers that had received nightly melatonin injections exhibited the correct migratory orientation (Schneider, Thalau, Semm, & Wiltschko, 1994). Binding of radiolabeled melatonin was decreased in the pars tuberalis of hibernating ground squirrels (Stanton, Siuciak, Dubocovich, & Krause, 1991), most likely due to a decrease in receptor number, suggesting that the melatonin system is seasonally regulated.

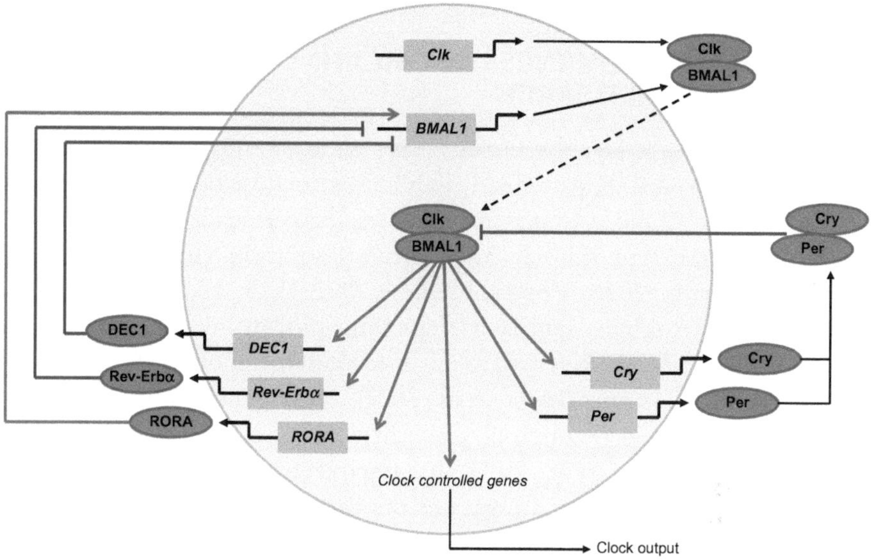

Figure 9.7 Mammalian circadian clock genes. Clk and Bmal1 dimerize and drive the expression of DEC1, Rev-erbα, RORA, Cry, and Per, along with clock controlled genes that drive rhythmic processes of the body. Per and Cry then dimerize and feed back to inhibit Clk and Bmal1. DEC1 and Rev-erbα inhibit transcription of Bmal1, while RORA drives transcription of Bmal1. The rhythmicity of these autoregulatory feedback loops drives the circadian rhythms in all cells. The master circadian clock is located in the suprachiasmatic nucleus of the hypothalamus and is entrained to the environmental photoperiod through a direct input from the retina. Bmal1, brain and muscle aryl hydrocarbon receptor nuclear translocator (ARNT)-like; Clk, Clock; Cry, Cryptochrome; DEC1, differentially expressed in chondrocytes 1; Per, Period; Rev-erbα, nuclear receptor subfamily 1; group D; member 1; RORA, retinoic acid receptor-related orphan receptor alpha. *Model is based on Kawamoto et al. (2004) and Ko and Takahashi (2006).* (For color version of this figure, the reader is referred to the online version of this chapter.)

Circadian clock genes could also be playing a role in seasonal timing. Clock genes were constantly expressed throughout hibernation in European hamsters (*Cricetus cricetus* L.) but their 24-h rhythm was eliminated (Revel et al., 2007). Specifically, *Per1* was elevated during hibernation to daytime levels of normothermic hamsters, while *Per2* was expressed at approximately nighttime levels. *Bmal1* expression was intermediate between day and night levels. In the thirteen-lined ground squirrel hypothalamus, two clock genes were differentially expressed: *BMAL1* (*ARNTL*) and *DEC1* (*BHLHE40*). *BMAL1* exhibits low expression in April compared to every other collection point, while *DEC1* is elevated during torpor (Schwartz et al., 2013).

6. RECENT ADVANCES IN CIRCANNUAL TIMING RESEARCH

Advances in molecular biology have allowed substantial progress in seasonal timing research, particularly in hibernation. Use of two dimensional gel electrophoresis and mass spectrometry has uncovered changes in the proteome of hibernating thirteen-lined ground squirrels (Epperson, Rose, Carey, & Martin, 2010; Epperson, Rose, Russel, et al., 2010; Russeth et al., 2006). An *in vivo* approach employing proton magnetic resonance spectroscopy (^{1}H MRS) was used to measure brain metabolites during hibernation in ground squirrels (Henry et al., 2007). Next-generation sequencing, including Roche 454 (Hampton et al., 2011) and more recently Illumina HiSeq 2000 (Schwartz et al., 2013), has uncovered transcriptional changes during hibernation in several tissues of the thirteen-lined ground squirrel. Additionally, microarrays have been recently utilized to analyze seasonal gene expression changes in song control regions in sparrows (Mukai et al., 2009; Thompson et al., 2012).

Further advancing research in circannual timing, the Broad Institute of MIT and Harvard (http://www.broad.mit.edu) has sequenced the genomes of four hibernating mammals over the last decade, including *Ictidomys* (formerly *Spermophilus*) *tridecemlineatus* (thirteen-lined ground squirrel), *Myotis lucifugus* (little brown bat), *Eptesicus fuscus* (big brown bat), and *Erinaceus europaeus* (common hedgehog). There are few genomic resources available for migrating bird species, although recently a genome-wide single nucleotide polymorphism set was developed for the barnacle goose (*Branta leucopsis*) (Jonker et al., 2012). The genome of the zebra finch (*Taeniopygia guttata*) has been sequenced as well (Warren et al., 2010), and while this is not a seasonal breeding/singing species, it does exhibit plasticity associated with song learning in the brain during juvenile development (Brenowitz & Beecher, 2005), which could provide clues about the pathways underlying seasonal plasticity in songbird brains.

In addition to recent technological advancements and approaches making it easier to examine molecular changes in nonmodel organisms, there have also been advances in genetic manipulations of nonmodel organisms used in seasonal studies. Recently, we published a study using adenovirus to overexpress the metabolic hormone *FGF21* at several different points during the year in thirteen-lined ground squirrels (Nelson et al., 2013). This transgenic approach sets the stage for similar studies in nonmodel organisms where generating a knockout animal or obtaining a specific cell line has been difficult or impossible.

Finally, advancements in husbandry resources have made keeping nonmodel experimental animals in the laboratory a less risky venture. For example, a wealth of information exists on caring for a captive colony of thirteen-lined ground squirrels, along with information on wild capture and captive breeding (Merriman, Lahvis, Jooss, Gesicki, & Schill, 2012; Vaughan, Gruber, Michalski, Seidling, & Schlink, 2006). Recently published husbandry practices for bats are also available (Giannini, 2011). Similarly, husbandry information exists for passerine birds (order Passeriformes), which includes many migrating species and seasonal singers (Bateson & Feenders, 2010). These resources provide an important foundation to help scientists focus on using these valuable species to address research questions without having to devote as much time to establishing protocols for animal care and colony maintenance.

7. CONCLUSIONS: FUTURE DIRECTIONS AND POTENTIAL APPLICATIONS

As the previous section details, advances in molecular tools, particularly for nonmodel organisms, will greatly improve and enhance our ability to study seasonal behaviors and the circannual timing underlying them. The value of nonmodel organisms to the study of seasonal phenotypes needs to be better understood and even heralded, particularly because these organisms can find answers to questions ill-suited to traditional model organisms (Bolker, 2012). Taking advantage of unique phenotypes like hibernation could greatly benefit several areas related to human health (e.g., Klein, Wendroth, Drewes, & Andrews, 2010; Mulier, Lexcen, Luzcek, Greenberg, & Beilman, 2011). Uncovering some of the genes and pathways underlying the induction and maintenance of hibernation in particular could benefit research in obesity, neuroprotection, cardiovascular disease, muscle disuse atrophy, sleep, and seasonal disorders like SAD. Similar benefits could arise from studying the molecular genetics of migrating birds or seasonally singing birds. In the words of Jessica Bolker: There's more to life than rats and flies (Bolker, 2012).

ACKNOWLEDGMENTS

During the writing of this chapter, the authors were supported by NIH 1RC2HL101625-01 (to MTA) and 1F32NS077643-01A1 (to CS), and U.S. Army Medical Research and Materiel Command W81XWH-11-1-0409 (to MTA).

REFERENCES

Adams, M. (2010). The primary cilium: An orphan organelle finds a home. *Nature Education, 3*, 54.

Andrews, M. T., Russeth, K. P., Drewes, L. R., & Henry, P. (2009). Adaptive mechanisms regulate preferred utilization of ketones in the heart and brain of a hibernating mammal during arousal from torpor. *American Journal of Physiology. Regulatory, Integrative and Comparative Physiology, 296*, R383–R393.

Andrews, M. T., Squire, T. L., Bowen, C. M., & Rollins, M. B. (1998). Low-temperature carbon utilization is regulated by novel gene activity in the heart of a hibernating mammal. *Proceedings of the National Academy of Sciences of the United States of America, 95*, 8392–8397.

Aoto, J., Nam, C. I., Poon, M. M., Ting, P., & Chen, L. (2008). Synaptic signaling by all-trans retinoic acid in homeostatic synaptic plasticity. *Neuron, 60*, 308–320.

Badano, J. L., Mitsuma, N., Beales, P. L., & Katsanis, N. (2006). The ciliopathies: An emerging class of human genetic disorders. *Annual Review of Genomics and Human Genetics, 7*, 125–148.

Barnes, B. M. (1989). Freeze avoidance in a mammal: Body temperatures below 0°C in an Arctic hibernator. *Science, 244*, 1593–1595.

Barnes, D., & Mrosovsky, N. (1974). Body weight regulation in ground squirrels and hypothalamically lesioned rats: Slow and sudden set point changes. *Physiology & Behavior, 12*, 251–258.

Barrett, P., Ivanova, E., Graham, E. S., Ross, A. W., Wilson, D., Plé, H., et al. (2006). Photoperiodic regulation of cellular retinoic acid-binding protein 1, GPR50 and nestin in tanycytes of the third ventricle ependymal layer of the Siberian hamster. *Journal of Endocrinology, 191*, 687–698.

Barrett, P., Ross, A. W., Balik, A., Littlewood, P. A., Mercer, J. G., Moar, K. M., et al. (2005). Photoperiodic regulation of histamine H3 receptor and VGF messenger ribonucleic acid in the arcuate nucleus of the Siberian hamster. *Endocrinology, 146*, 1930–1939.

Bateson, M., & Feenders, G. (2010). The use of passerine bird species in laboratory research: Implications of basic biology for husbandry and welfare. *ILAR Journal, 51*, 394–408.

Bitting, L., Sutin, E. L., Watson, F. L., Leard, L. E., O'Hara, B. F., Craig Heller, H., et al. (1994). *C-fos* mRNA increases in the ground squirrel suprachiasmatic nucleus during arousal from hibernation. *Neuroscience Letters, 165*, 117–121.

Bolker, J. (2012). Model organisms: There's more to life than rats and flies. *Nature, 491*, 31–33.

Bozdagi, O., Rich, E., Tronel, S., Sadahiro, M., Patterson, K., Shapiro, M. L., et al. (2008). The neurotrophin-inducible gene Vgf regulates hippocampal function and behavior through a brain-derived neurotrophic factor-dependent mechanism. *Journal of Neuroscience, 28*, 9857–9869.

Bratincsak, A., McMullen, D., Miyake, S., Toth, Z. E., Hallenbeck, J., & Palkovits, M. (2007). Spatial and temporal activation of brain regions in hibernation: *c-fos* expression during the hibernation bout in thirteen-lined ground squirrel. *The Journal of Comparative Neurology, 505*, 443–458.

Brenowitz, E. A., & Beecher, M. D. (2005). Song learning in birds: Diversity and plasticity, opportunities and challenges. *Trends in Neurosciences, 28*, 127–132.

Brenowitz, E. A., Margoliash, D., & Nordeen, K. W. (1997). An introduction to birdsong and the avian song system. *Journal of Neurobiology, 33*, 495–500.

Bruni, J. (1998). Ependymal development, proliferation, and functions: A review. *Microscopy Research and Technique, 41*, 2–13.

Buck, M. J., Squire, T. L., & Andrews, M. T. (2002). Coordinate expression of the PDK4 gene: A means of regulating fuel selection in a hibernating mammal. *Physiological Genomics, 8*, 5–13.

Caprioli, M., Ambrosini, R., Boncoraglio, G., Gatti, E., Romano, A., Romano, M., et al. (2012). Clock gene variation is associated with breeding phenology and maybe under directional selection in the migratory barn swallow. *PLoS One*, *7*, e35140.

Carey, H. V., Andrews, M. T., & Martin, S. L. (2003). Mammalian hibernation: Cellular and molecular responses to depressed metabolism and low temperature. *Physiological Reviews*, *83*, 1153–1181.

Cassone, V. M., Bartell, P. A., Earnest, B. J., & Kumar, V. (2008). Duration of melatonin regulates seasonal changes in song control nuclei of the house sparrow, *Passer domesticus*: Independence from gonads and circadian entrainment. *Journal of Biological Rhythms*, *23*, 49–58.

Chambon, P. (1996). A decade of molecular biology of retinoic acid receptors. *The FASEB Journal*, *10*, 940–954.

Chiang, M. Y., Misner, D., Kempermann, G., Schikorski, T., Giguère, V., Sucov, H. M., et al. (1998). An essential role for retinoid receptors RARβ and RXRγ in long-term potentiation and depression. *Neuron*, *21*, 1353–1361.

Davenport, J. R., Watts, A. J., Roper, V. C., Croyle, M. J., van Groen, T., Wyss, J. M., et al. (2007). Disruption of intraflagellar transport in adult mice leads to obesity and slow-onset cystic kidney disease. *Current Biology*, *17*, 1586–1594.

Dingle, H., & Drake, V. A. (2007). What is migration? *Bioscience*, *57*, 113–121.

Drew, K., Rice, M., Kuhn, T., & Smith, M. (2001). Neuroprotective adaptations in hibernation: Therapeutic implications for ischemia-reperfusion, traumatic brain injury and neurodegenerative diseases. *Free Radical Biology & Medicine*, *31*, 563–573.

Drickamer, L., Vessey, S., & Jakob, E. (2002). *Animal behavior: Mechanisms, ecology, evolution*. New York: McGraw-Hill.

Egevang, C., Stenhouse, I. J., Phillips, R. A., Petersen, A., Fox, J. W., & Silk, J. R. D. (2010). Tracking of Arctic terns *Sterna paradisaea* reveals longest animal migration. *Proceedings of the National Academy of Sciences of the United States of America*, *107*, 2078–2081.

Epperson, L. E., Dahl, T. A., & Martin, S. L. (2004). Quantitative analysis of liver protein expression during hibernation in the golden-mantled ground squirrel. *Molecular & Cellular Proteomics*, *3*, 920–933.

Epperson, L. E., & Martin, S. L. (2002). Quantitative assessment of ground squirrel mRNA levels in multiple stages of hibernation. *Physiological Genomics*, *10*, 93–102.

Epperson, L. E., Rose, J. C., Carey, H. V., & Martin, S. L. (2010). Seasonal proteomic changes reveal molecular adaptations to preserve and replenish liver proteins during ground squirrel hibernation. *American Journal of Physiology. Regulatory, Integrative and Comparative Physiology*, *298*, R329–R340.

Epperson, L. E., Rose, J. C., Russel, R. L., Nikrad, M. P., Carey, H. V., & Martin, S. L. (2010). Seasonal protein changes support rapid energy production in hibernator brainstem. *Journal of Comparative Physiology B*, *180*, 599–617.

Geiser, F. (2004). Metabolic rate and body temperature reduction during hibernation and daily torpor. *Annual Review of Physiology*, *66*, 239–274.

Giannini, N. P. (2011). How to study bats: Second edition of a classic some twenty years later. *Journal of Mammalian Evolution*, *18*, 219–221.

Gillis, E. A., Morrison, S. F., Zazula, G. D., & Hik, D. S. (2005). Evidence for selective caching by arctic ground squirrels living in alpine meadows in the Yukon. *Arctic*, *58*, 354–360.

Guglielmo, C. G., Haunerland, N. H., Hochachka, P. W., & Williams, T. D. (2002). Seasonal dynamics of flight muscle fatty acid binding protein and catabolic enzymes in a migratory shorebird. *American Journal of Physiology. Regulatory, Integrative and Comparative Physiology*, *282*, R1405–R1413.

Guglielmo, C., Williams, T., Zwingelstein, G., Brichon, G., & Weber, J. M. (2002). Plasma and muscle phospholipids are involved in the metabolic response to long-distance migration in a shorebird. *Journal of Comparative Physiology B*, *172*, 409–417.

Hahm, S., Fekete, C., Mizuno, T. M., Windsor, J., Yan, H., Boozer, C. N., et al. (2002). VGF is required for obesity induced by diet, gold thioglucose treatment, and agouti and is differentially regulated in pro-opiomelanocortin-and neuropeptide Y-containing arcuate neurons in response to fasting. *Journal of Neuroscience*, *22*, 6929–6938.

Hampton, M., Melvin, R. G., Kendall, A. H., Kirkpatrick, B. R., Peterson, N., & Andrews, M. T. (2011). Deep sequencing the transcriptome reveals seasonal adaptive mechanisms in a hibernating mammal. *PLoS One*, *6*, e27021.

Hampton, M., Nelson, B. T., & Andrews, M. T. (2010). Circulation and metabolic rates in a natural hibernator: An integrative physiological model. *American Journal of Physiology. Regulatory, Integrative and Comparative Physiology*, *299*, R1478–R1488.

Hannibal, J., Ding, J. M., Chen, D., Fahrenkrug, J., Larsen, P. J., Gillette, M. U., et al. (1997). Pituitary adenylate cyclase-activating peptide (PACAP) in the retinohypothalamic tract: A potential daytime regulator of the biological clock. *Journal of Neuroscience*, *17*, 2637–2644.

Heller, H. C. (1979). Hibernation: Neural aspects. *Annual Review of Physiology*, *41*, 305–321.

Henry, P., Russeth, K. P., Tkac, I., Drewes, L. R., Andrews, M. T., & Gruetter, R. (2007). Brain energy metabolism and neurotransmission at near-freezing temperatures: *In vivo* ^{1}H MRS study of a hibernating mammal. *Journal of Neurochemistry*, *101*, 1505–1515.

Hissa, R., Hohtola, E., Tuomala-Saramaki, T., Laine, T., & Kallio, H. (1998). Seasonal changes in fatty acids and leptin contents in the plasma of the European brown bear (*Ursus arctos arctos*). *Annales Zoologici Fennici*, *35*, 215–224.

Jameson, E., Jr., & Mead, R. A. (1964). Seasonal changes in body fat, water and basic weight in Citellus lateralis, Eutamias speciosus and E. amoenus. *Journal of Mammalogy*, *45*, 359–365.

Jethwa, P. H., & Ebling, F. J. P. (2008). Role of VGF-derived peptides in the control of food intake, body weight and reproduction. *Neuroendocrinology*, *88*, 80–87.

Jonker, R. M., Zhang, Q., Van Hooft, P., Loonen, M. J. J. E., Van der Jeugd, H. P., Crooijmans, R. P. M. A., et al. (2012). The development of a genome side SNP set for the barnacle goose *Branta leucopsis*. *PLoS One*, 7, e38412.

Kasus-Jacobi, A., Ou, J., Birch, D. G., Locke, K. G., Shelton, J. M., Richardson, J. A., et al. (2005). Functional characterization of mouse RDH11 as a retinol dehydrogenase involved in dark adaptation in vivo. *Journal of Biological Chemistry*, *280*, 20413–20420.

Kawamoto, T., Noshiro, M., Sato, F., Maemura, K., Takeda, N., Nagai, R., et al. (2004). A novel autofeedback loop of Dec1 transcription involved in circadian rhythm regulation. *Biochemical and Biophysical Research Communications*, *313*, 117–124.

Kilduff, T. S., Miller, J. D., Radeke, C. M., Sharp, F. R., & Heller, H. C. (1990). ^{14}C-2-deoxyglucose uptake in the ground squirrel brain during entrance to and arousal from hibernation. *Journal of Neuroscience*, *10*, 2463–2475.

Kilduff, T. S., Sharp, F. R., & Heller, H. C. (1982). [^{14}C]2-deoxyglucose uptake in ground squirrel brain during hibernation. *Journal of Neuroscience*, *2*, 143–157.

Klein, A. H., Wendroth, S. M., Drewes, L. R., & Andrews, M. T. (2010). Small-volume D-β-hydroxybutyrate solution infusion increases survivability of lethal hemorrhagic shock in rats. *Shock*, *34*, 565–572.

Ko, C. H., & Takahashi, J. S. (2006). Molecular components of the mammalian circadian clock. *Human Molecular Genetics*, *15*, R271–R277.

Larkin, R. P., Griffin, D. R., Torre-Bueno, J. R., & Teal, J. (1979). Radar observations of bird migration over the western North Atlantic Ocean. *Behavioral Ecology and Sociobiology*, *4*, 225–264.

Louvi, A., & Grove, E. A. (2011). Cilia in the CNS: The quiet organelle claims center stage. *Neuron*, *69*, 1046–1060.

Lowery, G. H. (1946). Evidence of trans-Gulf migration. *Auk*, *63*, 175–211.

Magariños, A. M., McEwen, B. S., Saboureau, M., & Pevet, P. (2006). Rapid and reversible changes in intrahippocampal connectivity during the course of hibernation in European hamsters. *Proceedings of the National Academy of Sciences of the United States of America, 103*, 18775–18780.

Mak, H. Y., Nelson, L. S., Basson, M., Johnson, C. D., & Ruvkun, G. (2006). Polygenic control of *Caenorhabditis elegans* fat storage. *Nature Genetics, 38*, 363–368.

Merriman, D. K., Lahvis, G., Jooss, M., Gesicki, J. A., & Schill, K. (2012). Current practices in a captive breeding colony of 13-lined ground squirrels (*Ictidomys tridecemlineatus*). *Lab Animal, 41*, 315.

Misner, D., Jacobs, S., Shimizu, Y., De Urquiza, A., Solomin, L., Perlmann, T., et al. (2001). Vitamin A deprivation results in reversible loss of hippocampal long-term synaptic plasticity. *Proceedings of the National Academy of Sciences of the United States of America, 98*, 11714–11719.

Moore, F. R., Woodrey, M. S., Buler, J. J., Woltmann, S., & Simons, T. R. (2005). Understanding the stopover of migratory birds: A scale dependent approach. In *Bird conservation implementation and integration in the Americas: Proceedings of the third international partners in flight conference, Asilomar, CA*.

Morgan, P. J., Barrett, P., Howell, H. E., & Helliwell, R. (1994). Melatonin receptors: Localization, molecular pharmacology and physiological significance. *Neurochemistry International, 24*, 101–146.

Mueller, J. C., Pulido, F., & Kempenaers, B. (2011). Identification of a gene associated with avian migratory behaviour. *Proceedings of the Royal Society B, 278*, 2848–2856.

Mukai, M., Replogle, K., Drnevich, J., Wang, G., Wacker, D., Band, M., et al. (2009). Seasonal differences of gene expression profiles in song sparrow (*Melospiza melodia*) hypothalamus in relation to territorial aggression. *PLoS One, 4*, e8182.

Mulier, K. E., Lexcen, D. R., Luzcek, E., Greenberg, J. J., & Beilman, G. J. (2011). Treatment with beta-hydroxybutyrate and melatonin is associated with improved survival in a porcine model of hemorrhagic shock. *Resuscitation, 83*, 253–258.

Mutch, D. M., & Clément, K. (2006). Unraveling the genetics of human obesity. *PLoS Genetics, 2*, e188.

Nagy, A. D., & Csernus, V. J. (2007). The role of PACAP in the control of circadian expression of clock genes in the chicken pineal gland. *Peptides, 28*, 1767–1774.

Nakao, N., Ono, H., Yamamura, T., Anraku, T., Takagi, T., Higashi, K., et al. (2008). Thyrotrophin in the pars tuberalis triggers photoperiodic response. *Nature, 452*, 317–322.

Nakao, N., Ono, H., & Yoshimura, T. (2008). Thyroid hormones and seasonal reproductive neuroendocrine interactions. *Reproduction, 136*, 1–8.

Nelson, B. T., Ding, X., Gerard, R. D., Boney-Montoya, J., Kliewer, S. A., & Andrews, M. T. (2013). Metabolic hormone FGF21 is induced in ground squirrels during hibernation but its overexpression is not sufficient to cause torpor. *PLoS One, 8*, e53574.

Nottebohm, F. (1980). Testosterone triggers growth of brain vocal control nuclei in adult female canaries. *Brain Research, 189*, 429–436.

Nottebohm, F. (1981). A brain for all seasons: Cyclical anatomical changes in song control nuclei of the canary brain. *Science, 214*, 1368–1370.

Nottebohm, F. (2005). The neural basis of birdsong. *PLoS Biology, 3*, e164.

Odum, E. P. (1960). Premigratory hyperphagia in birds. *American Journal of Clinical Nutrition, 8*, 621–629.

Ono, H., Hoshino, Y., Yasuo, S., Watanabe, M., Nakane, Y., Murai, A., et al. (2008). Involvement of thyrotropin in photoperiodic signal transduction in mice. *Proceedings of the National Academy of Sciences of the United States of America, 105*, 18238–18242.

Piersma, T. (1998). Phenotypic flexibility during migration: Optimization of organ size contingent on the risks and rewards of fueling and flight? *Journal of Avian Biology, 29*, 511–520.

Popov, V., Bocharova, L., & Bragin, A. (1992). Repeated changes of dendritic morphology in the hippocampus of ground squirrels in the course of hibernation. *Neuroscience, 48*, 45–51.

Ramenofsky, M., Savard, R., & Greenwood, M. (1999). Seasonal and diel transitions in physiology and behavior in the migratory dark-eyed junco. *Comparative Biochemistry and Physiology. Part A, Molecular & Integrative Physiology, 122*, 385–397.

Revel, F. G., Herwig, A., Garidou, M. L., Dardente, H., Menet, J. S., Masson-Pévet, M., et al. (2007). The circadian clock stops ticking during deep hibernation in the European hamster. *Proceedings of the National Academy of Sciences of the United States of America, 104*, 13816–13820.

Ross, A. W., Bell, L. M., Littlewood, P. A., Mercer, J. G., Barrett, P., & Morgan, P. J. (2005). Temporal changes in gene expression in the arcuate nucleus precede seasonal responses in adiposity and reproduction. *Endocrinology, 146*, 1940–1947.

Ross, A. W., Webster, C. A., Mercer, J. G., Moar, K. M., Ebling, F. J., Schuhler, S., et al. (2004). Photoperiodic regulation of hypothalamic retinoid signaling: Association of retinoid X receptor γ with body weight. *Endocrinology, 145*, 13–20.

Ruby, N. F., Dark, J., Heller, H. C., & Zucker, I. (1996). Ablation of suprachiasmatic nucleus alters timing of hibernation in ground squirrels. *Proceedings of the National Academy of Sciences of the United States of America, 93*, 9864–9868.

Russeth, K. P., Higgins, L., & Andrews, M. T. (2006). Identification of proteins from non-model organisms using mass spectrometry: Application to a hibernating mammal. *Journal of Proteome Research, 5*, 829–839.

Satinoff, E. (1967). Disruption of hibernation caused by hypothalamic lesions. *Science, 155*, 1031–1033.

Schaub, M., Jenni, L., & Bairlein, F. (2008). Fuel stores, fuel accumulation, and the decision to depart from a migration stopover site. *Behavioral Ecology, 19*, 657–666.

Schneider, T., Thalau, H. P., Semm, P., & Wiltschko, W. (1994). Melatonin is crucial for the migratory orientation of Pied Flycatchers (*Ficedula hypoleuca Pallas*). *Journal of Experimental Biology, 194*, 255–262.

Schwartz, C., Hampton, M., & Andrews, M. T. (2013). Seasonal and regional differences in gene expression in the brain of a hibernating mammal. *PLoS One, 8*, e58427.

Seo, S., Guo, D. F., Bugge, K., Morgan, D. A., Rahmouni, K., & Sheffield, V. C. (2009). Requirement of Bardet–Biedl syndrome proteins for leptin receptor signaling. *Human Molecular Genetics, 18*, 1323–1331.

Serkova, N. J., Rose, J. C., Epperson, L. E., Carey, H. V., & Martin, S. L. (2007). Quantitative analysis of liver metabolites in three stages of the circannual hibernation cycle in 13-lined ground squirrels by NMR. *Physiological Genomics, 31*, 15–24.

Shearer, K. D., Goodman, T. H., Ross, A. W., Reilly, L., Morgan, P. J., & McCaffery, P. J. (2010). Photoperiodic regulation of retinoic acid signaling in the hypothalamus. *Journal of Neurochemistry, 112*, 246–257.

Simonneaux, V., & Ouichou, A. (1993). Pituitary adenylate cyclase-activating polypeptide (PACAP) stimulates melatonin synthesis from rat pineal gland. *Brain Research, 603*, 148–152.

Smith, G. T. (1996). Seasonal plasticity in the song nuclei of wild rufous-sided towhees. *Brain Research, 734*, 79–85.

Smith, G. T., Brenowitz, E. A., Beecher, M. D., & Wingfield, J. C. (1997). Seasonal changes in testosterone, neural attributes of song control nuclei, and song structure in wild songbirds. *Journal of Neuroscience, 17*, 6001–6010.

Smith, G. T., Brenowitz, E. A., Wingfield, J. C., & Baptista, L. F. (2004). Seasonal changes in song nuclei and song behavior in Gambel's white-crowned sparrows. *Journal of Neurobiology, 28*, 114–125.

Squire, T. L., Lowe, M. E., Bauer, V. W., & Andrews, M. T. (2003). Pancreatic triacylglycerol lipase in a hibernating mammal. II. Cold-adapted function and differential expression. *Physiological Genomics*, *16*, 131–140.

Stanton, T. L., Siuciak, J. A., Dubocovich, M. L., & Krause, D. N. (1991). The area of 2-[^{125}I] iodomelatonin binding in the pars tuberalis of the ground squirrel is decreased during hibernation. *Brain Research*, *557*, 285–288.

Tachibana, T., Oikawa, D., Adachi, N., Boswell, T., & Furuse, M. (2007). Central administration of vasoactive intestinal peptide and pituitary adenylate cyclase-activating polypeptide differentially regulates energy metabolism in chicks. *Comparative Biochemistry and Physiology. Part A, Molecular & Integrative Physiology*, *147*, 156–164.

Thompson, C. K., Meitzen, J., Replogle, K., Drnevich, J., Lent, K. L., Wissman, A. M., et al. (2012). Seasonal changes in patterns of gene expression in avian song control brain regions. *PLoS One*, *7*, e35119.

Torke, K. G., & Twente, J. W. (1977). Behavior of *Spermophilus lateralis* between periods of hibernation. *Journal of Mammalogy*, *58*, 385–390.

Tramontin, A. D., & Brenowitz, E. A. (2000). Seasonal plasticity in the adult brain. *Trends in Neurosciences*, *23*, 251–258.

Vaughan, D. K., Gruber, A. R., Michalski, M. L., Seidling, J., & Schlink, S. (2006). Capture, care, and captive breeding of 13-lined ground squirrels, *Spermophilus tridecemlineatus*. *Lab Animal*, *35*, 33–40.

von der Ohe, C. G., Darian-Smith, C., Garner, C., & Heller, H. (2006). Ubiquitous and temperature-dependent neural plasticity in hibernators. *Journal of Neuroscience*, *26*, 10590–10598.

von der Ohe, C. G., Garner, C. C., Darian-Smith, C., & Heller, H. C. (2007). Synaptic protein dynamics in hibernation. *Journal of Neuroscience*, *27*, 84–92.

Ware, J. V., Nelson, O. L., Robbins, C. T., & Jansen, H. T. (2012). Temporal organization of activity in the brown bear (*Ursus arctos*): Roles of circadian rhythms, light and food entrainment. *American Journal of Physiology. Regulatory, Integrative and Comparative Physiology*, *303*, R890–R902.

Warren, W. C., Clayton, D. F., Ellegren, H., Arnold, A. P., Hillier, L. D. W., Künstner, A., et al. (2010). The genome of a songbird. *Nature*, *464*, 757–762.

Watanabe, M., Yasuo, S., Watanabe, T., Yamamura, T., Nakao, N., Ebihara, S., et al. (2004). Photoperiodic regulation of type 2 deiodinase gene in Djungarian hamster: Possible homologies between avian and mammalian photoperiodic regulation of reproduction. *Endocrinology*, *145*, 1546–1549.

Watson, E., Fargali, S., Okamoto, H., Sadahiro, M., Gordon, R., Chakraborty, T., et al. (2009). Analysis of knockout mice suggests a role for VGF in the control of fat storage and energy expenditure. *BMC Physiology*, *9*, 19–39.

Whitfield-Rucker, M. G., & Cassone, V. M. (1996). Melatonin binding in the house sparrow song control system: Sexual dimorphism and the effect of photoperiod. *Hormones and Behavior*, *30*, 528–537.

Yoshimura, T., Yasuo, S., Watanabe, M., Iigo, M., Yamamura, T., Hirunagi, K., et al. (2003). Light-induced hormone conversion of T4 to T3 regulates photoperiodic response of gonads in birds. *Nature*, *426*, 178–181.

CHAPTER TEN

Unliganded Thyroid Hormone Receptor Regulates Metamorphic Timing via the Recruitment of Histone Deacetylase Complexes

Yun-Bo Shi[1]

Section on Molecular Morphogenesis, Program in Cellular Regulation and Metabolism, NICHD, NIH, Bethesda, Maryland, USA

[1]Corresponding author: e-mail address: Shi@helix.nih.gov

Contents

1. Introduction 276
2. TR Function During Frog Development 279
3. Unliganded TR Recruits Corepressor Complexes to Control Metamorphic Timing 282
4. Coactivator Recruitment for Metamorphic Progression 285
5. Temporal Regulation of Organ-Specific Metamorphosis 285
6. Conclusions 291

Acknowledgment 291

References 291

Abstract

Anuran metamorphosis involves a complex series of tissue transformations that change an aquatic tadpole to a terrestrial frog and resembles the postembryonic perinatal period in mammals. Thyroid hormone (TH) plays a causative role in amphibian metamorphosis and its effect is mediated by TH receptors (TRs). Molecular analyses during *Xenopus* development have shown that unliganded TR recruits histone deacetylase (HDAC)-containing N-CoR/SMRT complexes and causes histone deacetylation at target genes while liganded TR leads to increased histone acetylations and altered histone methylations at target genes. Transgenic studies involving mutant TR-cofactors have shown that corepressor recruitment by unliganded TR is required to ensure proper timing of the onset of metamorphosis while coactivator levels influence the rate of metamorphic progression. In addition, a number of factors that can influence cellular free TH levels appear to contribute the timing of metamorphic transformations of different organs by regulating the levels of unliganded vs. liganded TR in an organ-specific manner. Thus, the recruitment of HDAC-containing corepressor complexes by unliganded TR likely controls both the timing of the initiation of metamorphosis and the temporal regulation of organ-specific transformations. Similar mechanisms likely mediate TR function in mammals as the maturation of many organs during

Current Topics in Developmental Biology, Volume 105
ISSN 0070-2153
http://dx.doi.org/10.1016/B978-0-12-396968-2.00010-5

postembryonic development is dependent upon TH and resembles organ metamorphosis in amphibians.

1. INTRODUCTION

Thyroid hormone (TH) plays critical roles in vertebrate development and adult organ function (Atkinson, 1994; Franklyn & Gammage, 1996; Freake & Oppenheimer, 1995; Hetzel, 1989; Lazar, 1993; Shi, 1999; Silva, 1995; Tata, 1993; Yen, 2001). The most dramatic developmental process controlled by TH is the metamorphosis in anurans, when an aquatic, herbivorous tadpole is transformed into a terrestrial, carnivorous frog (Dodd & Dodd, 1976; Gilbert, Tata, & Atkinson, 1996; Shi, 1999). During anuran metamorphosis, different tissues undergo distinct changes (Dodd & Dodd, 1976; Gilbert et al., 1996; Shi, 1999). Some, like the tail and gills, are completely resorbed. Others, such as the limbs, are formed *de novo*. Most of the organs, however, undergo partial, often drastic remodeling involving both larval cell death and adult tissue development. For example, in the larval intestine, the major tissue characteristic of the organ, the larval epithelium, degenerates during metamorphosis by undergoing apoptosis (Dodd & Dodd, 1976; Shi, Hasebe, Fu, Fujimoto, & Ishizuya-Oka, 2011; Shi & Ishizuya-Oka, 1996). Concurrently, the adult epithelial stem cells are formed *de novo* through dedifferentiation of some larval epithelial cells through yet unknown mechanisms. These cells then proliferate and differentiate to form the adult intestinal epithelium with a self-renewing system resembling that in the mammalian adult intestine (Ishizuya-Oka & Shi, 2011; Sun & Shi, 2012).

Extensive studies as early as a century ago have firmly established that TH is the causative agent of amphibian metamorphosis (Allen, 1918; Gilbert et al., 1996; Gudernatch, 1912; Shi, 1999). There are two natural THs, T3 (3, 5, 3′-triiodothyronine) and T4 (3, 5, 3′,5′-tetraiodothyronine, or thyroxine). T4 is biologically less active and often considered as a prehormone, whose deiodination gives to the more active hormone T3. The plasma levels of both T3 and T4 are low or undetectable in embryos or premetamorphic tadpoles and rise to peak at the climax of metamorphosis (Fig. 10.1) (Leloup & Buscaglia, 1977).

Although other vertebrates do not undergo such a complete TH-dependent transformation, TH is known to also play similarly critical

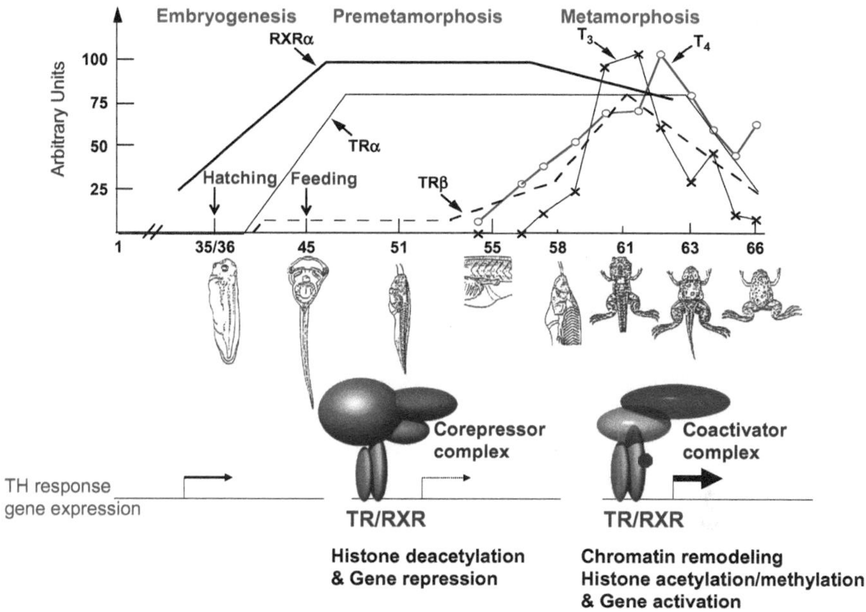

Figure 10.1 A dual function model of TR in frog development. Top: Correlation of the levels of endogenous TH and the mRNAs of TRα, TRβ, and RXRα genes with development in *X. laevis*. The embryos hatches around stage 35 and tadpole feeding begins around stage 45 (Nieuwkoop & Faber, 1956). The plasma concentrations (arbitrary units) of T_3 (crosses) and T_4 (open circles) are from Leloup and Buscaglia (1977). The mRNA levels for TRα (solid line), TRβ (broken line), and RXRα (bold line) are based on published data (Wong & Shi, 1995; Yaoita & Brown, 1990) and for clarity, plotted on arbitrary scales. Bottom: Regulation of TH-inducible genes during frog development. TH-inducible genes are expressed at the basal levels during embryogenesis when the levels of TH and TRs are low. After tadpole feeding begins at stage 45, TRα and RXRα expression increases dramatically. In the absence of the TH, the resulting TR-RXR heterodimers bind to the TREs in the target genes and recruit corepressor complexes to repress gene expression. After stage 55, the rise of endogenous TH concentration causes the release of the corepressor complexes and recruitment of coactivator complexes to the promoters due to its binding to TR. This leads to the activation of TH-inducible genes and metamorphosis. (For color version of this figure, the reader is referred to the online version of this chapter.)

roles in their development (Shi, 2013; Tata, 1993). The most obvious and earliest known abnormalities of human body and behavior associated with TH-deficiency are goiter and cretinism (Hetzel, 1989; Tata, 1993). As in amphibians, there is little TH in the embryonic plasma even though the thyroid gland is developed within the first 12 weeks of gestation in human (Hetzel, 1989; Howdeshell, 2002; Tata, 1993). Significant TH production probably does not occur until 20 weeks or later and in the fetal plasma, TH levels remain low until about 4 months prior to birth. The TH levels reach a peak around birth and remain high several months after birth. This post-embryonic/perinatal period in mammals such as human and mouse mimics anuran metamorphosis in many aspects (Atkinson, 1994; Shi, 1999; Tata, 1993). Among them are the transition from aquatic (living in the amniotic fluid) to terrestrial habitat with air-breathing after birth as the lung matures, the switching of the hemoglobin from the fetal to adult types, the increased albumin expression and the induction of urea cycle enzymes in the liver, the maturation of the intestine to the adult form, skin keratinization, and the developmental progression and restructuring of the central and peripheral nervous systems, etc. Thus, it may not be surprising that TH deficiency during neonatal period leads to irreversible, profound neurological deficit, and mental retardation as well as other defects in human (Hetzel, 1989; Hsu & Brent, 1998; Porterfield & Hendrich, 1993). While the importance of TH in mammalian development has long been established, it is difficult to manipulate the uterus-enclosed mammalian embryos for studying the mechanisms of TH action. This contrasts sharply with the total control of anuran metamorphosis by TH and easiness to manipulate this process in tadpoles or organ/primary cell cultures for functional and mechanistic analyses.

Natural metamorphosis in anurans takes place at precise developmental stages, which are defined based on animal morphology, including the presence or absence of the limb or tail or their morphology, etc. (Dodd & Dodd, 1976; Shi, 1999). In *Xenopus*, the onset of metamorphosis is around stage 55 when endogenous TH becomes detectable (Fig. 10.1) (Dodd & Dodd, 1976; Nieuwkoop & Faber, 1956). Thus, TH synthesis is the primary determinant for the timing of the onset of metamorphosis. TH is synthesized in the thyroid gland as either T3 or T4. In *Xenopus laevis*, the gland first appears in the embryo as a median thickening of the pharyngeal epithelium at the time of tadpole hatching (stages 35/36) (Dodd & Dodd, 1976). This rudiment then develops into a functional larval thyroid gland around stage 53. Subsequently, the plasma TH concentrations increase, reaching peak levels around stage 60, the climax of metamorphosis. Currently, the mechanisms underlying the developmental regulation of the plasma TH levels are unknown. However, as in mammals, many factors appear to be involved. As have been reviewed extensively, these factors include those that influence the growth and maturation of the thyroid gland and hormonal clues that regulate TH synthesis and release from the gland (Denver, 2013; Dodd & Dodd, 1976; Gilbert et al., 1996; Shi, 1999). Regardless of the regulation mechanisms, the synthesis of endogenous TH appears to dictate the timing of metamorphosis initiation.

On the other hand, metamorphosis can be precociously induced in premetamorphic tadpoles as early as at stage 41 for *X. laevis*, only 3 days after fertilization, by the treatment with exogenous TH, while younger tadpoles or embryos are resistant to TH (Shi, 1999; Tata, 1968). Thus, *Xenopus* tadpoles gain competence to respond to TH as early as stage 41, much earlier than the onset of metamorphosis at stage 55. In addition, different organs undergo metamorphosis at distinct developmental stages even though they are exposed to the same circulating plasma TH. Studies over the last two decades have demonstrated a critical role of TH receptor (TR) in both mediating the metamorphic effect of TH and ensuring proper growth and development of different tadpole organs prior to metamorphosis.

2. TR FUNCTION DURING FROG DEVELOPMENT

TH affects development mainly by regulating gene expression through TRs, although non-genomic effects of TH have also been documented (Buchholz, Paul, Fu, & Shi, 2006; Davis & Davis, 1996; Evans, 1988; Lazar, 1993; Tsai & O'Malley, 1994; Yen, 2001). There are two evolutionary conserved TR genes, TRα and TRβ, in all vertebrate species and both are members of the nuclear hormone receptor superfamily (Laudet & Gronemeyer, 2002). TRs can bind to TH response elements (TREs) as monomers, homodimers, as well as heterodimers formed with 9-cis-retinoic acid receptors (RXRs), also members of the superfamily. TH can both activate and repress gene expression. TH-inducible genes, TR-RXR heterodimers bind to TREs in/around the promoters of target genes even in the context of chromatin and repress their expression in the absence of TH. TH-binding to TR leads to chromatin disruption and the activation of these promoters (Hsia & Shi, 2002; Hsia, Wang, & Shi, 2001; Shi, Matsuura, Fujimoto, Wen, & Fu, 2012; Wong et al., 1998; Wong, Shi, & Wolffe, 1995, 1997).

TR regulates transcription by recruiting cofactor complexes to target genes. In the absence of TH, TR recruits corepressor complexes, such as those containing histone deacetylase histone deacetylase 3 (HDAC3) and the highly related TR-binding proteins N-CoR and SMRT (Burke & Baniahmad, 2000; Glass & Rosenfeld, 2000; Guenther et al., 2000; Ishizuka & Lazar, 2003; Jones, Sachs, Rouse, Wade, & Shi, 2001; Jones & Shi, 2003; Li, Wang, et al., 2000; Shi et al., 2012; Yoon et al.,

2003; Zhang, Kalkum, Chait, & Roeder, 2002; Zhang & Lazar, 2000). This leads to local deacetylation of histones H3 and H4 transcriptional inhibition (Fig. 10.1). The binding by TH induces a conformational change in TR, leading to the release of corepressor complexes and the binding of coactivator complexes and gene activation. Among the coactivator complexes include those involved in chromatin disruption/remodeling as well as histone modifications such as those containing the related TR-binding coactivators SRC-1, 2, and 3, which complex with among other proteins CBP/p300 and protein arginine methyltransferase 1 (PRMT1) (Chen et al., 1997, 1999; Demarest et al., 2002; Heimeier, Hsia, & Shi, 2008; Huang, Li, Sachs, Cole, & Wong, 2003; Ito & Roeder, 2001; Koh, Chen, Lee, & Stallcup, 2001; Li, O'Malley, & Wong, 2000; Matsuda, Paul, Choi, Hasebe, & Shi, 2009; Matsuda, Paul, Choi, & Shi, 2007; Matsuura et al., 2012; McKenna & O'Malley, 2001; Onate, Tsai, Tsai, & O'Malley, 1995; Rachez & Freedman, 2001; Sheppard, Harries, Hussain, Bevan, & Heery, 2001; Shi et al., 2012; Torchia et al., 1997; Wong et al., 1995, 1997; Yen, 2001; Zhang & Lazar, 2000).

As the mediators of the effects of TH (see below), TRs and RXRs are expected to be present during amphibian metamorphosis. Indeed, their mRNAs are highly expressed during amphibian metamorphosis. In the pseudo-tetraploid *X. laevis*, there are two TRα and two TRβ genes. The TRα genes are activated as early as stage 41, shortly after tadpole hatching at stage 35/36, and reached high levels by stage 45 (Fig. 10.1) (Kanamori & Brown, 1992; Yaoita & Brown, 1990), the onset of tadpole feeding (Nieuwkoop & Faber, 1956), and their expression is maintained at high levels throughout metamorphosis (Fig. 10.1). The TRβ genes have little expression in premetamorphic tadpoles but are highly activated during metamorphosis, coinciding with the rise of endogenous TH (Fig. 10.1) (Kanamori & Brown, 1992; Shi, Yaoita, & Brown, 1992; Wang & Brown, 1993; Yaoita & Brown, 1990). Similarly, RXRα and RXRγ are also expressed in premetamorphic as well as metamorphosing tadpoles (Fig. 10.1) (Wong & Shi, 1995). It has been shown that the TRβ genes are activated directly by TR via the binding to the promoter region (Ranjan, Wong, & Shi, 1994; Shi et al., 1992), which explains the correlation of the mRNA levels with plasma TH (Fig. 10.1). The mechanisms governing the developmental expression patterns of TRα and RXR genes remain to be determined.

The expression studies led us to propose a dual function model for TR during the frog development (Fig. 10.1) (Sachs et al., 2000; Shi, Wong,

Puzianowska-Kuznicka, & Stolow, 1996). During embryogenesis, many TH-inducible genes are expressed at basal levels sufficient for embryonic development. Once organogenesis is completed, that is, by stage 45 when feeding begins (Nieuwkoop & Faber, 1956), these genes are no longer needed. The increased expression of TR and RXR genes (e.g., RXRα, Fig. 10.1) leads to the formation of TR-RXR heterodimers that bind to the TREs of the TH-inducible genes. In the absence of TH, the unliganded TR recruits corepressor complexes to the genes, resulting in the repression of these genes. Since exogenous TH can induce metamorphosis as early as stage 41, the model hypothesizes that this early expression of TR is likely responsible for the tadpole competence to undergo TH-induced metamorphosis. At the same time, gene repression by the unliganded TR ensures proper tadpole growth and prevents premature metamorphic organ transformations, thus regulating the timing of metamorphosis.

After stage 55, the rise of endogenously synthesized TH leads to the binding of TH to the chromatin-bound TR, which releases the corepressor complexes from the target genes with concurrent recruitment of the coactivator complexes. This results in gene activation and metamorphic transformations in the tadpoles.

Over the years, we and others have carried out a series *in vivo* studies, including transgenic analyses with mutant TRs to test this dual function model (Buchholz, Hsia, Fu, & Shi, 2003; Buchholz, Paul, & Shi, 2005; Buchholz, Tomita, Fu, Paul, & Shi, 2004; Das, Schreiber, Huang, & Brown, 2002; Grimaldi, Buisine, Bilesimo, & Sachs, 2013; Hasebe, Buchholz, Shi, & Ishizuya-Oka, 2011; Nakajima & Yaoita, 2003; Puzianowska-Kuznicka, Damjanovski, & Shi, 1997; Sachs & Shi, 2000; Schreiber & Brown, 2003; Schreiber, Das, Huang, Marsh-Armstrong, & Brown, 2001). In particular, overexpression of a dominant negative TR that lacks T3 binding ability but retains the ability to heterodimerize with RXR and repress gene expression just like unliganded TR inhibits TH-induced gene activation and metamorphosis in *X. laevis* (Fig. 10.2) (Buchholz et al., 2003; Das et al., 2002; Nakajima & Yaoita, 2003; Schreiber & Brown, 2003; Schreiber et al., 2001), indicating an essential role of TH-bound TR in gene regulation and metamorphosis. Conversely, overexpression of a dominant positive TR that cannot bind to TH but constitutively activate TH target gene transcription leads to the upregulation of TH response genes and tadpole metamorphosis (Buchholz et al., 2004; Hasebe et al., 2011). These studies have thus demonstrated that TR is

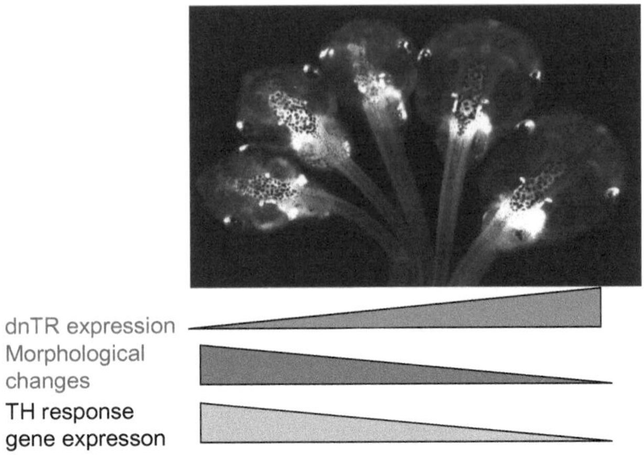

Figure 10.2 Dose-dependent inhibition of T3-induced metamorphosis by the expression of a dnTR. Transgenic tadpoles were generated to overexpress a dnTR with GFP fused at the N terminus. The premetamorphic tadpoles were treated with T3 to induce metamorphosis. Note that the degree of inhibition of T3-induced metamorphosis correlates with the level of dnTR overexpression as judged from the GFP expression levels in the tadpoles (DR Buchholz and Y-B Shi, unpublished). See Buchholz et al. (2003) for more details. (For color version of this figure, the reader is referred to the online version of this chapter.)

not only necessary but also sufficient to mediate the effects of TH during *Xenopus* metamorphosis.

3. UNLIGANDED TR RECRUITS COREPRESSOR COMPLEXES TO CONTROL METAMORPHIC TIMING

The dual function model for TR (Fig. 10.1) predicts that gene repression by unliganded TR during premetamorphosis is important for proper tadpole growth and development before undergoing metamorphic transformations. As summarized above, unliganded TR can recruit corepressor to target promoters to repress their expression. Chromatin immunization (ChIP) analyses have shown that during *Xenopus* development, the TR-binding corepressors N-CoR and SMRT are recruited by TR in premetamorphic tadpoles to endogenous target genes (Sachs et al., 2002; Tomita, Buchholz, & Shi, 2004). These corepressors are released from the target promoters during natural metamorphosis or upon TH treatment of premetamorphic tadpoles. Both N-CoR and SMRT form large complexes containing other proteins such as TBLR1 and HDAC3 (Guenther et al., 2000; Ishizuka & Lazar, 2003; Jones et al., 2001; Jones & Shi, 2003;

Li, Wang, et al., 2000; Shi et al., 2012; Yoon et al., 2003; Zhang et al., 2002; Zhang & Lazar, 2000). Consistently, TBLR1 is also recruited by unliganded TR to endogenous target genes in premetamorphic tadpoles and released during metamorphosis, accompanied by local increase in histone acetylation (Sachs et al., 2002; Tomita et al., 2004), suggesting that such corepressor complexes and histone deacetylation are important for the repression of TH target genes and premetamorphic animal development.

To investigate the role of corepressors in premetamorphic tadpoles, we generated two dominant negative forms of the *X. laevis* N-CoR (dnN-CoRs) that contained only the receptor interacting domain (ID) near the C-terminus of N-CoR (Fig. 10.3A). The first dnN-CoR, myc-ID monomer, comprised just the ID fused to an N-terminal peptide containing the myc tag and a nuclear localizing sequence (NLS). The second dnN-CoR, myc-ID dimer, is similar to myc-ID monomer except with two copies of the ID in the same directions connected by a linker sequence. In the reconstituted frog oocyte transcription system where one can study transcription of a reporter gene in the context of chromatin, both dnN-CoRs were able to compete against wild type N-CoR for binding to unliganded TR (Sato et al., 2007). More importantly, they reduced the repression of the TH target promoter by TR in the absence of TH with the myc-ID dimer more effective, while having no effect on the activation by TR in the presence of TH (Fig. 10.3B) (Sato et al., 2007).

To investigate the role of corepressor complex in TR function during development, we carried out transgenesis to express the myc-ID under the control of a heat shock-inducible promoter. In the absence of heat shock, the dnN-CoR was not expressed and the transgenic tadpoles and frogs were apparently normal. As this dominant N-CoR binds only to unliganded TR, it is expected to affect TR function only during the developmental stages 45–55 when TR expression is high while TH levels are low (Fig. 10.1), a window period of about 30 days (Nieuwkoop & Faber, 1956). Thus, we subjected transgenic and their wild type siblings to daily heat shock treatment, starting at stage 46. Heat shock treatment led to overexpression of the dnN-CoR and upregulation of most of the TH-response genes analyzed (Sato et al., 2007). More importantly, transgenic animals developed faster, reaching stage 55, the onset of metamorphosis, by as much as 7 days earlier than the wild type siblings within the 30-day experiment (Sato et al., 2007) (Fig. 10.3C). These observations demonstrated that the recruitment of HDAC-containing corepressor complexes is indeed important for gene repression by unliganded TR in

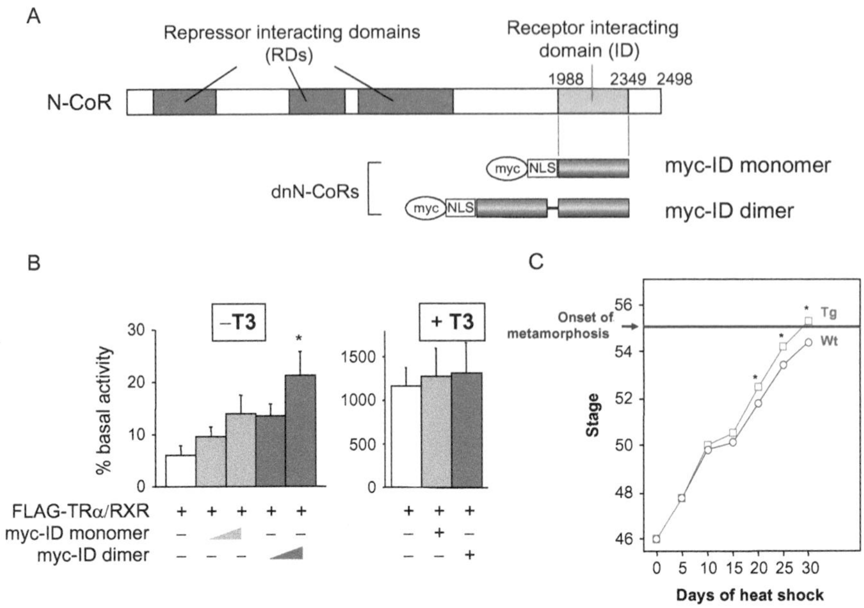

Figure 10.3 Transgenic expression of a dominant negative corepressor leads to premature metamorphosis. (A) Schematic representation of dnN-CoRs. *X. laevis* N-CoR interacts with unliganded TR via the receptor ID near the C-terminus. The repressor interacting domains (RDs) located in the N-terminal part are required for the recruitment of other co-repressors such as TBL1/TBLR1 and HDAC3. One dnN-CoRs, myc-ID monomer, comprises the ID (amino acids 1988–2349) fused to an N-terminal peptide containing a myc tag and a NLS. The other dnN-CoR, myc-ID dimer, consists of two IDs in the same directions connected by a linker sequence. (B) Of the two dnN-CoRs, myc-ID dimer relieves reporter gene repression by unliganded TR more efficiently than myc-ID monomer. The mRNAs for FLAG-tagged TR and RXR with or without the mRNA for myc-ID dimer or monomer were injected into *X. laevis* oocyte cytoplasm. The TRE-containing firefly luciferase reporter vector together with the control Renilla luciferase plasmid was then injected into the nucleus. After overnight incubation with or without T3, the oocytes were lysed and assayed for luciferase activities. The ratio of firefly luciferase activity to Renilla luciferase activity was determined as a measure of the reporter gene transcription level. The result was expressed as a percentage of the basal transcription level that was obtained from the oocytes without TR/RXR mRNA injection. Note that both dnN-CoRs relieved the unliganded TR-induced repression, but the derepression was more potent with the myc-ID dimer (left graph, $^{*}p<0.05$). Neither of dnN-CoRs affected the liganded-TR induced reporter gene activation (right graph). (C) Transgenic (Tg) tadpoles expressing the dnN-CoR myc-ID dimer under the control of a heat shock-inducible promoter initiate metamorphosis earlier than wild type (Wt) siblings. Wild type and transgenic sibling animals at stage 46 were treated daily with heat shock every day to induce dnN-CoR expression and their developmental stages were determined every 5 days for 30 days. Note that at the end of the 30-day experiments, Tg animals had initiated metamorphosis (passed stage 55) while Wt ones were one stage behind (a delay of about 7 days). $^{*}p$ values <0.05 (Tg vs. Wt). See for Sato, Buchholz, Paul, and Shi (2007) more details. (For color version of this figure, the reader is referred to the online version of this chapter.)

premetamorphic tadpoles as the dual function model predicted, and more importantly, plays a role in regulating the timing for the initiation of metamorphosis, thereby ensuring proper growth of the tadpole before its transformation into a frog, an important measure for the growth and survival of the postmetamorphic animal.

4. COACTIVATOR RECRUITMENT FOR METAMORPHIC PROGRESSION

Transcriptional activation by TH involves both the relief of the repression caused by unliganded TR and additional activation due to liganded TR. Expression and ChIP analyses have shown that a number of coactivators, including histone acetyltransferases SRC3 and p300 and histone methyltransferase PRMT1, are expressed and/or recruited by TR to endogenous target genes during *X. laevis* metamorphosis (Havis, Sachs, & Demeneix, 2003; Matsuda et al., 2009, 2007; Paul, Buchholz, Fu, & Shi, 2005, 2007; Paul, Fu, Buchholz, & Shi, 2005; Paul & Shi, 2003). More importantly, transgenic overexpression of a dominant negative form of *Xenopus* SRC3 that contains only the TR-ID (Paul, Fu, et al., 2005), or a dominant negative form of p300 that contains only the SRC-binding domain and is expected to bind to SRCs to disrupt the function of SRC-p300/CBP complexes (Paul et al., 2007), inhibits TH-induced gene activation and TH-metamorphosis as well as natural metamorphosis. These results demonstrate an essential role of the recruitment of coactivator complexes, in particular those containing SRC-p300/CBP, in gene regulation by TR and metamorphosis.

Interestingly, when we overexpressed the full length *X. laevis* PRMT1, another component of the SRC-p300 complexes, in transgenic tadpoles under the control of a heat shock-inducible promoter, heat shock treatment led to increased levels of PRMT1 in tadpoles. The resulting tadpoles had enhanced target gene expression in the presence of TH and accelerated natural as well as TH-induced metamorphosis (Fig. 10.4) (Matsuda et al., 2009). Thus, coactivator levels appear to also influence the rate of the progression of metamorphosis.

5. TEMPORAL REGULATION OF ORGAN-SPECIFIC METAMORPHOSIS

Aside from metamorphosis initiation, the timing of metamorphosis is also reflected at the level of individual organs as different organs undergo

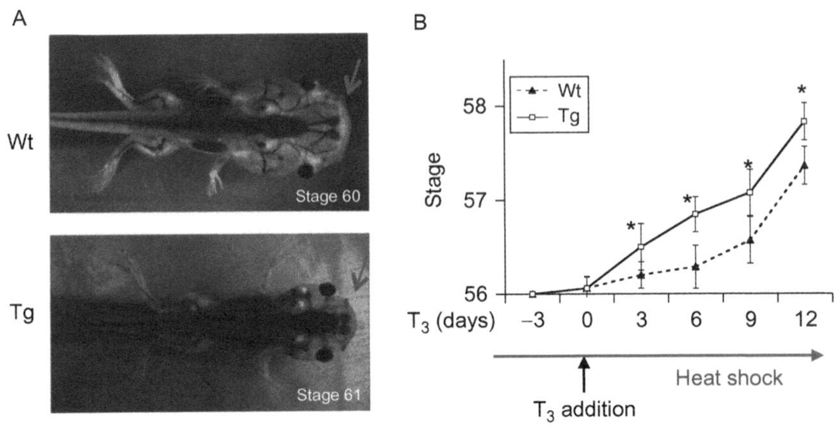

Figure 10.4 Transgenic expression of PRMT1 accelerates metamorphosis. (A) Overexpression of PRMT1 accelerates natural metamorphosis. A wild type (Wt) and transgenic (Tg) tadpole overexpression PRMT1 under a heat shock-inducible promoter control developed to stages 60 and 61 (note the pointier head of the transgenic animal, arrow), respectively, after 6 days of heat shock treatment starting at stage 58, the onset of metamorphic climax. (B) PRMT1 overexpression also leads to faster development during T3-induced metamorphosis. Wt and Tg animals were heat-shocked daily starting at stage 56. Three days after the first heat-shock treatment, all the animals were treated with T3 and continued to be subjected to heat shock treatment daily. The average developmental stages for the Wt and Tg animals at indicated time points were plotted. $^{*}p$ values < 0.05 when compared to the Wt animals. See Matsuda et al. (2009) for more details. (For color version of this figure, the reader is referred to the online version of this chapter.)

their unique metamorphic changes at specific developmental stages. For example, in *Xenopus*, the hindlimb bud begins to grow around stage 48 and the most dramatic morphological differentiation takes place around stages 53–58 (Fig. 10.5). Subsequently, the hindlimb undergoes mostly growth with little morphological changes (Fig. 10.5) (Nieuwkoop & Faber, 1956). On the other hand, tail metamorphosis, that is, total resorption, is one of the last changes to occur. Although some changes in the tail, such as fin resorption, take place earlier, the reduction in tail length does not occur until around stage 62 and its completion marks the end of metamorphosis (Fig. 10.5) (Dodd & Dodd, 1976). Finally, the intestine begins its remodeling around stage 58 when its connective tissue and muscle layers increase in thickness (Ishizuya-Oka & Shimozawa, 1987; McAvoy & Dixon, 1977; Shi & Ishizuya-Oka, 1996). This is followed by the degeneration of the larval epithelium through programmed cell death (apoptosis) around stages 60–62 and the formation of the adult epithelium toward the end of metamorphosis

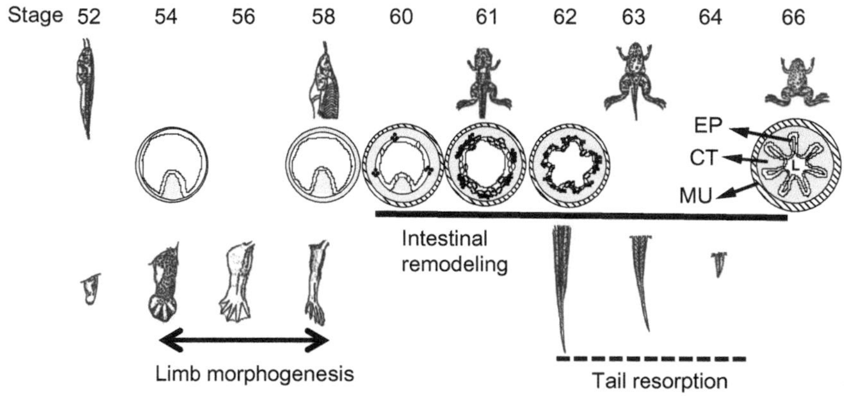

Figure 10.5 Temporal regulation of tissue specific metamorphosis. The developmental stages are based on Nieuwkoop and Faber (1956). The tails at stages 62–66 are drawn to the same scale to show the resorption (no tail at stage 66), while the tadpoles, intestinal cross-sections and hindlimbs at different stages are drawn in arbitrary sizes to highlight the morphological differences. Tadpole small intestine has a single epithelial fold, where connective tissue (CT) is abundant, while a frog has a multiply folded intestinal epithelium (EP), with elaborate connective tissue and muscle (MU). Dots, proliferating adult intestinal epithelial cells; open circles, apoptotic primary or larval intestinal epithelial cells; L, intestinal lumen.

(Fig. 10.5). A number of factors appear to contribute to this temporal regulation of organ-specific metamorphosis. These factors affect two aspects of TH signaling: TR-RXR heterodimer levels and availability of intracellular TH.

Expression analyses of TR and RXR have shown that the mRNAs for TRα, TRβ, and RXRs are highly expressed during amphibian metamorphosis (Davey, Schneider, & Galton, 1994; Eliceiri & Brown, 1994; Fairclough & Tata, 1997; Helbing, Gergely, & Atkinson, 1992; Kawahara, Baker, & Tata, 1991; Schneider & Galton, 1991; Wang & Brown, 1993; Wang, Matsuda, & Shi, 2008; Wong & Shi, 1995; Yaoita & Brown, 1990). Importantly, the expression of TR and RXR genes in individual organs during *Xenopus* metamorphosis strongly correlates with the temporal regulation of organ-specific transformations (Eliceiri & Brown, 1994; Wang & Brown, 1993; Wang et al., 2008; Wong & Shi, 1995; Yaoita & Brown, 1990). In general, TR and RXR genes are coordinately regulated during metamorphosis and their expression is high in a given organ when metamorphosis takes place and low before or after metamorphosis (Table 10.1). Thus, the mRNA levels for these receptor genes (except for TRβ genes) are high in the hindlimb around stages 54–56 when hindlimb morphogenesis, for example, digit formation, takes place (Wang &

Table 10.1 Correlations of the relative expression levels of TR, RXR, and some factors affecting cellular free T3 with organ transformations during *Xenopus* metamorphosis.

	Intestine	**Hindlimb**		**Tail**	
Stages	**58–66**	**54–56**	**58–66**	**54–60**	**62–66**
Plasma T_3	High	Low	High	Low–high	High
CTHBP/PKM2 mRNA	Low–moderate	Low	Increasing	High	Low
5-Deiodinase mRNA	Low	Low	Increasing	Increasing	Decreasing
Estimated free cellular T_3	High	Moderate	Low	Low	High
TR, RXR mRNA	Moderate–high	Very high	Low	Increasing	High
Developmental event	Remodeling	Morphogenesis	Growth	Few changes	Resorption

Brown, 1993; Wang et al., 2008; Wong & Shi, 1995). Similarly, in the tail, the genes are expressed at low levels until stage 62 when they are up-regulated, coinciding with the rapid tail resorption, while in the intestine, high levels of TR and RXR mRNAs are present during tissue remodeling (stages 58–66) (Table 10.1) (Wang & Brown, 1993; Wong & Shi, 1995). Thus, the levels of TR-RXR heterodimers likely help to determine temporal regulation of the changes in individual organs.

In the absence of TH, the presence of TR-RXR in a given organ can act to prevent precocious metamorphosis of the organ, thus ensuring its function in the developing tadpoles. TH levels in an organ is affected several mechanisms. The first is the circulating plasma levels of TH, governed mainly by the production of TH in the thyroid gland (Table 10.1 and Fig. 10.1) (Leloup & Buscaglia, 1977). Secondly, the levels of cellular TH could be affected by cellular proteins that bind TH, that is, cytosolic thyroid hormone binding proteins (CTHBPs). Finally, the metabolic conversion of T_4 to T_3 and/or the inactivation of both hormones may be an important step in regulating the levels of functional TH within an organ.

While all organs are exposed to the same circulating plasma TH, once TH enters the cytoplasm, it is likely to CTHBPs (Ashizawa & Chen, 1992; Barsano & Groot, 1983; Shi, Liang, Parkison, & Cheng, 1994; Yamauchi & Tata, 1994). In general, CTHBPs are multi-functional proteins that serve one or more other roles (e.g., pyruvate kinase, myosin light chain kinase, and disulfide isomerase) and bind TH with 10–100-fold weaker affinity than the nuclear TRs. The exact roles of these CTHBPs in TH signal transduction are still unclear. These proteins could participate in TH import from the extracellular medium, intracellular TH metabolism, and transport to the nucleus, or serve as a buffer to modulate intracellular free TH concentrations. Studies on the human M2 pyruvate kinase (PKM2) have shown that over-expression of the protein leads to an inhibition of TH-dependent transcriptional activation by TR in a tissue culture cell line (Ashizawa & Chen, 1992). Interestingly, analysis of the expression of the frog PKM2 in the tail, limb, and intestine revealed that the intestine had very low but relatively constant levels of its mRNA during development (Shi et al., 1994). In contrast, the PKM2 expression in the tail and hindlimb was drastically altered during metamorphosis. Low levels of PKM2 mRNA were present in either the hindlimb during morphogenesis (stages 54–56) or tail during resorption (stages 62–64) (Table 10.1). Interestingly, PKM2 mRNA levels increased dramatically in the growing hindlimb and decreased precipitously in the resorbing tail. Thus, while recent studies suggest that PKM2 may affect cell fate via multiple mechanisms (Lu, 2012; Luo & Semenza, 2012), these findings support a potential role of PKM2 in the regulation of free cellular TH to affect the metamorphic transition of these two organs.

Another factor affecting local TH levels is deiodination. TH is synthesized initially as T_4 or thyroxine in the thyroid gland (St Germain et al., 1994). Some T_4 in turn is converted in the thyroid gland into T_3 by 5′-deiodinases (Jorgensen, 1978; St Germain et al., 1994). Both T_3 and T_4 are then secreted into the serum and carried to target tissues (Hennemann & Visser, 1997; Oppenheimer et al., 1987; Ritchie et al., 2003; Robbins, 1992; Shi, Ritchie, & Taylor, 2002; Visser, Friesema, & Visser, 2011). There are two families of deiodinases, 5′- and 5-deiodinases (Arrojo & Bianco, 2011; Gereben et al., 2008; St Germain et al., 1994). The 5′-deiodinases convert T4 to T3 and thus enhance TH signaling while the 5-deiodinases inactivate both T4 and T3, thus reducing TH signaling. Analysis of the expression of the 5-deiodinase gene in *X. laevis* showed that it was regulated in a organ-specific manner (Table 10.1) (St Germain et al., 1994; Wang & Brown, 1993). It had little expression in the intestine. In the hindlimb, it is not expressed until after limb morphogenesis is completed, that is, around stage 60, when its mRNA level is up-regulated, paralleling the increase in limb size. In contrast, relatively high levels of its mRNA are present in the tail during premetamorphic stages. Immediately prior to rapid tail resorption (stages 58–61), the expression of the 5-deiodinase gene is suddenly up-regulated several fold. Its mRNA levels then return to the premetamorphic levels as tail resorption takes place after stage 61. While no detailed analyses have been done on 5′-deiodinase during *X. laevis* metamorphosis, studies in *Rana catesbeiana* argue that 5′-deiodinase expression is also temporally regulated in different organs such that the enzyme likely functions to increase T3 level within the organ as metamorphosis begins (Becker, Stephens, Davey, Schneider, & Galton, 1997; Davey, Becker, Schneider, St Germain, & Galton, 1995). These findings suggest that 5 and 5′-deiodinases function to regulate the levels of TH within an organ to regulate its metamorphic transformation.

While there are clearly many other factors contributing to the regulation of TH levels within an organ, the brief summary above suggests that the levels of both receptor and cellular free TH in individual organs during metamorphosis contribute to the timing of tissue-specific metamorphosis. At the mechanistic level, low levels of TH in a given organ will lead to a high fraction of TR-RXR heterodimers in the unliganded state. This will lead to the recruitment of HDAC-containing corepressor complexes to the TH target genes to repress their expression and thus prevent organ metamorphosis. Thus, there is likely a tissue-specific critical ratio of liganded to unliganded TR that determines the timing of organ metamorphosis.

6. CONCLUSIONS

Anuran metamorphosis is one of the most dramatic events regulated by TH. It is tightly controlled during development both on the timing of the initiation of this process at whole animal level and on the temporal coordination of organ-specific transformations. The biosynthesis of endogenous TH and the regulation of cellular free TH levels are the key determinants controlling the timing of metamorphosis. Mechanistically, the metamorphic effect of TH is mediated by TR through regulating the transcription of TH-target genes. Low levels of TH ensure a high fraction of TR in the unliganded state at the target genes. This leads to the recruitment of HDAC-containing corepressor complexes to repress the genes, thus preventing the onset of animal metamorphosis and organ transformations. During metamorphosis, high levels of TH lead to liganded TR that activates target gene transcription to induce metamorphosis. Such a ligand switch behavior controls the impressive morphological development of the frog.

ACKNOWLEDGMENT

The work in the author's laboratory has been supported by the Intramural Research Program of NICHD, NIH.

REFERENCES

Allen, B. M. (1918). The results of thyroid removal in the larvae of rana pipiens. *Journal of Experimental Zoology*, *24*, 499–519.

Arrojo, E. D. R., & Bianco, A. C. (2011). Type 2 deiodinase at the crossroads of thyroid hormone action. *International Journal of Biochemistry & Cell Biology*, *43*, 1432–1441.

Ashizawa, K., & Chen, S.-Y. (1992). Regulation of thyroid hormone receptor-mediated transcription by a cytosol protein. *Proceedings of the National Academy of Sciences of the United States of America*, *89*, 9277–9281.

Atkinson, B. G. (1994). Metamorphosis: Model systems for studying gene expression in postembryonic development. *Developmental Genetics*, *15*, 313–319.

Barsano, C. P., & Groot, L. (1983). Nuclear cytoplasmic interrelationships. In J. Oppenheimer & H. Samuels (Eds.), *Molecular basis of thyroid hormone action* (pp. 139–177). New York: Academic Press.

Becker, K. B., Stephens, K. C., Davey, J. C., Schneider, M. J., & Galton, V. A. (1997). The type 2 and type 3 iodothyronine deiodinases play important roles in coordinating development in *Rana catesbeiana* tadpoles. *Endocrinology*, *138*, 2989–2997.

Buchholz, D. R., Hsia, V. S.-C. , Fu, L., & Shi, Y.-B. (2003). A dominant negative thyroid hormone receptor blocks amphibian metamorphosis by retaining corepressors at target genes. *Molecular and Cellular Biology, 23*, 6750–6758.

Buchholz, D. R., Paul, B. D., Fu, L., & Shi, Y. B. (2006). Molecular and developmental analyses of thyroid hormone receptor function in *Xenopus laevis*, the African clawed frog. *General and Comparative Endocrinology, 145*, 1–19.

Buchholz, D. R., Paul, B. D., & Shi, Y. B. (2005). Gene-specific changes in promoter occupancy by thyroid hormone receptor during frog metamorphosis. Implications for developmental gene regulation. *Journal of Biological Chemistry, 280*, 41222–41228.

Buchholz, D. R., Tomita, A., Fu, L., Paul, B. D., & Shi, Y.-B. (2004). Transgenic analysis reveals that thyroid hormone receptor is sufficient to mediate the thyroid hormone signal in frog metamorphosis. *Molecular and Cellular Biology, 24*, 9026–9037.

Burke, L. J., & Baniahmad, A. (2000). Co-repressors 2000. *The FASEB Journal, 14*, 1876–1888.

Chen, H., Lin, R. J., Schiltz, R. L., Chakravarti, D., Nash, A., Nagy, L., et al. (1997). Nuclear receptor coactivator ACTR is a novel histone acetyltransferase and forms a multimeric activation complex with P/CAF and CBP/p300. *Cell, 90*, 569–580.

Chen, D., Ma, H., Hong, H., Koh, S. S., Huang, S. M., Schurter, B. T., et al. (1999). Regulation of transcription by a protein methyltransferase. *Science, 284*, 2174–2177.

Das, B., Schreiber, A. M., Huang, H., & Brown, D. D. (2002). Multiple thyroid hormone-induced muscle growth and death programs during metamorphosis in *Xenopus laevis*. *Proceedings of the National Academy of Sciences of the United States of America, 99*, 12230–12235.

Davey, J. C., Becker, K. B., Schneider, M. J., St Germain, D. L., & Galton, V. A. (1995). Cloning of a cDNA for the type II iodothyronine deiodinase. *Journal of Biological Chemistry, 270*, 26786–26789.

Davey, J. C., Schneider, M. J., & Galton, V. A. (1994). Cloning of a thyroid hormone-responsive *Rana catesbeiana* c-erbA-beta gene. *Developmental Genetics, 15*, 339–346.

Davis, P. J., & Davis, F. B. (1996). Nongenomic actions of thyroid hormone. *Thyroid, 6*, 497–504.

Demarest, S. J., Martinez-Yamout, M., Chung, J., Chen, H., Xu, W., Dyson, H. J., et al. (2002). Mutual synergistic folding in recruitment of CBP/p300 by p160 nuclear receptor coactivators. *Nature, 415*, 549–553.

Denver, R. J. (2013). Neuroendocrinology of amphibian metamorphosis. *Current Topics in Developmental Biology, 103*, 195–227.

Dodd, M. H. I., & Dodd, J. M. (1976). The biology of metamorphosis. In B. Lofts (Ed.), *Physiology of the amphibia* (pp. 467–599). New York: Academic Press.

Eliceiri, B. P., & Brown, D. D. (1994). Quantitation of endogenous thyroid hormone receptors alpha and beta during embryogenesis and metamorphosis in *Xenopus laevis*. *Journal of Biological Chemistry, 269*, 24459–24465.

Evans, R. M. (1988). The steroid and thyroid hormone receptor superfamily. *Science, 240*, 889–895.

Fairclough, L., & Tata, J. R. (1997). An immunocytochemical analysis of the expression of thyroid hormone receptor alpha and beta proteins during natural and thyroid hormone-induced metamorphosis in Xenopus. *Development, Growth & Differentiation, 39*, 273–283.

Franklyn, J. A., & Gammage, M. D. (1996). Thyroid disease: Effects on cardiovascular function. *Trends in Endocrinology and Metabolism, 7*, 50–54.

Freake, H. C., & Oppenheimer, J. H. (1995). Thermogenesis and thyroid function. *Annual Review of Nutrition, 15*, 263–291.

Gereben, B., Zavacki, A. M., Ribich, S., Kim, B. W., Huang, S. A., Simonides, W. S., et al. (2008). Cellular and molecular basis of deiodinase-regulated thyroid hormone signaling. *Endocrine Reviews, 29*, 898–938.

Gilbert, L. I., Tata, J. R., & Atkinson, B. G. (1996). *Metamorphosis: Post-embryonic reprogramming of gene expression in amphibian and insect cells*. New York: Academic Press.

Glass, C. K., & Rosenfeld, M. G. (2000). The coregulator exchange in transcriptional functions of nuclear receptors. *Genes & Development, 14*, 121–141.

Grimaldi, A. G., Buisine, N., Bilesimo, P., & Sachs, L. M. (2013). High-throughput sequencing will metamorphose the analysis of thyroid hormone receptor function during amphibian development. *Current Topics in Developmental Biology, 103*, 277–303.

Gudernatch, J. F. (1912). Feeding experiments on tadpoles. I. The influence of specific organs given as food on growth and differentiation: A contribution to the knowledge of organs with internal secretion. *Archiv für Entwicklungsmechanik der Organismen, 35*, 457–483.

Guenther, M. G., Lane, W. S., Fischle, W., Verdin, E., Lazar, M. A., & Shiekhattar, R. (2000). A core SMRT corepressor complex containing HDAC3 and TBL1, a WD40-repeat protein linked to deafness. *Genes & Development, 14*, 1048–1057.

Hasebe, T., Buchholz, D. R., Shi, Y. B., & Ishizuya-Oka, A. (2011). Epithelial-connective tissue interactions induced by thyroid hormone receptor are essential for adult stem cell development in the *Xenopus laevis* intestine. *Stem Cells, 29*, 154–161.

Havis, E., Sachs, L. M., & Demeneix, B. A. (2003). Metamorphic T3-response genes have specific co-regulator requirements. *EMBO Reports, 4*, 883–888.

Heimeier, R. A., Hsia, V. S.-C. , & Shi, Y.-B. (2008). Participation of BAF57 and BRG1-containing chromatin remodeling complexes in thyroid hormone-dependent gene activation during vertebrate development. *Molecular Endocrinology, 22*, 1065–1077.

Helbing, C. C., Gergely, C., & Atkinson, B. G. (1992). Sequential up-regulation of thyroid hormone ß receptor, ornithine transcarbamylase, and carbamyl phosphate synthetase mRNAs in the liver of Rana Catesbeiana tadpoles during spontaneous and thyroid hormone-induced metamorphosis. *Developmental Genetics, 13*, 289–301.

Hennemann, G., & Visser, T. J. (1997). Thyroid hormone syntesis, pasma membrane transport and metabolism. Pharmacotherapeutics of thyroid gland. In A. P. Weetman & A. Grossman (Eds.), *Handbook of experimental pharmacology: Vol. 128*. (pp. 75–117). Berlin/New York: Springer-Verlag.

Hetzel, B. S. (1989). *The story of iodine deficiency: An international challenge in nutrition*. Oxford: Oxford University Press.

Howdeshell, K. L. (2002). A model of the development of the brain as a construct of the thyroid system. *Environmental Health Perspectives, 110*, 337–348.

Hsia, V. S.-C., & Shi, Y.-B. (2002). Chromatin disruption and histone acetylation in the regulation of HIV-LTR by thyroid hormone receptor. *Molecular and Cellular Biology, 22*, 4043–4052.

Hsia, V. S.-C., Wang, H., & Shi, Y.-B. (2001). Involvement of chromatin and histone acetylation in the regulation of HIV-LTR by thyroid hormone receptor. *Cell Research, 11*, 8–16.

Hsu, J. H., & Brent, G. A. (1998). Thyroid hormone receptor gene knockouts. *Trends in Endocrinology and Metabolism, 9*, 103–112.

Huang, Z.-Q., Li, J., Sachs, L. M., Cole, P. A., & Wong, J. (2003). A role for cofactor–cofactor and cofactor–histone interactions in targeting p300, SWI/SNF and Mediator for transcription. *EMBO Journal, 22*, 2146–2155.

Ishizuka, T., & Lazar, M. A. (2003). The N-CoR/histone deacetylase 3 complex is required for repression by thyroid hormone receptor. *Molecular and Cellular Biology, 23*, 5122–5131.

Ishizuya-Oka, A., & Shi, Y. B. (2011). Evolutionary insights into postembryonic development of adult intestinal stem cells. *Cell & Bioscience, 1*, 37.

Ishizuya-Oka, A., & Shimozawa, A. (1987). Development of the connective tissue in the digestive tract of the larval and metamorphosing *Xenopus laevis*. *Anatomischer Anzeiger, 164*, 81–93.

Ito, M., & Roeder, R. G. (2001). The TRAP/SMCC/Mediator complex and thyroid hormone receptor function. *Trends in Endocrinology and Metabolism, 12*, 127–134.

Jones, P. L., Sachs, L. M., Rouse, N., Wade, P. A., & Shi, Y. B. (2001). Multiple N-CoR complexes contain distinct histone deacetylases. *Journal of Biological Chemistry, 276*, 8807–8811.

Jones, P. L., & Shi, Y.-B. (2003). N-CoR-HDAC corepressor complexes: Roles in transcriptional regulation by nuclear hormone receptors. In J. L. Workman (Ed.), *Current topics in microbiology and immunology: Protein complexes that modify chromatin* (pp. 237–268). Berlin: Springer-Verlag.

Jorgensen, E. C. (1978). Thyroid hormones and analogs. II. Structure–activity relationships. In C. H. Li (Ed.), *Hormonal proteins and peptides* (pp. 107–199). New York: Academic Press.

Kanamori, A., & Brown, D. D. (1992). The regulation of thyroid hormone receptor beta genes by thyroid hormone in *Xenopus laevis*. *Journal of Biological Chemistry, 267*, 739–745.

Kawahara, A., Baker, B. S., & Tata, J. R. (1991). Developmental and regional expression of thyroid hormone receptor genes during *Xenopus metamorphosis*. *Development, 112*, 933–943.

Koh, S. S., Chen, D. G., Lee, Y. H., & Stallcup, M. R. (2001). Synergistic enhancement of nuclear receptor function by p160 coactivators and two coactivators with protein methyltransferase activities. *Journal of Biological Chemistry, 276*, 1089–1098.

Laudet, V., & Gronemeyer, H. (2002). *The nuclear receptor FactsBook*. San Diego: Academic Press, 462 pp.

Lazar, M. A. (1993). Thyroid hormone receptors: Multiple forms, multiple possibilities. *Endocrine Reviews, 14*, 184–193.

Leloup, J., & Buscaglia, M. (1977). La triiodothyronine: Hormone de la métamorphose des amphibiens. *Academy of Sciences of the Czech Republic, 284*, 2261–2263.

Li, J., O'Malley, B. W., & Wong, J. (2000). p300 requires its histone acetyltransferase activity and SRC-1 interaction domain to facilitate thyroid hormone receptor activation in chromatin. *Molecular and Cellular Biology, 20*, 2031–2042.

Li, J., Wang, J., Wang, J., Nawaz, Z., Liu, J. M., Qin, J., et al. (2000). Both corepressor proteins SMRT and N-CoR exist in large protein complexes containing HDAC3. *EMBO Journal, 19*, 4342–4350.

Lu, Z. (2012). PKM2 functions as a histone kinase. *Cell Cycle, 11*, 4101–4102.

Luo, W., & Semenza, G. L. (2012). Emerging roles of PKM2 in cell metabolism and cancer progression. *Trends in Endocrinology and Metabolism, 23*, 560–566.

Matsuda, H., Paul, B. D., Choi, C. Y., Hasebe, T., & Shi, Y.-B. (2009). Novel functions of protein arginine methyltransferase 1 in thyroid hormone receptor-mediated transcription and in the regulation of metamorphic rate in *Xenopus laevis*. *Molecular and Cellular Biology, 29*, 745–757.

Matsuda, H., Paul, B. D., Choi, C. Y., & Shi, Y.-B. (2007). Contrasting effects of two alternative splicing forms of coactivator-associated arginine methyltransferase 1 on thyroid hormone receptor-mediated transcription in *Xenopus laevis*. *Molecular Endocrinology, 21*, 1082–1094.

Matsuura, K., Fujimoto, K., Das, B., Fu, L., Lu, C. D., & Shi, Y. B. (2012). Histone H3K79 methyltransferase Dot1L is directly activated by thyroid hormone receptor during *Xenopus metamorphosis*. *Cell & Bioscience, 2*, 25.

McAvoy, J. W., & Dixon, K. E. (1977). Cell proliferation and renewal in the small intestinal epithelium of metamorphosing and adult *Xenopus laevis*. *The Journal of Experimental Zoology, 202*, 129–138.

McKenna, N. J., & O'Malley, B. W. (2001). Nuclear receptors, coregulators, ligands, and selective receptor modulators: Making sense of the patchwork quilt. *Annals of the New York Academy of Sciences, 949*, 3–5.

Nakajima, K., & Yaoita, Y. (2003). Dual mechanisms governing muscle cell death in tadpole tail during amphibian metamorphosis. *Developmental Dynamics*, *227*, 246–255.

Nieuwkoop, P. D., & Faber, J. (1956). *Normal table of Xenopus laevis*. Amsterdam: North Holland Publishing, 252 pp.

Onate, S. A., Tsai, S. Y., Tsai, M. J., & O'Malley, B. W. (1995). Sequence and characterization of a coactivator for the steroid hormone receptor superfamily. *Science*, *270*, 1354–1357.

Oppenheimer, J. H., Schwartz, H. L., Mariash, C. N., Kinlaw, W. B., Wong, N. C., & Freake, H. C. (1987). Advances in our understanding of thyroid hormone action at the cellular level. *Endocrine Reviews*, *8*, 288–308.

Paul, B. D., Buchholz, D. R., Fu, L., & Shi, Y.-B. (2005). Tissue- and gene-specific recruitment of steroid receptor coactivator-3 by thyroid hormone receptor during development. *Journal of Biological Chemistry*, *280*, 27165–27172.

Paul, B. D., Buchholz, D. R., Fu, L., & Shi, Y.-B. (2007). SRC-p300 coactivator complex is required for thyroid hormone induced amphibian metamorphosis. *Journal of Biological Chemistry*, *282*, 7472–7481.

Paul, B. D., Fu, L., Buchholz, D. R., & Shi, Y.-B. (2005). Coactivator recruitment is essential for liganded thyroid hormone receptor to initiate amphibian metamorphosis. *Molecular and Cellular Biology*, *25*, 5712–5724.

Paul, B. D., & Shi, Y.-B. (2003). Distinct expression profiles of transcriptional coactivators for thyroid hormone receptors during *Xenopus laevis* metamorphosis. *Cell Research*, *13*, 459–464.

Porterfield, S. P., & Hendrich, C. E. (1993). The role of thyroid hormones in prenatal and neonatal neurological development–current perspectives. *Endocrine Reviews*, *14*, 94–106.

Puzianowska-Kuznicka, M., Damjanovski, S., & Shi, Y.-B. (1997). Both thyroid hormone and 9-cis retinoic acid receptors are required to efficiently mediate the effects of thyroid hormone on embryonic development and specific gene regulation in *Xenopus laevis*. *Molecular and Cellular Biology*, *17*, 4738–4749.

Rachez, C., & Freedman, L. P. (2001). Mediator complexes and transcription. *Current Opinion in Cell Biology*, *13*, 274–280.

Ranjan, M., Wong, J., & Shi, Y. B. (1994). Transcriptional repression of Xenopus TR beta gene is mediated by a thyroid hormone response element located near the start site. *Journal of Biological Chemistry*, *269*, 24699–24705.

Ritchie, J. W. A., Shi, Y.-B. , Hayashi, Y., Baird, F. E., Muchekehu, R. W., Christie, G. R., et al. (2003). A role for thyroid hormone transporters in transcriptional regulation by thyroid hormone receptors. *Molecular Endocrinology*, *17*, 653–661.

Robbins, J. (1992). Thyroxine transport and the free hormone hypothesis. *Endocrinology*, *131*, 546–547.

Sachs, L. M., Damjanovski, S., Jones, P. L., Li, Q., Amano, T., Ueda, S., et al. (2000). Dual functions of thyroid hormone receptors during Xenopus development. *Comparative Biochemistry and Physiology. Part B, Biochemistry & Molecular Biology*, *126*, 199–211.

Sachs, L. M., Jones, P. L., Havis, E., Rouse, N., Demeneix, B. A., & Shi, Y.-B. (2002). N-CoR recruitment by unliganded thyroid hormone receptor in gene repression during *Xenopus laevis* development. *Molecular and Cellular Biology*, *22*, 8527–8538.

Sachs, L. M., & Shi, Y.-B. (2000). Targeted chromatin binding and histone acetylation in vivo by thyroid hormone receptor during amphibian development. *Proceedings of the National Academy of Sciences of the United States of America*, *97*, 13138–13143.

Sato, Y., Buchholz, D. R., Paul, B. D., & Shi, Y.-B. (2007). A role of unliganded thyroid hormone receptor in postembryonic development in *Xenopus laevis*. *Mechanisms of Development*, *124*, 476–488.

Schneider, M. J., & Galton, V. A. (1991). Regulation of c-erbA-alpha messenger RNA species in tadpole erythrocytes by thyroid hormone. *Molecular Endocrinology*, *5*, 201–208.

Schreiber, A. M., & Brown, D. D. (2003). Tadpole skin dies autonomously in response to thyroid hormone at metamorphosis. *Proceedings of the National Academy of Sciences of the United States of America, 100*, 1769–1774.

Schreiber, A. M., Das, B., Huang, H., Marsh-Armstrong, N., & Brown, D. D. (2001). Diverse developmental programs of *Xenopus laevis* metamorphosis are inhibited by a dominant negative thyroid hormone receptor. *Proceedings of the National Academy of Sciences of the United States of America, 98*, 10739–10744.

Sheppard, H. M., Harries, J. C., Hussain, S., Bevan, C., & Heery, D. M. (2001). Analysis of the steroid receptor coactivator 1 (SRC1)–CREB binding protein interaction interface and its importance for the function of SRC1. *Molecular and Cellular Biology, 21*, 39–50.

Shi, Y.-B. (1999). *Amphibian metamorphosis: From morphology to molecular biology*. New York: John Wiley & Sons Inc, 288 pp.

Shi, Y. B. (2013). Current topics in developmental biology animal metamorphosis preface. *Current Topics in Developmental Biology, 103*, xv–xvi.

Shi, Y. B., Hasebe, T., Fu, L., Fujimoto, K., & Ishizuya-Oka, A. (2011). The development of the adult intestinal stem cells: Insights from studies on thyroid hormone-dependent amphibian metamorphosis. *Cell & Bioscience, 1*, 30.

Shi, Y.-B., & Ishizuya-Oka, A. (1996). Biphasic intestinal development in amphibians: Embryogensis and remodeling during metamorphosis. *Current Topics in Developmental Biology, 32*, 205–235.

Shi, Y.-B., Liang, V. C.-T., Parkison, C., & Cheng, S.-Y. (1994). Tissue-dependent developmental expression of a cytosolic thyroid hormone protein gene in Xenopus: Its role in the regulation of amphibian metamorphosis. *FEBS Letters, 355*, 61–64.

Shi, Y. B., Matsuura, K., Fujimoto, K., Wen, L., & Fu, L. (2012). Thyroid hormone receptor actions on transcription in amphibia: The roles of histone modification and chromatin disruption. *Cell & Bioscience, 2*, 42.

Shi, Y.-B. , Ritchie, J., & Taylor, P. M. (2002). Complex regulation of thyroid hormone action: Multiple opportunities for pharmacological intervention. *Pharmacology & Therapeutics, 94*, 235–251.

Shi, Y.-B., Wong, J., Puzianowska-Kuznicka, M., & Stolow, M. (1996). Tadpole competence and tissue-specific temporal regulation of amphibian metamorphosis: Roles of thyroid hormone and its receptors. *BioEssays, 18*, 391–399.

Shi, Y.-B., Yaoita, Y., & Brown, D. D. (1992). Genomic organization and alternative promoter usage of the two thyroid hormone receptor ß genes in *Xenopus laevis*. *Journal of Biological Chemistry, 267*, 733–788.

Silva, J. E. (1995). Thyroid hormone control of thermogenesis and energy balance. *Thyroid, 5*, 481–492.

St Germain, D. L., Schwartzman, R. A., Croteau, W., Kanamori, A., Wang, Z., Brown, D. D., et al. (1994). A thyroid hormone-regulated gene in *Xenopus laevis* encodes a type III iodothyronine 5-deiodinase. *Proceedings of the National Academy of Sciences of the United States of America, 91*, 7767–7771.

Sun, G., & Shi, Y.-B. (2012). Thyroid hormone regulation of adult intestinal stem cell development: Mechanisms and evolutionary conservations. *International Journal of Biological Sciences, 8*, 1217–1224.

Tata, J. R. (1968). Early metamorphic competence of *Xenopus larvae*. *Developmental Biology, 18*, 415–440.

Tata, J. R. (1993). Gene expression during metamorphosis: An ideal model for post-embryonic development. *BioEssays, 15*, 239–248.

Tomita, A., Buchholz, D. R., & Shi, Y.-B. (2004). Recruitment of N-CoR/SMRT-TBLR1 corepressor complex by unliganded thyroid hormone receptor for gene repression during frog development. *Molecular and Cellular Biology, 24*, 3337–3346.

Torchia, J., Rose, D. W., Inostroza, J., Kamei, Y., Westin, S., Glass, C. K., et al. (1997). The transcriptional co-activator p/CIP binds CBP and mediates nuclear-receptor function. *Nature, 387*, 677–684.

Tsai, M. J., & O'Malley, B. W. (1994). Molecular mechanisms of action of steroid/thyroid receptor superfamily members. *Annual Review of Biochemistry, 63*, 451–486.

Visser, W. E., Friesema, E. C., & Visser, T. J. (2011). Minireview: Thyroid hormone transporters: The knowns and the unknowns. *Molecular Endocrinology, 25*, 1–14.

Wang, Z., & Brown, D. D. (1993). Thyroid hormone-induced gene expression program for amphibian tail resorption. *Journal of Biological Chemistry, 268*, 16270–16278.

Wang, X., Matsuda, H., & Shi, Y.-B. (2008). Developmental regulation and function of thyroid hormone receptors and 9-cis retinoic acid receptors during Xenopus tropicalis metamorphosis. *Endocrinology, 149*, 5610–5618.

Wong, J., Patterton, D., Imhof, D., Guschin, D., Shi, Y.-B. , & Wolffe, A. P. (1998). Distinct requirements for chromatin assembly in transcriptional repression by thyroid hormone receptor and histone deacetylase. *EMBO Journal, 17*, 520–534.

Wong, J., & Shi, Y.-B. (1995). Coordinated regulation of and transcriptional activation by Xenopus thyroid hormone and retinoid X receptors. *Journal of Biological Chemistry, 270*, 18479–18483.

Wong, J., Shi, Y. B., & Wolffe, A. P. (1995). A role for nucleosome assembly in both silencing and activation of the Xenopus TR beta A gene by the thyroid hormone receptor. *Genes & Development, 9*, 2696–2711.

Wong, J., Shi, Y.-B., & Wolffe, A. P. (1997). Determinants of chromatin disruption and transcriptional regulation instigated by the thyroid hormone receptor: Hormone-regulated chromatin disruption is not sufficient for transcriptinal activation. *EMBO Journal, 16*, 3158–3171.

Yamauchi, K., & Tata, J. R. (1994). Purification and characterization of a cytosolic thyroid-hormone-binding protein (CTBP) in Xenopus liver. *European Journal of Biochemistry, 225*, 1105–1112.

Yaoita, Y., & Brown, D. D. (1990). A correlation of thyroid hormone receptor gene expression with amphibian metamorphosis. *Genes & Development, 4*, 1917–1924.

Yen, P. M. (2001). Physiological and molecular basis of thyroid hormone action. *Physiological Reviews, 81*, 1097–1142.

Yoon, H.-G., Chan, D. W., Huang, Z. Q., Li, J., Fondell, J. D., Qin, J., et al. (2003). Purification and functional characterization of the human N-CoR complex: The roles of HDAC3, TBL1 and TBLR1. *EMBO Journal, 22*, 1336–1346.

Zhang, J., Kalkum, M., Chait, B. T., & Roeder, R. G. (2002). The N-CoR-HDAC3 nuclear receptor corepressor complex inhibits the JNK pathway through the integral subunit GPS2. *Molecular Cell, 9*, 611–623.

Zhang, J., & Lazar, M. A. (2000). The mechanism of action of thyroid hormones. *Annual Review of Physiology, 62*, 439–466.

CHAPTER ELEVEN

Keeping Puberty on Time: *Novel* Signals and Mechanisms Involved

Manuel Tena-Sempere[*,†,‡,1]

[*]Department of Cell Biology, Physiology and Immunology, University of Córdoba, Córdoba, Spain
[†]CIBER Fisiopatología de la Obesidad y Nutrición, Instituto de Salud Carlos III, Córdoba, Spain
[‡]Instituto Maimónides de Investigaciones Biomédicas (IMIBIC)/Hospital Universitario Reina Sofía, Córdoba, Spain
[1]Corresponding author: e-mail address: fi1tesem@uco.es

Contents

1. Puberty is a Key Developmental Transition 300
2. Neurobiology of Puberty: Essential Roles of GnRH and Kiss1 Neurons 301
3. Putative Roles of Kisspeptin Partners in the Control of Puberty: The Case of NKB 307
4. Early Development and the Timing of Puberty: Kiss1 Neurons in Brain Sex Differentiation 309
5. Metabolic Control of Puberty: Leptin, Kisspeptins, and Cellular Energy Sensors 312
6. Novel Mechanisms in the Control of Puberty: Roles of Epigenetics and microRNAs 317
7. Concluding Remarks 320

Acknowledgments 321
References 321

Abstract

Puberty is the culmination of a complex series of maturational events that lead to the completion of sexual and somatic maturation and the acquisition of reproductive competence. This key developmental transition, which defines the boundary between immaturity and adulthood, is under the control of sophisticated regulatory networks that impinge upon the brain centers governing the reproductive axis. These networks are sensitive to earlier maturational events, such as brain sex differentiation, and dynamically regulated by a plethora of hormonal factors and environmental signals, which are essential for the fine-tuning of the tempo of puberty. While much knowledge on mammalian puberty had been gleaned during the last decades, important recent developments have substantially expanded our understanding of the neuroendocrine and molecular mechanisms governing puberty onset. We will provide here a synoptic account of some of these important advancements, including the identification of the essential roles of hypothalamic kisspeptin signaling, and some of its putative partners, in pubertal maturation, the characterization of novel mechanisms involved in the metabolic regulation of puberty, and the recognition of the potential roles of epigenetics and miRNA-related pathways in the central control of puberty. It is expected that further progress in these and related areas will follow in the coming years. This will

Current Topics in Developmental Biology, Volume 105
ISSN 0070-2153
http://dx.doi.org/10.1016/B978-0-12-396968-2.00011-7

permit a better understanding of the physiological mechanisms responsible for pubertal timing and will help to decipher the pathophysiological basis for pubertal alterations in humans and wildlife species.

1. PUBERTY IS A KEY DEVELOPMENTAL TRANSITION

Puberty is an essential event in the lifespan of any individual, in which sexual and somatic maturation is achieved, and reproductive capacity attained (Parent et al., 2003). In addition, important growth, behavioral, and psychological changes occur at puberty, thus allowing the acquisition of a complete adult phenotype. Because of its substantial importance as a life-changing phenomenon, the physiology of puberty and the pathophysiology of its alterations have been subjected to thorough analyses; yet, while much progress has taken place in the field in the last decades, our understanding of the molecular and neurohormonal basis of mammalian puberty is still incomplete. This is partially due to the multifaceted and transitional nature of puberty, which requires the concerted action of different homeostatic systems that become precisely activated (or inactivated) during pubertal maturation.

Indeed, rather than a specific time-point in postnatal development, puberty is the culmination of a complex series of maturational events that start *in utero* and progress throughout early postnatal, infantile, and juvenile ages. These result in a diversity of phenotypic indices, whose timing is shaped by the dynamic interplay between our gene load and the environment. Thus, while the age of puberty is subjected to considerable genetic determination (Gajdos, Henderson, Hirschhorn, & Palmert, 2010), environmental cues may account for a substantial variation in the age of puberty even within homogeneous populations (Parent et al., 2003). Accordingly, puberty has been regarded as a biological *sensor* and the trends for changes in the timing of puberty, as apparently detected in recent years in different wildlife species and human populations (see below), are considered a potential biomarker for the suspected deterioration of the environmental influences on reproductive health.

Although the age limits for normal human puberty are somewhat ample and dependent on different ethnic and geographical factors (Parent et al., 2003), several epidemiological studies published in recent years have documented a trend for an earlier initiation of puberty, as estimated by the beginning of breast development (namely, thelarche) (Aksglaede, Juul, Olsen, & Sorensen, 2009; Aksglaede, Sorensen, Petersen, Skakkebaek, & Juul, 2009).

This phenomenon, which requires further validation and demonstration in other populations, has been mainly identified in girls and has caused considerable concerns among scientists and even lay public. This is mainly due to the potential adverse outcomes of early puberty, which include increased risks of disordered behavior, cardiovascular and metabolic disease, hypertension, obesity, and diabetes, as well as lower adult height (Carter, Jaccard, Silverman, & Pina, 2009; Ibanez, Diaz, Lopez-Bermejo, & Marcos, 2009; Labayen et al., 2009; Lakshman et al., 2008, 2009; Taeymans et al., 2008). Of note, a very recent study has suggested that a similar trend for earlier puberty can be occurring also in US boys (Herman-Giddens et al., 2012). Such a *recent* decline in the age of puberty might be related to the influence of different modifiers, such as the higher prevalence of childhood obesity and/or the increased exposure to different environmental pollutants (Aksglaede, Juul, et al., 2009; Aksglaede, Sorensen, et al., 2009); earlier puberty just being a sign of disturbed homeostasis with potential serious consequences in the long term. Yet, as mentioned above, conclusive demonstration of a universal trend for earlier puberty in humans is still missing and epidemiological data must be taken with caution until such validation is provided.

In any event, the above evidence stresses the need for a deepening of our knowledge of the physiological mechanisms of normal puberty, and of the regulatory networks whereby puberty and other essential body functions, such as energy homeostasis, are jointly controlled. Indeed, in the last decade, important developments have taken place in this field, which have permitted a significant expansion of our understanding of the basis of normal and altered puberty (Tena-Sempere, 2012). In this chapter, we will provide a synoptic overview of some of these recent developments, with special emphasis on the identification of novel neuropeptide systems involved in the control of the pubertal awakening of the reproductive axis, and how these are modulated by metabolic signals and novel regulatory elements, such as epigenetics and miRNA pathways.

2. NEUROBIOLOGY OF PUBERTY: ESSENTIAL ROLES OF GnRH AND Kiss1 NEURONS

From a reproductive perspective, puberty involves the full activation of the activity of the so-called hypothalamic–pituitary–gonadal (HPG) or gonadotropic axis, which is responsible for the completion of gonadal development and the attainment of sexual phenotypic maturity (Tena-Sempere,

2012; Tena-Sempere & Huhtaniemi, 2003). The function of this neurohormonal system primarily relies on the dynamic interaction of three major groups of signals, generated at the three major levels of the axis: (a) the hypothalamus, where a relatively scarce population of neurons episodically release the decapeptide gonadotropin-releasing hormone (GnRH) into the portal circulation; (b) the anterior pituitary, that secretes gonadotropins, luteinizing hormone (LH), and follicle-stimulating hormone (FSH); and (c) the gonads, which are responsible not only for the generation of gametes but also for the release of sex steroids and peptides (Tena-Sempere & Huhtaniemi, 2003). These major components of the HPG axis are connected via feedforward and feedback loops (see Fig. 11.1). In this system, GnRH neurons hold a major hierarchical role and operate as the final output pathway for different regulatory signals, including central neuropeptides and neurotransmitters, as well as peripheral hormones (Fink, 2000).

A hallmark of puberty onset is the increase of the neurosecretory activity of GnRH neurons (Ojeda et al., 2006; Ojeda & Skinner, 2006). Episodic secretion of GnRH, which is mandatory for proper stimulation of gonadotropin release and, hence, of gonadal function, is the result of the interplay between the intrinsic oscillatory nature of GnRH neurons and a wide array of excitatory and inhibitory afferents that integrate at the so-called GnRH pulse generator (Ojeda & Skinner, 2006). On the later, it has been documented that pubertal changes in pulsatile GnRH secretion are caused by the concerted modifications in transsynaptic and glial inputs to the GnRH neuronal network (Ojeda et al., 2006; Ojeda & Skinner, 2006). Within this complex circuitry, neuronal afferents to GnRH neurons likely operate as ultimately responsible for the triggering of puberty. While the nature of such a network of neuronal transmitters has been partially elucidated in recent years, our understanding of the whole set of regulatory signals that project onto GnRH neurons, as well as of their effects and major mechanisms of action, remains incomplete. Similarly, the molecular mechanisms whereby these signals are integrated at the level of GnRH neurons to define specific patterns of pulsatile secretion of the decapeptide are still poorly understood. Of note, system biology approaches have recently allowed the identification of sets of genes/proteins that become activated at the time of puberty. This has led to the hypothesis that, rather than the consequence of the action of a single trigger, puberty is the end point of the concerted and hierarchical activation/inactivation of excitatory and inhibitory networks (Ojeda, Dubay, et al., 2010; Ojeda et al., 2006; Ojeda, Lomniczi, Sandau, & Matagne, 2010; Ojeda & Skinner, 2006). The timed regulation of those networks seemingly requires the operation of precise and multifaceted control

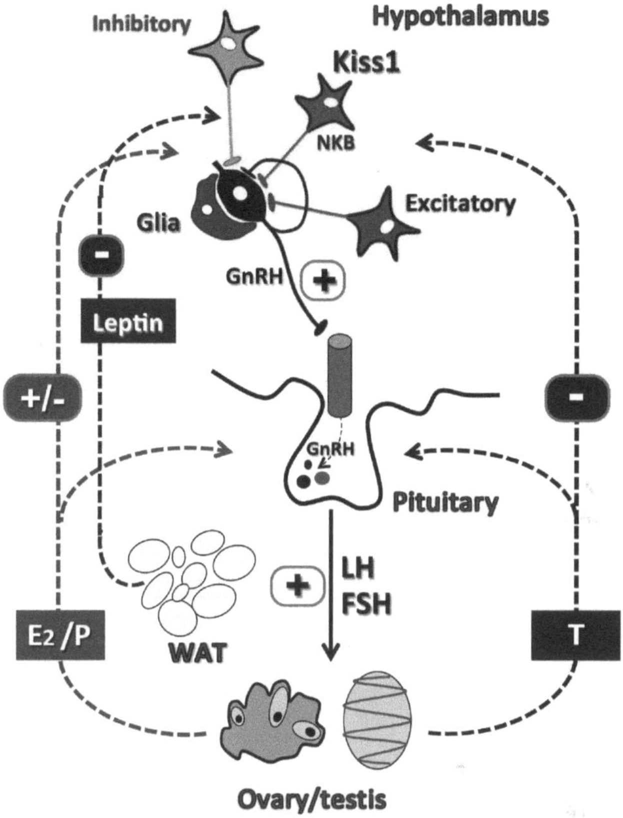

Figure 11.1 The hypothalamic–pituitary–gonadal (HPG) axis. A scheme is presented of the HPG axis, as master neurohormonal network for the control of reproduction. Hypothalamic GnRH neurons receive transsynaptic inputs (of inhibitory and excitatory nature) and glial-derived signals, which cooperate to modulate the pulsatile release of GnRH. This neuropeptide drives the secretion of gonadotropins, LH, and FSH, from the anterior pituitary, which in turn stimulate the maturation and function of the gonads; note that both the ovary and testis are presented in the figure. These major hormonal elements are connected via feedforward and feedback regulatory loops, in which gonadal steroids play a prominent role: testicular testosterone (T) inhibits GnRH/gonadotropin secretion (negative feedback), while ovarian steroids, mainly estradiol (E2) and progesterone (P), can carry out both negative and positive feedback actions depending on the stage of the ovarian cycle. Other peripheral regulators of the HPG axis mentioned in this review, such as the metabolic hormone, leptin, produced by the white adipose tissue (WAT), are also presented in the scheme. Note that for the sake of simplicity, central inhibitory and excitatory inputs to GnRH neurons are not itemized; only Kiss1 neurons are highlighted. *Adapted from references Pinilla, Aguilar, Dieguez, Millar, and Tena-Sempere (2012) and Roa, Aguilar, Dieguez, Pinilla, and Tena-Sempere (2008), with modifications.* (See color plate.)

mechanisms, whose nature has only recently begun to be elucidated (Lomniczi et al., 2013; Sangiao-Alvarellos et al., 2013).

Among the different neuropeptide regulators of GnRH neurons, kisspeptins have been recognized in recent years as essential signals for the control of puberty onset and reproductive function (Navarro & Tena-Sempere, 2012; Pinilla et al., 2012; Roa, Aguilar, et al., 2008; Tena-Sempere, 2008). Kisspeptins are a family of structurally related peptides of different amino acid length (e.g., Kp-54 and Kp-10) that share the C-terminal region. Kisspeptins are encoded by the *Kiss1* gene and operate through the G protein-coupled receptor, Gpr54, which is also termed kisspeptin receptor or Kiss1R (Oakley, Clifton, & Steiner, 2009; Pinilla et al., 2012; Roa, Aguilar, et al., 2008). While kisspeptins were initially identified as metastasis suppressors, in late 2003 two independent publications first reported the presence of inactivating mutations of their receptor in patients with isolated hypogonadotropic hypogonadism (HH); a rare condition of lack of puberty and infertility linked to low gonadotropin and sex steroid levels (de Roux et al., 2003; Seminara et al., 2003). These initial findings have been recently completed by the demonstration of the first inactivating mutation of the *KISS1* gene in humans with HH (Topaloglu et al., 2012). In good agreement, mice with inactivation of *Gpr54* or *Kiss1* genes display a phenotype roughly similar to that of affected humans (d'Anglemont de Tassigny et al., 2007; Seminara et al., 2003), thus supporting the conserved role of kisspeptin signaling in the control of the reproductive axis. This contention has been substantiated in the last years by a large number of molecular, neuroanatomical, and physiological studies in different species, which have jointly demonstrated that kisspeptins play an essential role in the direct (and probably indirect) control of GnRH neurons (Navarro & Tena-Sempere, 2012; Pinilla et al., 2012).

Based on the initial findings that humans and mice lacking functional Gpr54 or *Kiss1* do not display normal pubertal progression, different studies have addressed the putative physiological roles of kisspeptin signaling in the timing of puberty; detailed recapitulation of the results of such analyses can be found elsewhere (Navarro & Tena-Sempere, 2012; Pinilla et al., 2012). These studies have documented a complex and multifaceted pattern of activation of the Kiss1 system during the pubertal transition, which seems to involve, at least, the following major components:

- A rise in the hypothalamic expression of Kiss1 and kisspeptin immunoreactive content, which suggests an elevation of the endogenous kisspeptin tone during the pubertal transition. As deduced from pharmacological data, this increase would be sufficient to induce a state of maximal secretory activity of the GnRH/gonadotropin system (Navarro, Fernandez-Fernandez, et al., 2004; Navarro, Castellano, et al., 2004; Plant, Ramaswamy, & Dipietro, 2006; Shahab et al., 2005);

- An increase in the sensitivity to the stimulatory effects of kisspeptin on GnRH/LH secretion, as mainly documented in rodents (Castellano, Navarro, Fernandez-Fernandez, Castano, et al., 2006; Han et al., 2005);
- A heightening of Gpr54 signaling efficiency that seems to be associated to a state of resistance to desensitization to continuous kisspeptin stimulation (Han et al., 2005; Roa, Vigo, et al., 2008); and
- An elevation in the number of kisspeptin-positive neurons in the hypothalamus and of their projections to GnRH neurons, which has been documented mainly in rodents (Bentsen et al., 2010; Clarkson & Herbison, 2006).

Notably, this complex developmental program seems to form a fail-safe mechanism for the precise timing of puberty, since the GnRH/LH axis becomes fully responsive to kisspeptin stimulation at much earlier stages of postnatal development (e.g., neonatal/infantile period) (Castellano, Navarro, Fernandez-Fernandez, Castano, et al., 2006; Nazian, 2006). In the absence of precise regulatory mechanisms, the latter feature would make the system especially prone to precocious activation, as very potent gonadotropic responses are elicited in infantile rodents by administration of even low doses of kisspeptin (Castellano, Navarro, Fernandez-Fernandez, Castano, et al., 2006). The relevance of proper kisspeptin signaling for the precise timing of puberty is reinforced by the observation that pharmacological blockade of Gpr54 delays puberty onset in rats (Pineda et al., 2010) and timed ablation of Kiss1 neurons in the juvenile period prevents the attainment of reproductive competence (Mayer & Boehm, 2011).

Despite this compelling evidence, some aspects of the pubertal roles of kisspeptins remain obscure. For instance, whether the activational program described above equally applies to the different hypothalamic populations of Kiss1 neurons is yet to be characterized. In this sense, detailed studies in rodents have identified two prominent populations of Kiss1 neurons in the hypothalamus: one located in the arcuate nucleus (ARC) and the other in the rostral periventricular area of the 3rd ventricle (RP3V) (Clarkson, Anglemont de Tassigny, Colledge, Caraty, & Herbison, 2009; Pinilla et al., 2012). In other mammals, including primates, Kiss1 neurons are consistently present in the ARC/infundibular region, but the existence and/or abundance of a population equivalent to that of the RP3V Kiss1 neurons in rodents are yet the matter of debate (Pinilla et al., 2012). Of note, ARC and RP3V Kiss1 neurons seem to play different roles in the control of various aspects of reproduction and are differentially controlled by key regulators (e.g., sex steroids) (Garcia-Galiano, Pinilla, & Tena-Sempere, 2012); the basis for this phenomenon is yet to be deciphered. However, interesting differences in the modes of estrogen action (classical vs. nonclassical, as pertaining to estrogen receptor-α signaling) have been pointed out between ARC and RP3V Kiss1 neurons in mice (Gottsch et al., 2009). In addition, differential expression of the nuclear receptor, liver receptor homolog-1 (LRH-1), has been very recently documented between ARC and RP3V Kiss1 neurons; LHR-1 is selectively expressed in the ARC population, where it enhances *Kiss1* gene expression (Atkin et al., 2013). In this context, it is interesting to note that several expression analyses have suggested different roles for the ARC and RP3V Kiss1 neurons in the timing of the initiation and progression of puberty (Mayer et al., 2010), but the nature of such differential actions and their regulatory mechanisms remain ill defined (Pinilla et al., 2012).

The molecular mechanisms responsible for the activational program of Kiss1 neurons during puberty have not been fully clarified and have been addressed (mainly) in rodents. Thus, mouse studies have suggested that estrogens are indispensable for the pubertal expansion, and possibly activation, of Kiss1 neurons, mainly in the RP3V. This would imply that the initial stages of puberty, and hence the early rise of circulating sex steroids, would take place independently of kisspeptin drive. Thus, rather than the trigger of puberty, the Kiss1 system would operate as an indispensable amplifier of the secretory activity of GnRH neurons during puberty (Clarkson, Han, Liu, Lee, & Herbison, 2010). In any event, it must be stressed that equivalent studies, addressing whether a similar mechanism applies to other species, including humans, have not been conducted to date. In this context, it is worthy to note that the increase of the pulsatile secretion of gonadotropins in primates can take place in the absence of gonadal function; a finding that may argue against an indispensable role of sex steroids in the pubertal activation of the Kiss1 system in primate species.

As a final note in this section, a recent study (apparently) provided experimental evidence against an indispensable role of kisspeptin signaling in puberty onset in rodents, as it showed that congenital ablation of Kiss1 neurons in female mice was not incompatible with the attainment of adult fertility (Mayer & Boehm, 2011). It can be argued, however, that this phenotype, which is at odds with the reported impuberism of humans and mice with genetic inactivation of *Kiss1* or *Gpr54*, could be due to incomplete neuronal ablation and/or (more probably) developmental compensation.

In fact, in the same study, it was demonstrated that timed elimination of Kiss1 neurons during the early juvenile period prevented normal pubertal maturation and caused infertility (Mayer & Boehm, 2011). In fact, a similar phenomenon has been previously proposed also for GnRH, that is, that a congenital residual set of GnRH neurons might be sufficient to drive puberty onset, because of the occurrence of compensatory fail-safe (Herbison, Porteous, Pape, Mora, & Hurst, 2008). In fact, those findings further document the complex pattern of developmental regulation of Kiss1 neurons that, despite potential compensatory mechanism, are essential for the timed activation of the HPG axis that leads to puberty.

3. PUTATIVE ROLES OF KISSPEPTIN PARTNERS IN THE CONTROL OF PUBERTY: THE CASE OF NKB

Besides the demonstration of the important roles of kisspeptins in the control of puberty and fertility, significant efforts have been devoted recently to elucidate the major regulatory signals whereby the different Kiss1 neuronal populations are specifically controlled in different developmental periods and functional states of the HPG axis (Lehman, Coolen, & Goodman, 2010). In this context, anatomical analysis have demonstrated that at least an important fraction of Kiss1 neurons in the ARC, but not in the RP3V, coexpress neurokinin B (NKB) and its receptor (NK3R) (Navarro & Tena-Sempere, 2012). The physiological (and translational) relevance of this phenomenon is suggested by the finding that humans with inactivating mutations of the NKB or NK3R gene display a state of HH similar to that of *GPR54* or *KISS1* mutants (Topaloglu et al., 2009; Young et al., 2010). In the same vein, mouse lines with defective NKB signaling have been shown to display reduced fertility and some degree of impairment of gonadotropin function (Yang, Caligioni, Chan, & Seminara, 2012), although the rodent phenotype seems to be less dramatic than in humans. Of note, the coexpression of NKB and kisspeptins in the ARC have been documented in numerous species, including sheep (Goodman et al., 2007), goat, mouse (Navarro, Gottsch, et al., 2009; Wakabayashi et al., 2010), rat (Burke, Letts, Krajewski, & Rance, 2006), and monkey (Ramaswamy et al., 2010). Yet, a recent report suggested that a fraction of kisspeptin and NKB immune reactivity does not appear to colocalize in the infundibular area of the hypothalamus of young men (Hrabovszky et al., 2012). Yet, even in that study, as much as 75% of kisspeptin-positive perikarya showed also distinctive NKB expression. To add further complexity to the system, the expression of dynorphin-A

(Dyn) has also been demonstrated in a subset of ARC Kiss1/NKB neurons (Navarro, Gottsch, et al., 2009). Hence, the term KNDy has been coined to rename the ARC population of Kiss1 neurons, in order to reflect the concerted expression of these three neuropeptides (Lehman et al., 2010).

From a functional standpoint, numerous hormonal studies in several mammalian species have documented potent stimulatory actions of the NKB agonist, senktide, on LH secretion (Billings et al., 2010; Garcia-Galiano, van Ingen Schenau, et al., 2012; Navarro, Castellano, et al., 2011; Ramaswamy et al., 2010); effects that were absent in Gpr54 null mice thus suggesting that these require the integrity of kisspeptin signaling (Garcia-Galiano, van Ingen Schenau, et al., 2012). This contention is also supported by similar findings in the monkey (Ramaswamy, Seminara, & Plant, 2011), and by electrophysiological data demonstrating that senktide can stimulate Kiss1, but not GnRH, neurons (Navarro, Gottsch, et al., 2011). Altogether, integration of the available data has crystallized in a working model in which KNDy neurons in the ARC operate as major regulatory node for the generation of GnRH pulses (Navarro & Tena-Sempere, 2012). According to this model, KNDy neurons, which form a profusely interconnected network, are able to finely regulate kisspeptin output to GnRH neurons, and eventually other afferents, thanks in part to the positive drive of NKB, which would operate to activate GnRH neurons in an indirect, kisspeptin-mediated manner (see Fig. 11.2). Notwithstanding this, evidence for null or even inhibitory LH responses to NKB agonists has also been presented (Kinsey-Jones et al., 2012; Sandoval-Guzman & Rance, 2004); a phenomenon whose basis is yet to be clarified.

In the above context, the putative roles of NKB signaling in the control of puberty have attracted some attention recently (Navarro et al., 2012; Nestor et al., 2012; Ruiz-Pino et al., 2012). Our recent data demonstrated that, as is the case in adult females, prepubertal female rats display potent LH responses to the NKB agonist, senktide (Navarro et al., 2012). In addition, we have documented that during puberty the genes encoding NKB and NK3R are abundantly expressed in hypothalamic areas, such as the ARC, with key roles in the central control of the HPG axis, and its hypothalamic expression is substantially augmented during the infantile-to-juvenile transition (Navarro et al., 2012). Of note, comparison of the expression profiles of *NKB* and *Kiss1* genes in the hypothalamus of female rats during postnatal development would suggest that the rise of NKB expression anticipates the elevation of *Kiss1* mRNA levels; a phenomenon that has been recently documented in the female mouse (Gill et al., 2012). The functional relevance of this phenomenon is indirectly supported by the fact that blockade of NKB signaling by the use of a specific antagonist induces a modest but

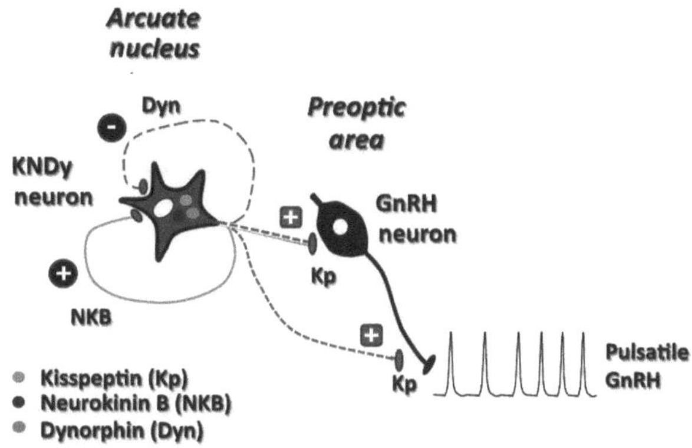

Figure 11.2 KNDy neurons and the regulation of GnRH secretion. Neurons in the arcuate nucleus (ARC) that coexpress kisspeptins (Kp), neurokinin B (NKB) and dynorphin (Dyn), also termed KNDy neurons, have been proposed to play an essential role in the control of the pulsatile secretory activity of GnRH neurons, mainly located in the preoptic area. In the scheme, a model for the functional roles of these KNDy transmitters in the dynamic regulation of Kp output and pulsatile GnRH secretion is presented. According to this model, NKB would operate as (auto)stimulatory signal for KNDy neurons, whereas Dyn would suppress Kp release. This model is supported by a wealth of pharmacological and neuroanatomical data, including the demonstration of the expression of NKB receptors in KNDy neurons and the dense network of fiber connections detected between ARC KNDy neurons (depicted in the scheme as auto-projections). However, the expression of the Dyn receptor in KNDy neurons is yet to be fully confirmed. In addition, some anatomical features of this network, for example, the precise site of projections of ARC KNDy neurons to GnRH neurons, remain unsolved and are depicted using *dotted lines. Adapted from references Navarro and Tena-Sempere (2012) and Roa, Aguilar, et al. (2008), with modifications.* (For color version of this figure, the reader is referred to the online version of this chapter.)

detectable delay in the timing of puberty in female rats (Navarro et al., 2012); admittedly, congenital NK3R null mice seem to display normal pubertal timing (Yang et al., 2012). In any event, a modulatory role of NKB signaling on puberty has been very recently proposed also in the female sheep (Nestor et al., 2012). Whether NKB signaling plays a similar role in male puberty is yet to be defined. In fact, our recent data suggest that LH responses to senktide are lower in male than in female rats before puberty, and males, but not females, become irresponsive to NKB agonist stimulation during the pubertal transition to adulthood (Ruiz-Pino et al., 2012).

4. EARLY DEVELOPMENT AND THE TIMING OF PUBERTY: Kiss1 NEURONS IN BRAIN SEX DIFFERENTIATION

Full activation of the gonadotropic system at puberty critically depends on the adequate functional organization of the hypothalamic

GnRH network at earlier stages of development (Gore, 2008); a phenomenon that takes place during the process of brain sex differentiation. Indeed, the timing of puberty and different adult reproductive traits are sexually dimorphic, and manifest as the late expression of the specific determination of the brain circuitries involved in the control of the HPG axis during critical developmental windows. In rodents, brain sex differentiation takes place during a period spanning from late embryonic to early postnatal age (Gore, 2008; Morris, Jordan, & Breedlove, 2004). Accordingly, experimental manipulations during this age-window that perturb the normal process of brain sex differentiation can alter the timing of puberty and/or disrupt the function of the adult gonadotropic system. For instance, neonatal androgenization of female rodents prevents the capacity of estrogen to induce positive feedback and hence to evoke the preovulatory surges of gonadotropins in adulthood that are the hormonal drive for ovulation.

Given its paramount importance in the timing of puberty and the regulation of adult reproductive function, the process of sexual differentiation of the hypothalamic Kiss1 system during early developmental periods has been the subject of analysis, especially in rodents. Initial studies already documented that the Kiss1 neuronal population in the RP3V in the rat is sexually dimorphic, with females having much greater numbers of Kiss1 neurons than males. Moreover, RP3V Kiss1 neurons appeared to be sensitive to the organizing actions of sex steroids during the window of sexual maturation of the brain (Kauffman et al., 2007). Thus, neonatal exposure of female rats to high doses of androgen caused the complete "masculinization" of the pattern of *Kiss1* mRNA expression at the RP3V in adulthood; namely, androgenized females showed much lower *Kiss1* mRNA levels than cyclic female rats (Kauffman et al., 2007). Moreover, exposure to estradiol as adults failed to increase *Kiss1* mRNA levels in the RP3V of neonatally androgenized females, in contrast to the clear stimulatory responses observed in cyclic females (Kauffman et al., 2007). Those observations were the first to demonstrate the sensitivity of the hypothalamic Kiss1 system to the organizing actions of endogenous (and eventually exogenous) sex steroids (Navarro & Tena-Sempere, 2008). In good agreement, later analyses in α-fetoprotein (AFP) knock-out mice, in which the congenital lack of this scavenger protein of circulating estrogens causes an excessive estrogenic input during early development, showed that sexual differentiation and function of the hypothalamic Kiss1 system is also severely disrupted in these mice (Gonzalez-Martinez, De Mees, Douhard, Szpirer, & Bakker, 2008). Furthermore, neonatal exposures to synthetic estrogens, known to alter the normal activation and function of the gonadotropic axis (Tena-Sempere, Pinilla, Gonzalez, & Aguilar, 2000), persistently suppressed the hypothalamic expression of *Kiss1* gene at the expected time of puberty and adulthood in the rat (Navarro, Fernandez-Fernandez, et al., 2004; Navarro, Sanchez-Garrido, et al., 2009).

Additional evidence for the influence of early sex steroid input for proper sexual organization of the hypothalamic Kiss1 system came from the studies involving neonatal gonadectomy (GNX) of male rats, a procedure that results in the elimination of androgen actions during the critical period of brain sex differentiation. This manipulation induced the *feminization* of *Kiss1*/kisspeptin expression at the RP3V, namely, males undergoing neonatal GNX displayed higher expression than control male rats in adulthood (Homma et al., 2009). Furthermore, neonatal GNX also caused the acquisition of positive feedback and surge-like LH responses to ovulatory doses of estradiol (Homma et al., 2009), which is a characteristic female trait. These findings would suggest that feminization of the RP3V Kiss1 population is merely a *default* pathway. However, more recent data from the hypogonadal *hpg* mouse, which is severely deficient in GnRH secretion and, hence, has very low sex steroids levels, evidenced that *hpg* female mice have lower numbers of kisspeptin neurons in the RP3V than controls and do not show the normal stimulatory response to estrogen at this nucleus, thus suggesting that some degree of estrogenic input is needed for the complete functional differentiation of RP3V population of Kiss1 neurons also in the female (Gill et al., 2010).

While compelling evidence supports that the RP3V population of Kiss1 neurons is sexually dimorphic, a feature that is likely to contribute to the differential timing of puberty between sexes (Clarkson et al., 2010; Kauffman, Navarro, Kim, Clifton, & Steiner, 2009), it remains debatable whether the same applies to the ARC population of Kiss1 neurons. Thus, initial studies, assessing *Kiss1* mRNA expression, suggested that the ARC population of Kiss1 neurons is not sexually dimorphic in the rat (Kauffman et al., 2007). However, more recent analyses have illustrated that, as is the case in the RP3V, Kiss1 neurons in the ARC are likely more numerous in the female, as documented recently by immunohistochemical and/or *in situ* hybridization studies in rodents, sheep, and primates (Cheng, Coolen, Padmanabhan, Goodman, & Lehman, 2010; Hrabovszky et al., 2010; Takumi, Iijima, & Ozawa, 2011). Moreover, *Kiss1* mRNA levels in the ARC seem to be higher in female rats already during the neonatal period, as reported recently (Cao & Patisaul, 2011). Interestingly, studies in the sheep suggest that prenatal exposure to testosterone does not alter kisspeptin content in the ARC of adult females, but blunts the expression of NKB (Cheng et al., 2010).

As final comment in this section, it is stressed that most of the information regarding the process of sexual differentiation of the hypothalamic Kiss1 system has been obtained in rodents, and some of the features described above might not be shared by other mammalian species. In fact, classical studies suggested decades ago that the mechanisms of sexual differentiation of gonadotropin control might be partially different between rodents and primates, as LH surges could be induced in male monkeys, even not subjected to hormonal manipulation at early stages of development, if provided with an appropriate ovarian (estrogen) signal as adults (Norman & Spies, 1986). In this context, additional studies are needed to elucidate the nature and physiological relevance of the process of sexual differentiation of the hypothalamic kisspeptin system in non-rodent mammalian species, including humans, and whether this phenomenon plays a significant role in the timing of puberty in such species.

5. METABOLIC CONTROL OF PUBERTY: LEPTIN, KISSPEPTINS, AND CELLULAR ENERGY SENSORS

It is well known since Old Ages that the amount of energy reserves of the organism, and by extension its metabolic status, are important factors in the modulation of the timing of puberty (Fernandez-Fernandez et al., 2006; Martos-Moreno, Chowen, & Argente, 2010). This is especially evident in the female, in which acquisition of reproductive capacity implies the need to face the eventual metabolic drainage of pregnancy and lactation. Hence, from a teleological perspective, it makes biological sense that puberty is only achieved when sufficient energy stores and optimal metabolic conditions are attained. Notwithstanding this, the influence of metabolic cues and energy homeostasis on puberty and reproductive function is also detected in males (Castellano et al., 2009), in which energy reserves are needed (at least in wildlife species) to cope with some key aspect of reproduction, such as territoriality and partner selection (Elias, 2012; Elias & Purohit, 2013). Furthermore, the tight connection between energy balance, puberty, and reproduction becomes evident not only in conditions of energy deficit, but also in situations of energy excess that, if of sufficient magnitude and persistent in time (e.g., early-onset morbid obesity), are associated to detectable alterations of the timing of puberty; for instance, precocious puberty seems to be more frequent in obese girls (Castellano et al., 2011). Moreover, alterations of the reproductive axis in adulthood have also been observed in conditions of persistent overweight; for example, hypogonadism of central origin has been described in obese men (Tajar et al., 2010). As a whole, these observations stress the close connection between the metabolic status and reproductive function, including puberty onset.

Compelling evidence gathered in the last two decades has documented the essential roles of the adipose hormone, leptin, in the metabolic control of puberty and fertility (Fernandez-Fernandez et al., 2006; Tena-Sempere, 2007). The circulating levels of leptin are proportional to the amount of body energy (fat) stores, so that leptin operates as signal of energy abundance for the reproductive system. However, despite active research in this area, the precise mechanisms and circuits whereby leptin conducts such biological function are yet to be completely elucidated. In any event, it has now been well demonstrated that leptin likely acts as a permissive (rather than trigger) factor that allows puberty to proceed only if sufficient body energy reserves are achieved (Castellano et al., 2009; Roa et al., 2010). The mechanisms for this permissive action are probably multifaceted, as illustrated by the capacity of leptin to operate at the different levels of the HPG axis (Tena-Sempere, 2007). Nonetheless, it is clear that the major site for the permissive effects of leptin is the hypothalamus, where the adipose hormone is able to modulate the GnRH neuronal system (Cunningham, Clifton, & Steiner, 1999). Yet, this action seems to be conducted in an indirect manner, as GnRH neurons do not physiologically express functional leptin receptors, as documented by expression studies in rodents and primates (Finn et al., 1998; Hakansson, Brown, Ghilardi, Skoda, & Meister, 1998), as well as by functional genomic analyses that showed that selective elimination of leptin receptors in GnRH neurons did not prevent puberty or fertility (Quennell et al., 2009).

Because of the indirect mode of leptin action on GnRH neurons, research efforts have been devoted to elucidate the neuronal circuit whereby leptin, and eventually other key metabolic signals, may transmit their regulatory actions to GnRH neurons. One pathway that has attracted considerable attention is that of Kiss1 neurons, whose role in the metabolic control of reproduction in general, and of puberty onset in particular, has been extensively analyzed. In this context, compelling evidence accumulated in recent years has demonstrated that Kiss1 neurons in the hypothalamus may sense different forms of metabolic stress and thus may convey the regulatory actions of metabolic cues to the centers that control puberty (Castellano et al., 2009; Tena-Sempere, 2010). The sensitivity of the hypothalamic Kiss1 system to changes in the metabolic status and energy homeostasis has been well characterized in conditions of negative energy balance. For instance, acute fasting is known to suppress of *Kiss1* mRNA expression and kisspeptin content in the hypothalamus of pubertal rats (Castellano, Bentsen, Mikkelsen, & Tena-Sempere, 2010; Castellano et al., 2005); similar observations have been made in adult rats subjected to 50% caloric restriction (True, Kirigiti, Kievit, Grove, & Susan Smith, 2011). In turn, the administration of kisspeptin can reverse the state of low gonadotropin levels and delayed puberty caused by chronic undernutrition in prepubertal female rats (Castellano et al., 2005). Analogous findings have been reported in adult models of metabolic stress (Castellano, Navarro, Fernandez-Fernandez, Roa, et al., 2006; Kalamatianos, Grimshaw, Poorun, Hahn, & Coen, 2008; Luque, Kineman, & Tena-Sempere, 2007; Smith, Acohido, Clifton, & Steiner, 2006), where the hypothalamic Kiss1 tone is supposedly inhibited, and the hypogonadotropic state rescued by exogenous kisspeptin. Altogether, these data stress the capacity of Kiss1 neurons to detect conditions of energy insufficiency that may suppress the functionality of the HPG axis, to a large extent, via inhibition of kisspeptin signaling.

Interestingly, the studies in models of metabolic stress in pubertal animals have recently revealed that, as is the case for Kiss1, the hypothalamic expression of NKB is also sensitive to situations of negative energy balance. Thus, acute fasting not only inhibited *Kiss1* mRNA expression in the ARC and RP3V, but also suppressed the hypothalamic expression of the genes encoding NKB and NK3R in pubertal female rats (Navarro et al., 2012). Moreover, as shown for exogenous kisspeptin administration, the delay of puberty caused by undernutrition was partially prevented by the administration of an NKB agonist, thus supporting that NKB signaling may cooperate with kisspeptins in the metabolic control of puberty (Navarro et al., 2012). Additional demonstration of the sensitivity of NKB expression to metabolic regulation during puberty was recently provided by studies on the impact of feeding a high fat diet to female rats from weaning onwards, which documented not only the occurrence of precocious puberty onset, but also of increased expression in the ARC of the gene encoding NKB, as well as of the *Kiss1* gene (Feng Li, Lin, Kinsey-Jones, & O'Byrne, 2012).

In fact, fragmentary evidence has now demonstrated that not only conditions of energy deficit but also persistent energy surplus has a discernible impact on the hypothalamic Kiss1 system. Admittedly, however, this phenomenon remains ill defined and the net impact of morbid obesity on central kisspeptin signaling likely depends on its duration and age of presentation.

Thus, conditions of long-term obesity in adulthood have been associated to some degree of suppression of the hypothalamic Kiss1 system. This contention was initially suggested by our studies in diet-induced obese male mice of the C57/B6 strain (Luque et al., 2007), and was later documented by subsequent analyses in DBA/2J mice, which are prone to develop obesity-induced infertility (Tortoriello, McMinn, & Chua, 2004), in which diet-induced obesity caused a significant suppression of kisspeptin expression in the ARC and RP3V of females (Quennell et al., 2011). On the other hand, early-onset obesity might cause opposite effects on the hypothalamic Kiss1 system, at least during puberty, as we have recently documented that postnatal overfeeding of female rats induces earlier occurrence of puberty in association with increased hypothalamic expression of *Kiss1* and a trend for higher numbers of kisspeptin fibers in the RP3V immediately prior the onset of puberty (Castellano et al., 2011). Admittedly, however, the magnitude of these changes seems modest and other studies in rats and mice have failed to detect clear-cut changes in hypothalamic Kiss1 expression following postnatal over-nutrition (Caron, Ciofi, Prevot, & Bouret, 2012; Smith & Spencer, 2012); a discrepancy that might be due to differences in the methodological approaches, age-windows of analysis and species used among those studies. In any event, the sensitivity of the developing Kiss1 system to early metabolic influences is stressed by findings in undernourished rats and mice during the postnatal period that display indices of suppression of the hypothalamic Kiss1 system, namely decreased *Kiss1* expression and kisspeptin-positive neurons/fibers, coupled to delayed puberty (Caron et al., 2012; Castellano et al., 2011).

In this context, the possibility that metabolic signals may act directly on Kiss1 neurons has been actively investigated. Again, most of the attention on this front has concentrated primarily on leptin. Yet, direct or indirect evidence for the potential actions of ghrelin and insulin on Kiss1 neurons has also been reported (Forbes, Li, Kinsey-Jones, & O'Byrne, 2009; Qiu et al., 2013). Regarding leptin effects, the detectable expression of the mRNA encoding the functional leptin receptor, LepRb (also termed Ob-Rb), at least in a subset of Kiss1 neurons in the ARC, as documented in the mouse and sheep (Backholer et al., 2010; Smith et al., 2006), and the capacity of leptin, at least at high doses, to enhance the hypothalamic expression of *Kiss1* gene in different models of severe metabolic stress, such as the leptin-deficient ob/ob mouse and the diabetic rat, led to the proposal that leptin acts directly on Kiss1 neurons to conduct at least part of its stimulatory/permissive effects on GnRH neurons (Castellano Navarro, Fernandez-Fernandez, Roa, et al., 2006; Luque et al., 2007; Smith et al., 2006). This possibility is further supported by *in vitro* work, showing that leptin increases *Kiss1* mRNA expression in suitable neuronal cell lines (Luque et al., 2007), and is able to induce the electrical firing of Kiss1 neurons via activation of TRPC channels in the guinea pig (Qiu, Fang, Bosch, Ronnekleiv, & Kelly, 2011).

Recent data, however, seem to suggest that the actions of leptin on the hypothalamic Kiss1 system might be predominantly indirect. Thus, selective congenital elimination of Ob-Rb in Kiss1 cells failed to prevent the normal occurrence of puberty onset and the attainment of fertility (Donato et al., 2011). Admittedly, however, the latter approach using the Cre-loxP technology cannot guarantee the selective ablation of Ob-Rb in postnatal Kiss1 neurons in the hypothalamus, as the *Kiss1* gene is widely expressed in other brain and other peripheral tissues at early stages of development. Furthermore, since knockout of Ob-Rb was induced congenitally in Kiss1 cells, the possibility of compensatory changes that may have masked the impact of the lack of leptin signaling in this neuronal population cannot be ruled out. In any event, an independent study mapping the distribution of the functional Ob-Rs in various hypothalamic areas failed to demonstrate substantial expression of leptin receptors in GnRH or Kiss1 neurons, except for a small fraction of Kiss1 cells in the ARC (Louis et al., 2011). Intriguingly, the latter study found an as yet uncharacterized population of ObRb-expressing neurons in the ARC and RP3V, in the proximity of Kiss1 neurons (Louis et al., 2011). Thus, it is possible that a part of the positive effects of leptin on the hypothalamic expression of *Kiss1*/kisspeptin is conducted indirectly and transmitted via intermediate circuits (Louis et al., 2011), whose nature and transmission await further clarification. Interestingly, a recent report has suggested that direct leptin signaling in Kiss1 neurons may arise after puberty (Cravo et al., 2013), suggesting an intriguing switch between indirect and direct effects of leptin on Kiss1 circuits during the time of pubertal maturation, whose physiological relevance is yet to be exposed. On the other hand, solid experimental evidence has documented that part of the reproductive effects of leptin are primarily transmitted via hypothalamic areas other than the major Kiss1-expressing nuclei, such as the ventral premammillary nucleus (PMV). Extensive description of the roles of PMV as a relay for the reproductive effects of leptin can be found elsewhere (Donato et al., 2011, 2009). Of important note, recent evidence strongly suggests that PMV projections participate in the regulation of Kiss1 neurons (Donato et al., 2013) and, hence, may contribute to the transmission of leptin effects onto Kiss1 circuits, as discussed above.

The central molecular mediators for the metabolic control of puberty in general, and of the Kiss1 system in particular, have also been studied, and partially characterized, in recent years. A paradigmatic example in this area was the identification of the putative roles of signaling pathways involving the cellular energy sensor, mammalian target of rapamycin (mTOR), in mediating the effects of leptin in the metabolic regulation of puberty and the hypothalamic expression of *Kiss1*. Detailed description of the physiological roles of mTOR in the control of cellular energy homeostasis and food intake can be found elsewhere (Roa & Tena-Sempere, 2010). Notably, central mTOR signaling has been proposed to be essential in keeping energy balance at the whole body level, among other mechanisms, by transmitting the anorectic effects of leptin (Cota et al., 2006). Studies conducted in pubertal female rats strongly suggested that central mTOR signaling also plays a relevant role in the control of puberty, as demonstrated by the impact of chronic blockade of mTOR by central injection of rapamycin, which significantly delayed the onset of puberty (Roa et al., 2009). In keeping with this action, inhibition of central mTOR blocked the permissive/stimulatory effect of leptin on pubertal onset, as illustrated by studies in pubertal female rats subjected to undernutrition in which the positive effects of leptin in terms of *rescue* of puberty onset against unfavorable metabolic conditions were prevented by central coadministration of rapamycin (Roa et al., 2009). The impact of mTOR blockade on puberty onset is seemingly mediated, at least partially, by the inhibition of the hypothalamic Kiss1 system, as suggested by the fact that central rapamycin treatment caused a significant suppression of *Kiss1* mRNA levels in key hypothalamic nuclei, such as the ARC (Roa et al., 2009). Altogether, the above observations would suggest that a leptin–mTOR–kisspeptin pathway plays a key role in the metabolic regulation of puberty (Roa & Tena-Sempere, 2010). Yet, the neuroanatomical features of such a circuitry remain unsolved. In fact, recent studies documented that the downstream target of mTOR, pS6, is not apparently expressed in Kiss1 neurons (Quennell et al., 2011). This would support the contention that leptin effects on *Kiss1* expression are mainly conducted via indirect afferents rather than through direct regulatory action of Kiss1 neurons; a phenomenon whose physiological importance during puberty and adulthood merits further analyses.

6. NOVEL MECHANISMS IN THE CONTROL OF PUBERTY: ROLES OF EPIGENETICS AND microRNAs

In previous sections, we have highlighted the complex nature of puberty onset, in which the time-course for the developmental maturation and full activation of the different elements of the HPG axis is tailored by the intricate interplay between central and peripheral signals, including numerous neuropeptides and hormones. Such an inherent complexity indirectly suggests the operation of multiple regulatory mechanisms, which would be needed to enable the proper timing of puberty. Recent evidence has suggested that such mechanisms likely include, in addition to the *classical* transcriptional regulation of sets of genes, nontranscriptional regulatory phenomena, such as epigenetics and microRNAs (miRNAs). Admittedly, the precise roles of such *nonclassical* mechanisms are yet to be fully exposed, as recognition of their participation in the control of puberty *escaped* conventional endocrine/molecular analyses until very recently.

In this context, evidence is mounting that the regulation of key components of the central pathways governing puberty onset involve epigenetic mechanisms. Epigenetics is defined as inheritable information that is not encoded by the nucleotide sequence of a given gene, but rather by modifications of the DNA or the protein component of chromatin, histones (Paulsen, Tierling, & Walter, 2008; Sims, Magazinnik, Houston, Wu, & Rice, 2008; Van Emburgh & Robertson, 2008). Incorporation of methyl groups to CpG islands in the DNA is the only known form of epigenetic modification of DNA, which is a general mark of gene repression (Paulsen et al., 2008; Van Emburgh & Robertson, 2008). In contrast, epigenetic changes on histones may include a diversity of modifications, such acetylation, methylation, and phosphorylation (Sims et al., 2008), whose net effect in terms of gene repression or activation depends on the histone and residue subjected to modification (Sims et al., 2008). Of note, epigenetic changes may participate in the control of long-term developmental modifications as well as in rapid modulatory events; both phenomena are seemingly involved in the regulation of key aspects of puberty (Ojeda, Dubay, et al., 2010). However, studies addressing the potential participation of epigenetics in the control of mammalian puberty have not been undertaken until very recently.

In a series of elegant studies, Ojeda and colleagues have very recently provided compelling evidence for the involvement of epigenetic regulatory mechanisms in the central control of female puberty. Thus, profound changes in the patterns of DNA methylation have been found to take place in the hypothalamus during puberty in female rodents (Lomniczi et al., 2013). In addition, the hypothalamic expression of two key transcriptional repressors of the Polycomb group (PcG), namely Eed and Cbx7, has been shown to decrease during pubertal progression due to enhanced methylation of their promoters (Lomniczi et al., 2013). Accordingly, systemic pharmacological blockade of DNA methylation in juvenile female rats was found to delay the timing of puberty (Lomniczi et al., 2013). Analogous results, albeit of more modest magnitude, have been obtained by our group in initial phenotypic and hormonal analyses of rat models of central pharmacological manipulation of methylation or deacetylation during the pubertal transition (Leon, Castellano, & Tena-Sempere, unpublished). The mechanism whereby methylation-induced suppression of PcG repressors induces the onset of puberty likely involves the activation of *Kiss1* expression, and prevention of this phenomenon resulted in delayed puberty (Lomniczi et al., 2013). Altogether, the above findings suggest that the tempo of puberty is finely controlled by inhibitory mechanisms that are lifted at the time of pubertal maturation by methylation changes, thus allowing the increase in hypothalamic *Kiss1* expression (and likely activity) that characterizes the onset of puberty (Lomniczi et al., 2013). In line with this phenomenon, recent observations strongly suggest that the differential methylation of the *Kiss1* gene might contribute to the dimorphism in the expression levels of *Kiss1* in the RP3V between male and female rodents (Semaan, Dhamija, Kim, Ku, & Kauffman, 2012), which arises in part during the pubertal transition. In addition, estrogen-induced changes in histone acetylation at the *Kiss1* promoter in this rostral area may play a role in the positive feedback responsible for the generation of the preovulatory surge (Tomikawa et al., 2012); a finding that stresses the importance of epigenetic regulatory mechanisms in the control of the Kiss1 system.

In addition to epigenetic changes, miRNA regulatory pathways might also be involved in the control of pubertal maturation; yet, the available evidence remains (mostly) circumstantial and needs to be thoroughly validated from a functional standpoint. In this context, in 2009, four independent genome-wide association studies identified an association of the age at menarche with variability at 6q21, in or near the *Lin28B* gene (He et al., 2009; Ong et al., 2009; Perry et al., 2009; Sulem et al., 2009). The robustness of this finding was later confirmed by an extensive meta-analysis (Elks et al., 2010), and the fact that Lin28B has also been linked to breast development and adult height. Furthermore, over-expression of the related Lin28A was shown to delay puberty in mice (Zhu et al., 2010), among other phenotypic features. While the mechanisms for such a pubertal role of Lin28 proteins is yet to be exposed, it was interesting to note that Lin28s are RNA-binding proteins whose major known function is to block the processing of miRNAs of the *let-7* family, via inhibition of the maturation of *let-7* precursors (Heo et al., 2008; Viswanathan, Daley, & Gregory, 2008). This was the first (indirect) suggestion of the potential involvement of miRNAs, of the *let-7* family and/or other related groups, in the regulation of puberty.

Departing from these initial observations, we have recently undertaken expression analyses to document the presence and potential changes in the expression levels of the members of the Lin28/*let-7* hub in the hypothalamus, as a key center for the control of puberty, as well as the gonads during the pubertal transition. These analyses have allowed us to document the robust expression of *Lin28A* and *Lin28B* mRNAs in rodent hypothalamus, whose relative levels markedly decreased from the neonatal to the pubertal period (Sangiao-Alvarellos et al., 2013). Such a developmental trend for decrease expression was not detected in the rat cerebral cortex (Sangiao-Alvarellos et al., 2013) or in the mouse testis (Gaytan et al., 2013), therefore suggesting tissue specificity. In addition, the expression of *let-7a* and *let-7b* miRNAs in the rat hypothalamus was also detected in our study; yet, in contrast to Lin28 mRNAs, their relative levels significantly increased during postnatal maturation (Sangiao-Alvarellos et al., 2013). These observations suggest an inverse relationship between *Lin28* and *let-7* expression levels in the hypothalamus, which is in keeping with the proposed role of Lin28 as putative repressor of *let-7* maturation. Of note, perturbation of early sex differentiation and puberty, by means of neonatal treatment with androgen or estrogen, resulted in significantly altered rations of *Lin28/let-7* at the time of puberty; a phenomenon that was also observed, albeit at lower magnitude, in other models of early perturbation of puberty onset (Sangiao-Alvarellos et al., 2013). While functional data to document the impact of changes in the hypothalamic *Lin28/let-7* ratios upon the onset of puberty are still missing (studies which are presently in progress in our laboratory), the available indirect evidence strongly suggests that developmental changes in hypothalamic *Lin28/let-7* expression might have a role in the mechanisms permitting/leading to puberty onset.

7. CONCLUDING REMARKS

Among the different developmental phenomena that lead to the acquisition of the adult phenotype, puberty is the culmination of a complex maturational program that is largely dictated by genetic factors but whose timing is shaped by the dynamic interplay of a plethora of endogenous factors and environmental cues. As illustrated in this chapter, much has been learnt in the last decade about the molecular and endocrine mechanisms responsible for the timing of puberty; a phenomenon whose relevance is highlighted by the diversity of maturational (reproductive, somatic, and psychological) events that take place during puberty, and the suspected impact that substantial changes in the tempo of puberty might have in various health parameters later on life. While the progress of our knowledge of the (patho) physiology of puberty has been astonishing, the occurrence of very recent, exciting developments, regarding the identification of novel neuropeptide pathways (such as kisspeptins and NKB) and molecular mechanisms (such as epigenetics and miRNA pathways) involved in the control of pubertal maturation, allows us to predict that much research activity will concentrate in these and related areas in the coming years; efforts that are expected to further clarify one of the most intriguing and complex developmental phenomena in the lifespan of any individual.

ACKNOWLEDGMENTS

The author is indebted with the members of the research team at the Physiology Section of the University of Cordoba, who actively participated in the generation of experimental data discussed herein. The work from the author's laboratory summarized in this chapter was supported by grants BFU 2008-00984 and BFU2011-25021 (Ministerio de Economía y Competitividad, Spain; co-funded with EU FEDER funds), Project PI042082 (Ministerio de Sanidad, Spain), Project P08-CVI-03788 (Junta de Andalucía, Spain), and EU research contract DEER FP7-ENV-2007-1. CIBER is an initiative of Instituto de Salud Carlos III (Ministerio de Sanidad, Spain).

Disclosure statement: The author has nothing to disclose.

REFERENCES

Aksglaede, L., Juul, A., Olsen, L. W., & Sorensen, T. I. (2009). Age at puberty and the emerging obesity epidemic. *PLoS One, 4*, e8450.

Aksglaede, L., Sorensen, K., Petersen, J. H., Skakkebaek, N. E., & Juul, A. (2009). Recent decline in age at breast development: The Copenhagen Puberty Study. *Pediatrics, 123*, e932–e939.

Atkin, S. D., Owen, B. M., Bookout, A. L., Cravo, R. M., Lee, C., Elias, C. F., et al. (2013). Nuclear receptor LRH-1 induces the reproductive neuropeptide kisspeptin in the hypothalamus. *Mol Endocrinol, 27*, 598–605.

Backholer, K., Smith, J. T., Rao, A., Pereira, A., Iqbal, J., Ogawa, S., et al. (2010). Kisspeptin cells in the ewe brain respond to leptin and communicate with neuropeptide Y and proopiomelanocortin cells. *Endocrinology, 151*, 2233–2243.

Bentsen, A. H., Ansel, L., Simonneaux, V., Tena-Sempere, M., Juul, A., & Mikkelsen, J. D. (2010). Maturation of kisspeptinergic neurons coincides with puberty onset in male rats. *Peptides, 31*, 275–283.

Billings, H. J., Connors, J. M., Altman, S. N., Hileman, S. M., Holaskova, I., Lehman, M. N., et al. (2010). Neurokinin B acts via the neurokinin-3 receptor in the retrochiasmatic area to stimulate luteinizing hormone secretion in sheep. *Endocrinology, 151*, 3836–3846.

Burke, M. C., Letts, P. A., Krajewski, S. J., & Rance, N. E. (2006). Coexpression of dynorphin and neurokinin B immunoreactivity in the rat hypothalamus: Morphologic evidence of interrelated function within the arcuate nucleus. *The Journal of Comparative Neurology, 498*, 712–726.

Cao, J., & Patisaul, H. B. (2011). Sexually dimorphic expression of hypothalamic estrogen receptors alpha and beta and kiss1 in neonatal male and female rats. *The Journal of Comparative Neurology*, *519*, 2954–2977.

Caron, E., Ciofi, P., Prevot, V., & Bouret, S. G. (2012). Alteration in neonatal nutrition causes perturbations in hypothalamic neural circuits controlling reproductive function. *Journal of Neuroscience*, *32*, 11486–11494.

Carter, R., Jaccard, J., Silverman, W. K., & Pina, A. A. (2009). Pubertal timing and its link to behavioral and emotional problems among 'at-risk' African American adolescent girls. *Journal of Adolescence*, *32*, 467–481.

Castellano, J. M., Bentsen, A. H., Mikkelsen, J. D., & Tena-Sempere, M. (2010). Kisspeptins: Bridging energy homeostasis and reproduction. *Brain Research*, *1364*, 129–138.

Castellano, J. M., Bentsen, A. H., Sanchez-Garrido, M. A., Ruiz-Pino, F., Romero, M., Garcia-Galiano, D., et al. (2011). Early metabolic programming of puberty onset: Impact of changes in postnatal feeding and rearing conditions on the timing of puberty and development of the hypothalamic kisspeptin system. *Endocrinology*, *152*, 3396–3408.

Castellano, J. M., Navarro, V. M., Fernandez-Fernandez, R., Castano, J. P., Malagon, M. M., Aguilar, E., et al. (2006). Ontogeny and mechanisms of action for the stimulatory effect of kisspeptin on gonadotropin-releasing hormone system of the rat. *Molecular and Cellular Endocrinology*, *257–258*, 75–83.

Castellano, J. M., Navarro, V. M., Fernandez-Fernandez, R., Nogueiras, R., Tovar, S., Roa, J., et al. (2005). Changes in hypothalamic KiSS-1 system and restoration of pubertal activation of the reproductive axis by kisspeptin in undernutrition. *Endocrinology*, *146*, 3917–3925.

Castellano, J. M., Navarro, V. M., Fernandez-Fernandez, R., Roa, J., Vigo, E., Pineda, R., et al. (2006). Expression of hypothalamic KiSS-1 system and rescue of defective gonadotropic responses by kisspeptin in streptozotocin-induced diabetic male rats. *Diabetes*, *55*, 2602–2610.

Castellano, J. M., Roa, J., Luque, R. M., Dieguez, C., Aguilar, E., Pinilla, L., et al. (2009). KiSS-1/kisspeptins and the metabolic control of reproduction: Physiologic roles and putative physiopathological implications. *Peptides*, *30*, 139–145.

Cheng, G., Coolen, L. M., Padmanabhan, V., Goodman, R. L., & Lehman, M. N. (2010). The kisspeptin/neurokinin B/dynorphin (KNDy) cell population of the arcuate nucleus: Sex differences and effects of prenatal testosterone in sheep. *Endocrinology*, *151*, 301–311.

Clarkson, J., Anglemont de Tassigny, X., Colledge, W. H., Caraty, A., & Herbison, A. E. (2009). Distribution of kisspeptin neurones in the adult female mouse brain. *Journal of Neuroendocrinology*, *21*, 673–682.

Clarkson, J., Han, S. K., Liu, X., Lee, K., & Herbison, A. E. (2010). Neurobiological mechanisms underlying kisspeptin activation of gonadotropin-releasing hormone (GnRH) neurons at puberty. *Molecular and Cellular Endocrinology*, *324*, 45–50.

Clarkson, J., & Herbison, A. E. (2006). Postnatal development of kisspeptin neurons in mouse hypothalamus; sexual dimorphism and projections to gonadotropin-releasing hormone neurons. *Endocrinology*, *147*, 5817–5825.

Cota, D., Proulx, K., Smith, K. A., Kozma, S. C., Thomas, G., Woods, S. C., et al. (2006). Hypothalamic mTOR signaling regulates food intake. *Science*, *312*, 927–930.

Cravo, R. M., Frazao, R., Perello, M., Osborne-Lawrence, S., Williams, K. W., Zigman, J. M., et al. (2013). Leptin signaling in kiss1 neurons arises after pubertal development. *PLoS One*, *8*, e58698.

Cunningham, M. J., Clifton, D. K., & Steiner, R. A. (1999). Leptin's actions on the reproductive axis: Perspectives and mechanisms. *Biology of Reproduction*, *60*, 216–222.

d'Anglemont de Tassigny, X., Fagg, L. A., Dixon, J. P., Day, K., Leitch, H. G., Hendrick, A. G., et al. (2007). Hypogonadotropic hypogonadism in mice lacking a functional Kiss1 gene. *Proceedings of the National Academy of Sciences of the United States of America, 104*, 10714–10719.

de Roux, N., Genin, E., Carel, J. C., Matsuda, F., Chaussain, J. L., & Milgrom, E. (2003). Hypogonadotropic hypogonadism due to loss of function of the KiSS1-derived peptide receptor GPR54. *Proceedings of the National Academy of Sciences of the United States of America, 100*, 10972–10976.

Donato, J., Jr., Cravo, R. M., Frazao, R., Gautron, L., Scott, M. M., Lachey, J., et al. (2011). Leptin's effect on puberty in mice is relayed by the ventral premammillary nucleus and does not require signaling in Kiss1 neurons. *Journal of Clinical Investigation, 121*, 355–368.

Donato, J., Jr., Lee, C., Ratra, D., Franci, C. R., Canteras, N. S., & Elias, C. F. (2013). Lesions of the ventral premammillary nucleus disrupt the dynamic changes in Kiss1 and GnRH expression characteristic of the proestrus-estrus transition. *Neuroscience, 241*, 67–79.

Donato, J., Jr., Silva, R. J., Sita, L. V., Lee, S., Lee, C., Lacchini, S., et al. (2009). The ventral premammillary nucleus links fasting-induced changes in leptin levels and coordinated luteinizing hormone secretion. *Journal of Neuroscience, 29*, 5240–5250.

Elias, C. F. (2012). Leptin action in pubertal development: Recent advances and unanswered questions. *Trends in Endocrinology and Metabolism, 23*, 9–15.

Elias, C. F., & Purohit, D. (2013). Leptin signaling and circuits in puberty and fertility. *Cellular and Molecular Life Sciences, 70*, 841–862.

Elks, C. E., Perry, J. R., Sulem, P., Chasman, D. I., Franceschini, N., He, C., et al. (2010). Thirty new loci for age at menarche identified by a meta-analysis of genome-wide association studies. *Nature Genetics, 42*, 1077–1085.

Feng Li, X., Lin, Y. S., Kinsey-Jones, J. S., & O'Byrne, K. T. (2012). High-fat diet increases LH pulse frequency and kisspeptin-neurokinin B expression in puberty-advanced female rats. *Endocrinology, 153*, 4422–4431.

Fernandez-Fernandez, R., Martini, A. C., Navarro, V. M., Castellano, J. M., Dieguez, C., Aguilar, E., et al. (2006). Novel signals for the integration of energy balance and reproduction. *Molecular and Cellular Endocrinology, 254–255*, 127–132.

Fink, G. (2000). Neuroendocrine regulation of pituitary function: General principles. In P. M. Conn & M. E. Freeman (Eds.), *Neuroendocrinology in physiology and medicine* (pp. 107–134). Totowa, NJ: Humana Press.

Finn, P. D., Cunningham, M. J., Pau, K. Y., Spies, H. G., Clifton, D. K., & Steiner, R. A. (1998). The stimulatory effect of leptin on the neuroendocrine reproductive axis of the monkey. *Endocrinology, 139*, 4652–4662.

Forbes, S., Li, X. F., Kinsey-Jones, J., & O'Byrne, K. (2009). Effects of ghrelin on Kisspeptin mRNA expression in the hypothalamic medial preoptic area and pulsatile luteinising hormone secretion in the female rat. *Neuroscience Letters, 460*, 143–147.

Gajdos, Z. K., Henderson, K. D., Hirschhorn, J. N., & Palmert, M. R. (2010). Genetic determinants of pubertal timing in the general population. *Molecular and Cellular Endocrinology, 324*, 21–29.

Garcia-Galiano, D., Pinilla, L., & Tena-Sempere, M. (2012). Sex steroids and the control of the Kiss1 system: Developmental roles and major regulatory actions. *Journal of Neuroendocrinology, 24*, 22–33.

Garcia-Galiano, D., van Ingen Schenau, D., Leon, S., Krajnc-Franken, M. A., Manfredi-Lozano, M., Romero-Ruiz, A., et al. (2012). Kisspeptin signaling is indispensable for neurokinin B, but not glutamate, stimulation of gonadotropin secretion in mice. *Endocrinology, 153*, 316–328.

Gaytan, F., Sangiao-Alvarellos, S., Manfredi-Lozano, M., Garcia-Galiano, D., Ruiz-Pino, F., Romero-Ruiz, A., et al. (2013). Distinct expression patterns predict differential roles of the miRNA-binding proteins, Lin28 and Lin28b, in the mouse testis: Studies during postnatal development and in a model of hypogonadotropic hypogonadism. *Endocrinology, 154*, 1321–1336.

Gill, J. C., Navarro, V. M., Kwong, C., Noel, S. D., Martin, C., Xu, S., et al. (2012). Increased neurokinin B (Tac2) expression in the mouse arcuate nucleus is an early marker of pubertal onset with differential sensitivity to sex steroid-negative feedback than Kiss1. *Endocrinology, 153*, 4883–4893.

Gill, J. C., Wang, O., Kakar, S., Martinelli, E., Carroll, R. S., & Kaiser, U. B. (2010). Reproductive hormone-dependent and -independent contributions to developmental changes in kisspeptin in GnRH-deficient hypogonadal mice. *PLoS One, 5*, e11911.

Gonzalez-Martinez, D., De Mees, C., Douhard, Q., Szpirer, C., & Bakker, J. (2008). Absence of gonadotropin-releasing hormone 1 and Kiss1 activation in alpha-fetoprotein knockout mice: Prenatal estrogens defeminize the potential to show preovulatory luteinizing hormone surges. *Endocrinology, 149*, 2333–2340.

Goodman, R. L., Lehman, M. N., Smith, J. T., Coolen, L. M., de Oliveira, C. V., Jafarzadehshirazi, M. R., et al. (2007). Kisspeptin neurons in the arcuate nucleus of the ewe express both dynorphin A and neurokinin B. *Endocrinology, 148*, 5752–5760.

Gore, A. C. (2008). Developmental programming and endocrine disruptor effects on reproductive neuroendocrine systems. *Frontiers in Neuroendocrinology, 29*, 358–374.

Gottsch, M. L., Navarro, V. M., Zhao, Z., Glidewell-Kenney, C., Weiss, J., Jameson, J. L., et al. (2009). Regulation of Kiss1 and dynorphin gene expression in the murine brain by classical and nonclassical estrogen receptor pathways. *Journal of Neuroscience, 29*, 9390–9395.

Hakansson, M. L., Brown, H., Ghilardi, N., Skoda, R. C., & Meister, B. (1998). Leptin receptor immunoreactivity in chemically defined target neurons of the hypothalamus. *Journal of Neuroscience, 18*, 559–572.

Han, S. K., Gottsch, M. L., Lee, K. J., Popa, S. M., Smith, J. T., Jakawich, S. K., et al. (2005). Activation of gonadotropin-releasing hormone neurons by kisspeptin as a neuroendocrine switch for the onset of puberty. *Journal of Neuroscience, 25*, 11349–11356.

He, C., Kraft, P., Chen, C., Buring, J. E., Pare, G., Hankinson, S. E., et al. (2009). Genome-wide association studies identify loci associated with age at menarche and age at natural menopause. *Nature Genetics, 41*, 724–728.

Heo, I., Joo, C., Cho, J., Ha, M., Han, J., & Kim, V. N. (2008). Lin28 mediates the terminal uridylation of let-7 precursor MicroRNA. *Molecular Cell, 32*, 276–284.

Herbison, A. E., Porteous, R., Pape, J. R., Mora, J. M., & Hurst, P. R. (2008). Gonadotropin-releasing hormone neuron requirements for puberty, ovulation, and fertility. *Endocrinology, 149*, 597–604.

Herman-Giddens, M. E., Steffes, J., Harris, D., Slora, E., Hussey, M., Dowshen, S. A., et al. (2012). Secondary sexual characteristics in boys: Data from the Pediatric Research in Office Settings Network. *Pediatrics, 130*, e1058–e1068.

Homma, T., Sakakibara, M., Yamada, S., Kinoshita, M., Iwata, K., Tomikawa, J., et al. (2009). Significance of neonatal testicular sex steroids to defeminize anteroventral periventricular kisspeptin neurons and the GnRH/LH surge system in male rats. *Biology of Reproduction, 81*, 1216–1225.

Hrabovszky, E., Ciofi, P., Vida, B., Horvath, M. C., Keller, E., Caraty, A., et al. (2010). The kisspeptin system of the human hypothalamus: Sexual dimorphism and relationship with gonadotropin-releasing hormone and neurokinin B neurons. *European Journal of Neuroscience, 31*, 1984–1998.

Hrabovszky, E., Sipos, M. T., Molnar, C. S., Ciofi, P., Borsay, B. A., Gergely, P., et al. (2012). Low degree of overlap between kisspeptin, neurokinin B, and dynorphin immunoreactivities in the infundibular nucleus of young male human subjects challenges the KNDy neuron concept. *Endocrinology, 153*, 4978–4989.

Ibanez, L., Diaz, R., Lopez-Bermejo, A., & Marcos, M. V. (2009). Clinical spectrum of premature pubarche: Links to metabolic syndrome and ovarian hyperandrogenism. *Reviews in Endocrine & Metabolic Disorders, 10*, 63–76.

Kalamatianos, T., Grimshaw, S. E., Poorun, R., Hahn, J. D., & Coen, C. W. (2008). Fasting reduces KiSS-1 expression in the anteroventral periventricular nucleus (AVPV): Effects of fasting on the expression of KiSS-1 and neuropeptide Y in the AVPV or arcuate nucleus of female rats. *Journal of Neuroendocrinology, 20*, 1089–1097.

Kauffman, A. S., Gottsch, M. L., Roa, J., Byquist, A. C., Crown, A., Clifton, D. K., et al. (2007). Sexual differentiation of Kiss1 gene expression in the brain of the rat. *Endocrinology, 148*, 1774–1783.

Kauffman, A. S., Navarro, V. M., Kim, J., Clifton, D., & Steiner, R. A. (2009). Sex differences in the regulation of Kiss1/NKB neurons in juvenile mice: Implications for the timing of puberty. *American Journal of Physiology, Endocrinology and Metabolism, 297*, E1212–E1221.

Kinsey-Jones, J. S., Grachev, P., Li, X. F., Lin, Y. S., Milligan, S. R., Lightman, S. L., et al. (2012). The inhibitory effects of neurokinin B on GnRH pulse generator frequency in the female rat. *Endocrinology, 153*, 307–315.

Labayen, I., Ortega, F. B., Moreno, L. A., Redondo-Figuero, C., Bueno, G., Gomez-Martinez, S., et al. (2009). The effect of early menarche on later body composition and fat distribution in female adolescents: Role of birth weight. *Annals of Nutrition and Metabolism, 54*, 313–320.

Lakshman, R., Forouhi, N., Luben, R., Bingham, S., Khaw, K., Wareham, N., et al. (2008). Association between age at menarche and risk of diabetes in adults: Results from the EPIC-Norfolk cohort study. *Diabetologia, 51*, 781–786.

Lakshman, R., Forouhi, N. G., Sharp, S. J., Luben, R., Bingham, S. A., Khaw, K. T., et al. (2009). Early age at menarche associated with cardiovascular disease and mortality. *Journal of Clinical Endocrinology and Metabolism, 94*, 4953–4960.

Lehman, M. N., Coolen, L. M., & Goodman, R. L. (2010). Minireview: Kisspeptin/neurokinin B/dynorphin (KNDy) cells of the arcuate nucleus: A central node in the control of gonadotropin-releasing hormone secretion. *Endocrinology, 151*, 3479–3489.

Lomniczi, A., Loche, A., Castellano, J. M., Ronnekleiv, O. K., Bosch, M., Kaidar, G., et al. (2013). Epigenetic control of female puberty. *Nature Neuroscience, 16*, 281–289.

Louis, G. W., Greenwald-Yarnell, M., Phillips, R., Coolen, L. M., Lehman, M. N., & Myers, M. G., Jr. (2011). Molecular mapping of the neural pathways linking leptin to the neuroendocrine reproductive axis. *Endocrinology, 152*, 2302–2310.

Luque, R. M., Kineman, R. D., & Tena-Sempere, M. (2007). Regulation of hypothalamic expression of KiSS-1 and GPR54 genes by metabolic factors: Analyses using mouse models and a cell line. *Endocrinology, 148*, 4601–4611.

Martos-Moreno, G. A., Chowen, J. A., & Argente, J. (2010). Metabolic signals in human puberty: Effects of over and undernutrition. *Molecular and Cellular Endocrinology, 324*, 70–81.

Mayer, C., Acosta-Martinez, M., Dubois, S. L., Wolfe, A., Radovick, S., Boehm, U., et al. (2010). Timing and completion of puberty in female mice depend on estrogen receptor alpha-signaling in kisspeptin neurons. *Proceedings of the National Academy of Sciences of the United States of America, 107*, 22693–22698.

Mayer, C., & Boehm, U. (2011). Female reproductive maturation in the absence of kisspeptin/GPR54 signaling. *Nature Neuroscience, 14*, 704–710.

Morris, J. A., Jordan, C. L., & Breedlove, S. M. (2004). Sexual differentiation of the vertebrate nervous system. *Nature Neuroscience, 7*, 1034–1039.

Navarro, V. M., Castellano, J. M., Fernandez-Fernandez, R., Barreiro, M. L., Roa, J., Sanchez-Criado, J. E., et al. (2004). Developmental and hormonally regulated messenger ribonucleic acid expression of KiSS-1 and its putative receptor, GPR54, in rat hypothalamus and potent luteinizing hormone-releasing activity of KiSS-1 peptide. *Endocrinology, 145*, 4565–4574.

Navarro, V. M., Castellano, J. M., McConkey, S. M., Pineda, R., Ruiz-Pino, F., Pinilla, L., et al. (2011). Interactions between kisspeptin and neurokinin B in the control of GnRH secretion in the female rat. *American Journal of Physiology, Endocrinology and Metabolism, 300*, E202–E210.

Navarro, V. M., Fernandez-Fernandez, R., Castellano, J. M., Roa, J., Mayen, A., Barreiro, M. L., et al. (2004). Advanced vaginal opening and precocious activation of the reproductive axis by KiSS-1 peptide, the endogenous ligand of GPR54. *Journal de Physiologie, 561*, 379–386.

Navarro, V. M., Gottsch, M. L., Chavkin, C., Okamura, H., Clifton, D. K., & Steiner, R. A. (2009). Regulation of gonadotropin-releasing hormone secretion by kisspeptin/dynorphin/neurokinin B neurons in the arcuate nucleus of the mouse. *Journal of Neuroscience, 29*, 11859–11866.

Navarro, V. M., Gottsch, M. L., Wu, M., Garcia-Galiano, D., Hobbs, S. J., Bosch, M. A., et al. (2011). Regulation of NKB pathways and their roles in the control of Kiss1 neurons in the arcuate nucleus of the male mouse. *Endocrinology, 152*, 4265–4275.

Navarro, V. M., Ruiz-Pino, F., Sanchez-Garrido, M. A., Garcia-Galiano, D., Hobbs, S. J., Manfredi-Lozano, M., et al. (2012). Role of neurokinin B in the control of female puberty and its modulation by metabolic status. *Journal of Neuroscience, 32*, 2388–2397.

Navarro, V. M., Sanchez-Garrido, M. A., Castellano, J. M., Roa, J., Garcia-Galiano, D., Pineda, R., et al. (2009). Persistent impairment of hypothalamic KiSS-1 system after exposures to estrogenic compounds at critical periods of brain sex differentiation. *Endocrinology, 150*, 2359–2367.

Navarro, V. M., & Tena-Sempere, M. (2008). The KiSS-1/GPR54 system: Putative target for endocrine disruption of reproduction at hypothalamic–pituitary unit? *International Journal of Andrology, 31*, 224–232.

Navarro, V. M., & Tena-Sempere, M. (2012). Neuroendocrine control by kisspeptins: Role in metabolic regulation of fertility. *Nature Reviews. Endocrinology, 8*, 40–53.

Nazian, S. J. (2006). Role of metastin in the release of gonadotropin-releasing hormone from the hypothalamus of the male rat. *Journal of Andrology, 27*, 444–449.

Nestor, C. C., Briscoe, A. M., Davis, S. M., Valent, M., Goodman, R. L., & Hileman, S. M. (2012). Evidence of a role for kisspeptin and neurokinin B in puberty of female sheep. *Endocrinology, 153*, 2756–2765.

Norman, R. L., & Spies, H. G. (1986). Cyclic ovarian function in a male macaque: Additional evidence for a lack of sexual differentiation in the physiological mechanisms that regulate the cyclic release of gonadotropins in primates. *Endocrinology, 118*, 2608–2610.

Oakley, A. E., Clifton, D. K., & Steiner, R. A. (2009). Kisspeptin signaling in the brain. *Endocrine Reviews, 30*, 713–743.

Ojeda, S. R., Dubay, C., Lomniczi, A., Kaidar, G., Matagne, V., Sandau, U. S., et al. (2010). Gene networks and the neuroendocrine regulation of puberty. *Molecular and Cellular Endocrinology, 324*, 3–11.

Ojeda, S. R., Lomniczi, A., Mastronardi, C., Heger, S., Roth, C., Parent, A. S., et al. (2006). Minireview: The neuroendocrine regulation of puberty: Is the time ripe for a systems biology approach? *Endocrinology, 147*, 1166–1174.

Ojeda, S. R., Lomniczi, A., Sandau, U., & Matagne, V. (2010). New concepts on the control of the onset of puberty. *Endocrine Development, 17*, 44–51.

Ojeda, S. R., & Skinner, M. K. (2006). Puberty in the rat. In J. D. Neill (Ed.), *The physiology of reproduction* (pp. 2061–2126). San Diego: Academic Press/Elsevier.

Ong, K. K., Elks, C. E., Li, S., Zhao, J. H., Luan, J., Andersen, L. B., et al. (2009). Genetic variation in LIN28B is associated with the timing of puberty. *Nature Genetics, 41*, 729–733.

Parent, A. S., Teilmann, G., Juul, A., Skakkebaek, N. E., Toppari, J., & Bourguignon, J. P. (2003). The timing of normal puberty and the age limits of sexual precocity: Variations around the world, secular trends, and changes after migration. *Endocrine Reviews, 24*, 668–693.

Paulsen, M., Tierling, S., & Walter, J. (2008). DNA methylation and the mammalian genome. In J. Tost (Ed.), *Epigenetics* (pp. 1–22). Norfolk: Caister Academic Press.

Perry, J. R., Stolk, L., Franceschini, N., Lunetta, K. L., Zhai, G., McArdle, P. F., et al. (2009). Meta-analysis of genome-wide association data identifies two loci influencing age at menarche. *Nature Genetics, 41*, 648–650.

Pineda, R., Garcia-Galiano, D., Roseweir, A., Romero, M., Sanchez-Garrido, M. A., Ruiz-Pino, F., et al. (2010). Critical roles of kisspeptins in female puberty and preovulatory gonadotropin surges as revealed by a novel antagonist. *Endocrinology, 151*, 722–730.

Pinilla, L., Aguilar, E., Dieguez, C., Millar, R. P., & Tena-Sempere, M. (2012). Kisspeptins and reproduction: Physiological roles and regulatory mechanisms. *Physiological Reviews, 92*, 1235–1316.

Plant, T. M., Ramaswamy, S., & Dipietro, M. J. (2006). Repetitive activation of hypothalamic G protein-coupled receptor 54 with intravenous pulses of kisspeptin in the juvenile monkey (Macaca mulatta) elicits a sustained train of gonadotropin-releasing hormone discharges. *Endocrinology, 147*, 1007–1013.

Qiu, X., Dowling, A. R., Marino, J. S., Faulkner, L. D., Bryant, B., Bruning, J. C., et al. (2013). Delayed puberty but normal fertility in mice with selective deletion of insulin receptors from kiss1 cells. *Endocrinology, 154*, 1337–1348.

Qiu, J., Fang, Y., Bosch, M. A., Ronnekleiv, O. K., & Kelly, M. J. (2011). Guinea pig kisspeptin neurons are depolarized by leptin via activation of TRPC channels. *Endocrinology, 152*, 1503–1514.

Quennell, J. H., Howell, C. S., Roa, J., Augustine, R. A., Grattan, D. R., & Anderson, G. M. (2011). Leptin deficiency and diet-induced obesity reduce hypothalamic kisspeptin expression in mice. *Endocrinology, 152*, 1541–1550.

Quennell, J. H., Mulligan, A. C., Tups, A., Liu, X., Phipps, S. J., Kemp, C. J., et al. (2009). Leptin indirectly regulates gonadotropin-releasing hormone neuronal function. *Endocrinology, 150*, 2805–2812.

Ramaswamy, S., Seminara, S. B., Ali, B., Ciofi, P., Amin, N. A., & Plant, T. M. (2010). Neurokinin B stimulates GnRH release in the male monkey (Macaca mulatta) and is colocalized with kisspeptin in the arcuate nucleus. *Endocrinology, 151*, 4494–4503.

Ramaswamy, S., Seminara, S. B., & Plant, T. M. (2011). Evidence from the agonadal juvenile male rhesus monkey (Macaca mulatta) for the view that the action of neurokinin B to trigger gonadotropin-releasing hormone release is upstream from the kisspeptin receptor. *Neuroendocrinology, 94*, 237–245.

Roa, J., Aguilar, E., Dieguez, C., Pinilla, L., & Tena-Sempere, M. (2008). New frontiers in kisspeptin/GPR54 physiology as fundamental gatekeepers of reproductive function. *Frontiers in Neuroendocrinology, 29*, 48–69.

Roa, J., Garcia-Galiano, D., Castellano, J. M., Gaytan, F., Pinilla, L., & Tena-Sempere, M. (2010). Metabolic control of puberty onset: New players, new mechanisms. *Molecular and Cellular Endocrinology, 324*, 87–94.

Roa, J., Garcia-Galiano, D., Varela, L., Sanchez-Garrido, M. A., Pineda, R., Castellano, J. M., et al. (2009). The mammalian target of rapamycin as novel central regulator of puberty onset via modulation of hypothalamic Kiss1 system. *Endocrinology, 150*, 5016–5026.

Roa, J., & Tena-Sempere, M. (2010). Energy balance and puberty onset: Emerging role of central mTOR signaling. *Trends in Endocrinology and Metabolism, 21*, 519–528.

Roa, J., Vigo, E., Garcia-Galiano, D., Castellano, J. M., Navarro, V. M., Pineda, R., et al. (2008). Desensitization of gonadotropin responses to kisspeptin in the female rat: Analyses of LH and FSH secretion at different developmental and metabolic states. *American Journal of Physiology, Endocrinology and Metabolism, 294*, E1088–E1096.

Ruiz-Pino, F., Navarro, V. M., Bentsen, A. H., Garcia-Galiano, D., Sanchez-Garrido, M. A., Ciofi, P., et al. (2012). Neurokinin B and the control of the gonadotropic axis in the rat: Developmental changes, sexual dimorphism, and regulation by gonadal steroids. *Endocrinology, 153*, 4818–4829.

Sandoval-Guzman, T., & Rance, N. E. (2004). Central injection of senktide, an NK3 receptor agonist, or neuropeptide Y inhibits LH secretion and induces different patterns of Fos expression in the rat hypothalamus. *Brain Research, 1026*, 307–312.

Sangiao-Alvarellos, S., Manfredi-Lozano, M., Ruiz-Pino, F., Navarro, V. M., Sanchez-Garrido, M. A., Leon, S., et al. (2013). Changes in hypothalamic expression of the Lin28/let-7 system and related microRNAs during postnatal maturation and after experimental manipulations of puberty. *Endocrinology, 154*, 942–955.

Semaan, S. J., Dhamija, S., Kim, J., Ku, E. C., & Kauffman, A. S. (2012). Assessment of epigenetic contributions to sexually-dimorphic Kiss1 expression in the anteroventral periventricular nucleus of mice. *Endocrinology, 153*, 1875–1886.

Seminara, S. B., Messager, S., Chatzidaki, E. E., Thresher, R. R., Acierno, J. S., Jr., Shagoury, J. K., et al. (2003). The GPR54 gene as a regulator of puberty. *The New England Journal of Medicine, 349*, 1614–1627.

Shahab, M., Mastronardi, C., Seminara, S. B., Crowley, W. F., Ojeda, S. R., & Plant, T. M. (2005). Increased hypothalamic GPR54 signaling: A potential mechanism for initiation of puberty in primates. *Proceedings of the National Academy of Sciences of the United States of America, 102*, 2129–2134.

Sims, J. K., Magazinnik, T., Houston, S. I., Wu, S., & Rice, J. C. (2008). Histone modifications and epigenetics. In J. Tost (Ed.), *Epigenetics* (pp. 127–152). Norfolk: Caister Academic Press.

Smith, J. T., Acohido, B. V., Clifton, D. K., & Steiner, R. A. (2006). KiSS-1 neurones are direct targets for leptin in the ob/ob mouse. *Journal of Neuroendocrinology, 18*, 298–303.

Smith, J. T., & Spencer, S. J. (2012). Preweaning over- and underfeeding alters onset of puberty in the rat without affecting kisspeptin. *Biology of Reproduction, 86*(145), 141–148.

Sulem, P., Gudbjartsson, D. F., Rafnar, T., Holm, H., Olafsdottir, E. J., Olafsdottir, G. H., et al. (2009). Genome-wide association study identifies sequence variants on 6q21 associated with age at menarche. *Nature Genetics, 41*, 734–738.

Taeymans, J., Hebbelinck, M., Borms, J., Clarys, P., Abidi, H., & Duquet, W. (2008). Tracking of adult adiposity in early, average and late maturing children: A thirty year longitudinal growth study. *The Journal of Sports Medicine and Physical Fitness, 48*, 326–334.

Tajar, A., Forti, G., O'Neill, T. W., Lee, D. M., Silman, A. J., Finn, J. D., et al. (2010). Characteristics of secondary, primary, and compensated hypogonadism in aging men: Evidence from the European Male Ageing Study. *Journal of Clinical Endocrinology and Metabolism, 95*, 1810–1818.

Takumi, K., Iijima, N., & Ozawa, H. (2011). Developmental changes in the expression of kisspeptin mRNA in rat hypothalamus. *Journal of Molecular Neuroscience, 43*, 138–145.

Tena-Sempere, M. (2007). Roles of ghrelin and leptin in the control of reproductive function. *Neuroendocrinology, 86*, 229–241.

Tena-Sempere, M. (2008). Timeline: The role of kisspeptins in reproductive biology. *Nature Medicine, 14*, 1196.

Tena-Sempere, M. (2010). Roles of kisspeptins in the control of hypothalamic–gonadotropic function: Focus on sexual differentiation and puberty onset. *Endocrine Development, 17*, 52–62.

Tena-Sempere, M. (2012). Deciphering puberty: Novel partners, novel mechanisms. *European Journal of Endocrinology, 167*, 733–747.

Tena-Sempere, M., & Huhtaniemi, I. (2003). Gonadotropins and gonadotropin receptors. In B. C. J. M. Fauser (Ed.), *Reproductive medicine – Molecular, cellular and genetic fundamentals* (pp. 225–244). New York: Parthenon Publishing.

Tena-Sempere, M., Pinilla, L., Gonzalez, L. C., & Aguilar, E. (2000). Reproductive disruption by exposure to exogenous estrogenic compounds during sex differentiation: Lessons from the neonatally estrogenized male rat. *Current Topics in Steroid Research, 3*, 23–37.

Tomikawa, J., Uenoyama, Y., Ozawa, M., Fukanuma, T., Takase, K., Goto, T., et al. (2012). Epigenetic regulation of Kiss1 gene expression mediating estrogen-positive feedback action in the mouse brain. *Proceedings of the National Academy of Sciences of the United States of America, 109*, E1294–E1301.

Topaloglu, A. K., Reimann, F., Guclu, M., Yalin, A. S., Kotan, L. D., Porter, K. M., et al. (2009). TAC3 and TACR3 mutations in familial hypogonadotropic hypogonadism reveal a key role for neurokinin B in the central control of reproduction. *Nature Genetics, 41*, 354–358.

Topaloglu, A. K., Tello, J. A., Kotan, L. D., Ozbek, M. N., Yilmaz, M. B., Erdogan, S., et al. (2012). Inactivating KISS1 mutation and hypogonadotropic hypogonadism. *The New England Journal of Medicine, 366*, 629–635.

Tortoriello, D. V., McMinn, J., & Chua, S. C. (2004). Dietary-induced obesity and hypothalamic infertility in female DBA/2J mice. *Endocrinology, 145*, 1238–1247.

True, C., Kirigiti, M. A., Kievit, P., Grove, K. L., & Susan Smith, M. (2011). Leptin is not the critical signal for kisspeptin or luteinising hormone restoration during exit from negative energy balance. *Journal of Neuroendocrinology, 23*, 1099–1112.

Van Emburgh, B. O., & Robertson, K. D. (2008). DNA methyltransferases and methyl-CpG binding proteins as multifunctional regulators of chromatin structure and development in mammalian cells. In J. Tost (Ed.), *Epigenetics* (pp. 23–62). Norfolk: Caister Academic Press.

Viswanathan, S. R., Daley, G. Q., & Gregory, R. I. (2008). Selective blockade of microRNA processing by Lin28. *Science, 320*, 97–100.

Wakabayashi, Y., Nakada, T., Murata, K., Ohkura, S., Mogi, K., Navarro, V. M., et al. (2010). Neurokinin B and dynorphin A in kisspeptin neurons of the arcuate nucleus participate in generation of periodic oscillation of neural activity driving pulsatile gonadotropin-releasing hormone secretion in the goat. *Journal of Neuroscience, 30*, 3124–3132.

Yang, J. J., Caligioni, C. S., Chan, Y. M., & Seminara, S. B. (2012). Uncovering novel reproductive defects in neurokinin B receptor null mice: Closing the gap between mice and men. *Endocrinology, 153*, 1498–1508.

Young, J., Bouligand, J., Francou, B., Raffin-Sanson, M. L., Gaillez, S., Jeanpierre, M., et al. (2010). TAC3 and TACR3 defects cause hypothalamic congenital hypogonadotropic hypogonadism in humans. *Journal of Clinical Endocrinology and Metabolism, 95*, 2287–2295.

Zhu, H., Shah, S., Shyh-Chang, N., Shinoda, G., Einhorn, W. S., Viswanathan, S. R., et al. (2010). Lin28a transgenic mice manifest size and puberty phenotypes identified in human genetic association studies. *Nature Genetics, 42*, 626–630.

INDEX

Note: Page numbers followed by "*f*" indicate figures, and "*t*" indicate tables.

A

ACOD. *See* Acyl Co-A desaturase (ACOD)
ACTH. *See* Adrenocorticotropic hormone (ACTH)
Acyl Co-A desaturase (ACOD), 256
Adrenocorticotropic hormone (ACTH), 5
α-Fetoprotein (AFP), 310–311
AFP. *See* α-Fetoprotein (AFP)
Agouti-related protein (AGRP), 255*f*, 256
AGRP. *See* Agouti-related protein (AGRP)
Amphibian metamorphosis
  CTHBPs, 289–290
  5'-deiodinases, 290
  morphological differentiation, 285–287
  mRNA levels, 287–289
  organ transformations, 288*t*, 289–290
  PKM2 mRNA, 289–290
  Rana catesbeiana, 290
  signaling, TH, 285–287
  temporal regulation, 285–287, 287*f*
  TH absence, 289
  TR and RXR expression, 287–289
  T_4/thyroxine, 290
Arabidopsis
  genetic analysis, 136
  heteroblastic species, 134
  miR156
    inactivation, 129–130
    inflorescence, 133
    overexpression, 136–137
    SPL9, 133–134
  seedlings, 141, 142–143
  trehalose-6-phosphate (T6P), 143
  vegetative phase change, 129
Araschnia levana (A. levana)
  "diapause-bound" morphs, 228–229, 229*f*
  larval instars, 228–229
ARC. *See* Arcuate nucleus (ARC)
Arcuate nucleus (ARC)
  infundibular region, 305–306
  Kiss1/NKB neurons, 307–308
  RP3V, 305–306
Austrofundulus limnaeus (A. limnaeus)
  developmental growth, 233–234, 234*f*
  diapause, 233–234

B

BAIAP2. *See* Brain-specific angiogenesis inhibitor 1-associated protein 2 (BAIAP2)
Bardet–Biedl syndrome (BBS), 261–263
BBS. *See* Bardet–Biedl syndrome (BBS)
BHB. *See* D-β-Hydroxybutyrate (BHB)
BmDH gene. *See* Bombyx DH *(BmDH)* gene
Bombyx DH *(BmDH)* gene, 219, 225
Bombyx mori (B. mori)
  diapause induction, 217–218, 218*f*
  diapause-linked diphenism and traits, 229–230, 233, 236
  dormant embryos, 214
  embryonic mesoderm segmentation, 216–217
  larval development, 225
  young "nondiapause" pupae, 229–230
Brain sex differentiation
  adult gonadotropic system, 309–310
  AFP knock-out mice, 310–311
  ARC population, 311–312
  GNX, 311
  hypothalamic Kiss1 system, 311
  mammalian species, 312
  rodents, 309–310
  RP3V population, 311–312
  sexual differentiation, 310–311
Brain-specific angiogenesis inhibitor 1-associated protein 2 (BAIAP2), 251

C

CA. *See Corpus allatum* (CA)
Caenorhabditis elegans (C. elegans)
  characterization, 4

Caenorhabditis elegans (C. elegans) (*Continued*)
DAF-12 regulation (*see* DAF-12 regulation, *C. elegans*)
dauer diapause (*see* Dauer diapause, *C. elegans*)
environmental cues, 203–205
heterochronic mutations, 3
larval stages
animals, 173
DAF-12, 167–169
development, 171–173
developmental transitions, 154
diverse timing pathways, 173–174
environmental conditions, 174
heterochronic (*see* Heterochronic, *C. elegans*)
interrupted development, 169–171
LIN-28, 160–161
LIN-42, 164–167
microRNA switches, 159–160
molecular mechanisms and regulatory networks, 154
molts, 162–164
oscillatory levels, 174
worm well-suited, 154–156
"lineage" genes, 4
longevity (*see* Longevity)
maturation, 182–183
CARTPT. *See* Cocaine- and amphetamine-regulated protein prepropeptide (CARTPT)
CBLN2. *See* Cerebellin precursor protein 2 (CBLN2)
CCAP. *See* Crustacean cardioactive peptide (CCAP)
CCAP efferent (CCAP-EN)
CCAP-PL neurons, 19–21
Dachshund marker, 19–21, 20*f*
EcR function, 21–22
larval and pupal ecdysis, 23
CCAP-EN. *See* CCAP efferent (CCAP-EN)
CCAP-IN. *See* CCAP interneuron (CCAP-IN)
CCAP interneuron (CCAP-IN), 19, 20*f*
Cellular energy sensors, 316–317
Cellular retinoic acid binding protein 1 (CRBP1), 263–264
Central nervous system (CNS)
Ca^{2+} spiking rhythms, 17
and CNS-RGs, 14–15
nuclear TIMELESS rhythms, 16
oscillations (*see* Prothoracic gland (PG))
PG components, 16–17
Cerebellin precursor protein 2 (CBLN2), 251
Circadian control, insect development
clock, PG, 8
Drosophila, 8
photoperiodic gate, 7–8
prothoracic glands harbor, 9
PTTH secretion, 7–8
Rhodnius harbors, 8
rhythms, 7
Circadian core clock, flies
CRYPTOCHROME (CRY), 9–11
Drosophila, 9–11, 10*f*
environmental changes, 9
fundamental characteristics, 9
heterodimer CLOCK/CYCLE binds, 9–11
metabolism and feeding behavior, 11–12
molecular components, 9–11
pacemaker neurons, *Drosophila* brain, 11
peripheral tissues, 11–12
timeless and *period* dimers, 9–11
Circadian rhythm pathways, miRNAs
bantam, 114
D. melanogaster, 113
identification, 114
miR-279, 115–116
miR-959-964 cluster, 116
molecular overview, 113
Circannual timing
adaptation, 248–249
Caenorhabditis elegans, 248–249
events, 248–249, 248*f*
genomes, hibernating mammals, 266
hibernation (*see* Hibernation)
husbandry resources, 267
melatonin and clock genes, 264–265
migration, 251–253
molecular biology, 266
neuronal cilia, 261–263
next-generation sequencing, 266
nonmodel organisms, 266–267

retinoic acid signaling, 263–264
seasonal courtship and territorial behavior, 249–251
Taeniopygia guttata (zebra finch), 266
thyroid hormone signaling, 259–260
VGF signaling, 260–261
9-Cis-retinoic acid receptors (RXRs)
amphibian metamorphosis, 280
repress gene expression, 279
RXRα and RXRγ, 280
TR-RXR heterodimers, 280–281
Xenopus metamorphosis, 288*t*
CNS. *See* Central nervous system (CNS)
Coactivator recruitment
PRMT1, 285, 286*f*
SRC-p300 complexes, 285
transcriptional activation, 285
Xenopus SRC3, 285
Cocaine- and amphetamine-regulated protein prepropeptide (CARTPT), 255*f*, 256
Corepressor complexes
dual function model, TR, 282–283
heat shock treatment, 283–285
myc-ID dimer, 283–285
N-CoR and SMRT, 282–283
premetamorphic tadpoles, 283
transgenic expression, 283, 284*f*
Corpus allatum (CA), 11–12
CRBP1. *See* Cellular retinoic acid binding protein 1 (CRBP1)
Crustacean cardioactive peptide (CCAP)
blocks pre-ecdysis and triggers ecdysis, 18*f*
BrdU incorporation, 21
βFTZ-F1 expression, 22
larval and pupal ecdysis, 23
late-emerging, 19–21
proper larval and pupal ecdysis, 19
puparium formation, 21–22
RNAi expression, 21–22
ventral nerve cord, 19
wing inflation, 19–21
CTHBPs. *See* Cytosolic thyroid hormone binding proteins (CTHBPs)
Cytosolic thyroid hormone binding proteins (CTHBPs)
cellular TH, 289
multi-functional proteins, 289–290

D

Dafachronic acids (DAs)
biosynthetic pathway, 186–189
DAF-12 regulation (*see* DAF-12 regulation, *C. elegans*)
production, 189–190
signal amplification enforces reproduction, 190–192
DAF-12 regulation, *C. elegans*
DAs (*see* Dafachronic acids (DAs))
developmental timing
dauer reprogramming, 198
heterochronic genes, 192
human cell culture, 195–196
identification, 195–196
let-7 microRNAs, 193–195, 196
lin-42/period, 196–198
microRNA paradigm, 192–193
molt cycle, 198–201
environmental and physiologic information, 182–183
germline removal, 202
gonadal longevity, 202
heterochronic circuit, 203
liver-X and vitamin-D receptors, 186
steroidal receptor, 203–205
steroidal signaling, 203
transcription factors, 201–202
Dauer diapause, *C.elegans*
DA (*see* Dafachronic acids (DAs))
environmental quality and nutrient availability, 183
hormone-regulated switch governs, 183–186
"prepubertal" quiescent state, 183
D-β-Hydroxybutyrate (BHB), 257–258
DH. *See* Diapause hormone (DH)
Diapause hormone (DH)
Gymnaephora groenlandica, 214–215
IGF-1 signaling, 237
induction, 214
insects (*see* Insects, embryonic diapause)
killifish embryo (*see Austrofundulus limnaeus*)
larval diapause, 219–221
lepidoptera species, 215
modulations, larval/pupal growth, 226–227
multiple polyphenic traits, 236

Diapause hormone (DH) (*Continued*)
organisms, 214–215
Papilio machaon, 214–215
phase, 214
photoperiodic changes, 236–237
pupal, 221–225
seasonal polyphenisms (*see* Seasonal polyphenisms, diapause)
trajectory, 236–237
Diatraea grandiosella (D. grandiosella)
larval diapause, 220*f*
"spotted" morph, 219
DILPs. *See Drosophila* insulin-like peptides (DILPs)
DLM. *See* Dorsal lateral nucleus (DLM)
Dorsal lateral nucleus (DLM), 249–251, 249*f*
Drosophila insulin-like peptides (DILPs)
circulating activity, 55–57
DILP2, DILP3 and *DILP5*, 42–43
DILP6 expression, 51–52
insulin/TOR, 43*f*
nutrient components, 47
signaling molecules, 38–39
Drosophila melanogaster (D. melanogaster)
adult brain harbors, 11
calories and growth, 41
CCAP neurons, 19
circadian clock neurons, 8
core clock, 9–11, 10*f*
corpora allata, 23–24
DILPS (*see* Drosophila insulin-like peptides (DILPs))
20E, 104–105
eclosion rhythms, 15–16, 17
embryo to adulthood, 44
Gce and Met double mutants, 25–26
Imp-L2, 55–56
insulin gene expression, 100
intrinsic clock, 113
JH titers, 6*f*
larval instars, 39
miR-33, 103
miR-34, 112–113
miRNAs (*see* MicroRNAs (miRNAs))
pathological conditions, 60–61
peripheral tissues, 11–12
PG cells, 14–15
primary classes, 104
PTTH neurons, 13
redundant JH receptors, 24–25
royalactin, 46–47

E

ECDPPase. *See* Ecdysteroid-phosphate phosphatase (ECDPPase)
Ecdysis triggering hormone (ETH), 17–19, 18*f*
Ecdysone (ECD)
applications, 215–216, 221
biosynthesis, 39–41
developmental transition, 39
diapause, 215–216
early L3 ecdysone pulse, 42
20-ECD injection, 228–229, 231
ECDPPase, 216–217
inhibition, 216
JHs pathway (*see* Juvenile hormones (JHs))
larval instars, 40*f*
Laspeyresia polmonella and *Ostrinia nubilalis*, 221
let-7-C miRNAs, biological roles
abrupt rescues, genetic depletion, 110–111
adult behaviors, 108–109
BTB-ZF, 109–110
Chinmo expression, 109–110
distinct neuronal subtypes, 109–110
20E-receptor, 110
GSC behavior, 111
mitotic programs, 108–109
regulation, dopaminergic neuron, 111
steroid hormone induction, 108–109
temporal wave, 109
mid-embryonic, 216–217
PG, 39–41
prolonged and intense, 216–217
PTTH signaling, 224–225
pupal diapause, 221–222
regulatory mechanism, 222
requirement, 215–216
RH5849, 222
synthesis and nutrition, 53
timing, 222–224, 228–229
Ecdysone receptor (EcR)
CCAP and bursicon expression, 21–22

circadian oscillations, insects
hormones act, 14
immunostains, 13–14
nuclear receptor-signaling pathways, 14
production, 13–14
REV-ERB/retinoid-related orphan receptor, 14
Rhodnius, 13–14
βFTZ-F1 expression, 22
Ecdysteroid-phosphate phosphatase (ECDPPase), 216–217
Ecdysteroids
nuclear receptors EcR, 13–14
signal transduction, 16–17
Eclosion hormone (EH), 17–19, 18*f*
EcR. *See* Ecdysone receptor (EcR)
EH. *See* Eclosion hormone (EH)
Endocrine regulation
dauer diapause (*see* Dauer diapause, *C.elegans*)
metazoan reproductive, 182–183
Epigenetics
central and peripheral signals, 317–318
CpG islands, 318
definition, 318
female puberty, 318–319
hypothalamic Kiss1 expression, 318–319
Lin28, 320
miRNA regulatory pathways, 319
PcG, 318–319
rat hypothalamus, 320
ETH. *See* Ecdysis triggering hormone (ETH)
Eucalyptus tenuiramis
branch matches, phase identity, 140
vegetative phase change, 139–140

F

FABP4. *See* Fatty acid binding protein (FABP4)
Fatty acid binding protein (FABP4), 256
Fine-tuning systemic insulin signaling
DILP activity, 55–57
ecdysone and insulin, 54–55
Frog development, TR function
chromatin disruption/remodeling, 279–280
dual function model, 280–281
HDAC3, 279–280
histones H3 and H4, 279–280
in vivo studies, 281–282
pseudo-tetraploid *X. laevis*, 280
T3-induced metamorphosis, 281–282, 282*f*
TRβ genes, 280
TREs, 279

G

GA. *See* Gibberellic acid (GA)
Ganglion mother cell (GMC)
division, 71, 88–89
embryonic NB lineages, 88–89
and INPs, 73
NB3-1 lineage, 75
NB7-1 lineage, 73–75
Prospero (Pros), 86–87
Gibberellic acid (GA), 136–137, 141–142
GMC. *See* Ganglion mother cell (GMC)
GnRH. *See* Gonadotropin-releasing hormone (GnRH)
GNX. *See* Gonadectomy (GNX)
Gonadectomy (GNX), 311
Gonadotropin-releasing hormone (GnRH)
and Kiss1 neurons, 301–307
KNDy neurons, 309*f*

H

HDAC3. *See* Histone deacetylase 3 (HDAC3)
Heteroblasty
ch1 gene, 142–143
description, 126–127
vegetative phase change
analysis, 128–129
environmental heterogeneity, 127–128
heterogeneous and quite subtle changes, 128–129
higher plants, 128–129
"ontogenetic maturation", 127–128
"physiological aging", 127–128
"seasonal heterophylly", 127–128
shoot morphology/physiology, 127–128
weakly heteroblastic species, 134

Heterochronic, *C. elegans*
animals
LIN41, 173
mammalian LIN28, 173
lin-42a over-expression, 166–167
mir-71 mutants, 172–173
mutants
"cell lineage abnormal" class, 156
cell lineages, 156, 157*f*
core pathway, 158, 159*f*
intermolt period, 156–157
lin-28 mutants, 158
lin-4 null mutants, 157
microRNAs, 158
null allele, 158
stage-specific sublineages, 156
temporal progress, 156
pathway
components, 168–169
developmental timing, 162
L1 diapause, 171–172
lin-28 and *hbl-1*, 161–162
lin-4 and *lin-14* comprise, 161
and molting control, 162
times larval development and paces, 174
Heterochronic genes
animal development, 192
DAF-12 targets, 195–196
epidermal seam and intestinal cells, 192
heterochronic loci, 192
LIN-28 regulation, 192
molt cycle, 198–201
phenotypes, 193–194
stage- and tissue-specific manner, 192
temporal selector genes, 192
HH. *See* Hypogonadotropic hypogonadism (HH)
Hibernation
central nervous system, 258–259
Class Mammalia, 253–254
description, 253–254
food consumption and satiety signaling, 253–254, 255*f*
fuel switch, 257–258
induction and maintenance, 267
mammals, 266
satiety and feeding behavior, 256–257
torpor, 254–256
Histone deacetylase 3 (HDAC3)
N-CoR and SMRT, 279–280
and TBL1/TBLR1, 284*f*
Histone modification
acetyltransferase, 285
dual function model, 277*f*, 279–280
H3 and H4 transcriptional inhibition, 279–280
HMGCS2. *See* Hydroxymethylglutaryl-CoA synthase 2 (HMGCS2)
HPG. *See* Hypothalamic–pituitary–gonadal (HPG)
Human M2 pyruvate kinase (PKM2)
and frog, 289–290
mRNA levels, 289–290
over-expression, protein, 289–290
tail and hindlimb, 289–290
Hydroxymethylglutaryl-CoA synthase 2 (HMGCS2), 257–258
Hypogonadotropic hypogonadism (HH), 304
Hypothalamic–pituitary–gonadal (HPG) axis
feedforward and feedback loops, 301–302
female mice, 311
neurohormonal network, 303*f*
sexual phenotypic maturity, 301–302

I

IBAs. *See* Interbout arousals (IBAs)
ICV injection. *See* Intracerebroventricular (ICV) injection
IGF1. *See* Insulin growth factor 1 (IGF1)
Immune response pathways, miRNAs
bioinformatics predicts, 118
description, 116–117
innate immunity, 116–117
miR-8, 117
miR-959-964 cluster, 117
INPs. *See* Intermediate neural progenitors (INPs)
Insect development
brain (*see* Central nervous system (CNS))
categories, 2
CCAP neurons (*see* Crustacean cardioactive peptide (CCAP))

circadian control (*see* Circadian control, insect development)
circadian core clock, flies, 9–12
circadian oscillations, EcR (*see* Ecdysone receptor (EcR))
circadian rhythm genes, 27–28
complex parameters, 28
developing organisms, 2
EcR circadian oscillations, 13–14
"hardwired", 2–3
hormonal control, 4–7
integrate environmental variables, 2–3
JHs function (*see* Juvenile hormones (JHs))
Manduca, 26–27
prothoracic glands, 12–13
timing, 3–4
transitions and circadian control, 7–9
Insects, embryonic diapause
"hatch-ready" larvae, 215–216
mid-embryonic arrest
autumn, 216–217
DH expression control, 219
diapause hormone, pupa, 217–218
dormant embryos break diapause, 216–217
ERK signaling, 216–217
Insulin growth factor 1 (IGF1), 235
Insulin signaling pathways, miRNAs and nutritional
dcr-1, 103
description, 99
genetic manipulation, 99
identification, 102
ILPs and insulin signaling, 100–101
miR-1, 104
miR-8, 102–103
miR-14, 102
miR-33, 103
miR-278, 102
and steroid, 111–112
Interbout arousals (IBAs), 250*f*, 254–256, 255*f*, 262–263
Intermediate neural progenitors (INPs)
asymmetrical division, 73
combinatorial temporal patterning, 77
Intracerebroventricular (ICV) injection, 253, 259–260

J

JHs. *See* Juvenile hormones (JHs)
Juvenile hormones (JHs)
description, 219–220
developmental hormone, 231–232
developmental timing
adult insects, 23
Aedes aegyti, 26
Bombyx kr-h1 gene, 25
Cecropia and *Cynthia* moths, 23–24
corpora allata, 23–24
Drosophila Met mutant, 24–25
Gce/Met function, 25
identification, 24–25
Kr-h1 and *Hairy*, 26
kruppel-homolog-1 (kr-h1) expression, 25
PTTH secretion, 23
receptor, 26, 27*f*
Rhodnius nymphs, 23–24
RNAi-mediated knockdown, 26
Tribolium castaneum, 25–26
USP, 23–24
diapausing/nondiapausing larvae, 219–220, 221
ECD signaling, 219–220
ecdysone titers, *Drosophila*, 6*f*
genetic manipulations, 4–5
hemolymphatic levels, 219–220
insect, 4–5
Ostrinia nubilalis larvae, 221
signaling pathway
anti-metamorphic hormone, 106
D. melanogaster molting hormone, 104–105, 105*f*
identification, 106
molecular analysis, 107–108
peripheral tissues, 104–105
processes, 106
twenty-hydroxyecdysone (20E), 104–105
"spotted" larvae, 219–220
stationary molts, 221

K

Kisspeptin/neurokinin B/dynorphin (KNDy)
ARC population, 307–308

Kisspeptin/neurokinin B/dynorphin (KNDy) (*Continued*)
- regulation, GnRH secretion, 309*f*

Kisspeptins (Kp)
- amino acid length, 304
- and ARC, 307–308
- GnRH neurons, 304
- Kiss1R, 304
- metabolic control, puberty, 312–317
- and NKB (*see* Neurokinin B (NKB))
- pubertal transition, 304–305
- reproductive function, 304

Kp. *See* Kisspeptins (Kp)

L

Leptin
- biological function, 313
- energy abundance, 313
- GnRH, 313
- HPG axis, 313
- Kiss1 neurons, 316
- mRNA encoding, 315–316
- mTOR–kisspeptin pathway, 316–317
- neuronal population, 316
- puberty and fertility, 313
- reproductive capacity, 312–313

Lipophilic hormone signaling pathways, miRNAs
- description, 104
- ECD and JH pathway (*see* Juvenile hormones (JHs))
- ecdysone regulated *let-7-C* miRNAs (*see* Ecdysone (ECD))
- *miR-34*, 112–113
- *miR-14* and *miR-8*, 111–112

Longevity
- description, 201
- gonadal pathway, 201–203

Long-term depression (LTD), 260–261, 263
Long-term potentiation (LTP), 260–261, 263
LTD. *See* Long-term depression (LTD)
LTP. *See* Long-term potentiation (LTP)

M

Mammalian target of rapamycin (mTOR)
- anorectic effects, 316–317
- cellular energy homeostasis, 316–317
- permissive/stimulatory effect, 316–317

Manduca sexta (M. sexta)
- "black" mutant, 231–232
- diapausing pupae, 225
- *MsPTTH* gene, 222–224
- PTTH injections, 222–224
- pupal diapause, 223*f*
- triggers pupal diapause, 222

MAPK pathway. *See* Mitogen-activated protein kinase (MAPK) pathway
MCT1. *See* Monocarboxylic acid transporter 1 (MCT1)
Melatonin and clock genes, 264–265
MicroRNAs (miRNAs)
- biogenesis and function, 99
- cell autonomous, 193
- circadian rhythm pathways (*see* Circadian rhythm pathways, miRNAs)
- DA/DAF-12 signaling, 202
- DAF-12 regulation
  - biogenesis and activity, 195–196
  - *Drosophila*, 195
  - environments global production, 195
  - HBL-1/hunchback, 194
  - heterochronic phenotypes, 193–194
  - LBD mutations, 193–194
  - *let-7* microRNAs, 196
  - L2/L3 temporal programs, 193–194
  - mammals, 195
  - *mir-84* and *mir-241* in human cell culture, 194
- developmental timing mechanism, 159
- and epigenetics (*see* Epigenetics)
- forward genetic analysis, 98
- heterochronic gene pathway, 173–174
- hormones activation, 98
- Hunchback-like transcription factor, 160
- insulin and nutritional pathways (*see* Insulin signaling pathways, miRNAs)
- L1 diapause, 172
- *let-7* family, 163–164, 168–169
- LIN-28, 160–161, 173–174
- *lin-4* encodes, 159–160, 171–172
- lipophilic hormone pathways (*see* Lipophilic hormone signaling pathways, miRNAs)
- L1/L2 and L2/L3 transitions, 192–193

mammalian *let-7* downregulation, 192–193
mammals, 203
pri-microRNA, 200–201
pubertal maturation, 319
stress induced and immune response pathways (*see* Immune response pathways, miRNAs)
switches, 160
systemic signaling pathways, 118
temporal control, 192–193

miR156
adult shoot apices, 141
Arabidopsis, 136–137
ch1 and wild-type plants, 142–143
controls reproductive competence, 136–137, 137*f*
divergent functions, 133–134
endogenous sugars, 143
GA-mediated flower, 141–142
GW-repeat proteins, 130–132
and miR157, 134–135
and miR172, 137–138
phylogenetic analysis, 133
plants overexpression, 141–142
precise relationship, 134–135
SBP/SPL genes, 130–132
SPL15, 132
spl2 and *spl9* mutations, 133–134
SPL3 protein, 130–132
SPL10 transcripts, 133–134
SPL3/4/5 transcripts flower, 138–139
targets, 138
tps1 mutants, 143
transgenic plants, 132
vegetative phase change (*see* Vegetative phase change, miR156)

miR157
measurement, 134–135
and miR156 (*see* miR156)
vegetative and reproductive development, 144

miR172
high and reduced levels, 137–138
MIR172B expression, 137–138
promote flowering, 137–138
vegetative phase, woody species, 131*f*

miRNAs. *See* MicroRNAs (miRNAs)

Mitogen-activated protein kinase (MAPK) pathway
ecdysone, 39–41
and Erk, 53

Molting cycle, *C. elegans*
adult population, 164
downstream components, 162–163
and *Drosophila*, 162
EcR and USP orthologs, 162–163
genome-wide searches, 163
heterochronic pathway and molting control, 162
let-7 mutants, 163–164
LIN-42, 164–167
NHR-25 function, 163
paced and aligned, 174
periodic regulation, 163
timing pathway components, 163–164

Monocarboxylic acid transporter 1 (MCT1), 258

mTOR. *See* Mammalian target of rapamycin (mTOR)

N

N-CoR/SMRT complexes
myc-ID, 283
TBLR1 and HDAC3, 282–283
transgenic expression, 284*f*
TR-binding proteins, 279–280

Neural stem cells, *Drosophila*. *See* Neuroblasts (NBs)

Neurobiology, puberty
ARC and RP3V, 305–306
GnRH (*see* Gonadotropin-releasing hormone (GnRH))
HPG/gonadotropic axis, 301–302
Kiss1/Gpr54, 306–307
mouse studies, 306

Neuroblasts (NBs)
Notch-dependent asymmetric divisions, 71
temporal patterning (*see* Temporal patterning, NBs)

Neurokinin B (NKB)
coexpression, 307–308
female rats, 308–309
GPR54/Kiss1 mutants, 307–308
hormonal studies, 308

Neurokinin B (NKB) (*Continued*)
KNDy neurons, 308, 309*f*
NKB and NK3R, 308–309
puberty and fertility, 307–308
Neuronal cilia modulation, 261–263
Niemann-pick type C (npc) genes, 45–46
NKB. *See* Neurokinin B (NKB)
npc genes. *See Niemann-pick type C (npc)* genes
Nutrient signaling
amino acids, sugar and fat
developmental time and growth, 44–45
diabetes, 48
dietary components, 44–45
DILP2 and *DILP5*, 44–45
Drosophila (*see Drosophila*)
endocrine system, 48
genetic perturbations, 46–47
hypoglycemia and disruption, 44–45
insulin/TOR signaling, 45–46
life span, *Drosophila*, 47–48
npc genes, 45–46
royalactin, 46–47
systemic growth and developmental timing, 47
based on tissues (*see* Tissues crosstalk, nutrient signaling)
Drosophila, larval instars, 39
ecdysone, 39–41
fine-tuning systems (*see* Fine-tuning systemic insulin signaling)
insulin-like peptides, 38–39
juvenile stages, 38–39
nutrient-dependent growth
DILP1–7, 42–43
FOXO activity, 42–43
internal nutrient status, 44
phosphorylation, 42–43
PI3K/AKT, 44
nutrient-independent development
developmental timing and body size, 39–41, 40*f*
ecdysone, 41–42
insulin-like peptides, 42
TGP, 41–42
organ growth and tissue regeneration, 57–60
PTTH activation, 39–41

O

Olfactory receptor neurons (ORNs), 72
Organ growth and tissue regeneration
DILP8, 60
DILP/InR and TOR pathways, 57–58
discs growth program, 58–60
growth perturbation and damage, 58
larval-specific organs, 57–58
mid-L3 checkpoint, 58–60
progression, adulthood, 58–60, 59*f*
size and dimensions, 57–58
Orgyia thyellina (O. thyellina), 230–231, 230*f*
ORNs. *See* Olfactory receptor neurons (ORNs)

P

Pancreatic triacylglycerol lipase (PTL), 257
PcG. *See* Polycomb group (PcG)
PDF neurons. *See* Pigment dispersing factor (PDF) neurons
PDK4. *See* Pyruvate dehydrogenase kinase isozyme 4 (PDK4)
PG. *See* Prothoracic gland (PG)
Pieris brassicae (P. brassicae)
developing pupae, 222
diapausing pupae, 224–225
noctuid moth, 231
PTTH, 224
Pieris napi (P. napi), 222, 228
Pigment dispersing factor (PDF) neurons, 8, 11
Polycomb group (PcG), 318–319
Polycomb repressor complexes (PRCs)
Drosophila, 83
NB competence, 83
repressive chromatin marks, 82–83
PRCs. *See* Polycomb repressor complexes (PRCs)
PRMT1. *See* Protein arginine methyltransferase 1 (PRMT1)
Pros-dependent cell-cycle exit, thoracic and central brain NBs
Cas expression, 86*f*, 87
GMCs, 86–87
sustained Grh expression, 87
Protein arginine methyltransferase 1 (PRMT1)
and CBP/p300, 279–280

transgenic expression, 286*f*
X. laevis, 285
Prothoracic gland (PG)
autonomous process, 224
block PTTH/ECD signaling, 224–225
cholesterol delivery, 45–46
diapausing embryos, 216
ecdysone, 39–41
endocrine system, 12
maintenance, 216
Manduca sexta, 224
nutrient sensor, 53
oscillations
Ca^{2+} spike frequencies, 17
CCAP neurons, 17–19
cryptochrome function, 15
cryptochrome mutants, 16–17
developmental timing, 15–16
Drosophila eclosion rhythms, 17
ecdysone concentration, 17–19
intracellular Ca^{2+} signaling, 16–17
neuro-endocrine network, 17–19
PERIOD, 16
period-luciferase transgene, 16
period transcript levels, CNS, 14–15
TIMELESS and PERIOD regulation, 16
Papilio xuthus, 224
prohormone, 5
PTTH, ECD synthesis, 12–13
Rhodnius larvae, 12–13
Samia cynthia, 224
s-LN_vs appearance, 11
Prothoracicotropic hormones (PTTHs)
brain, 224
brain-derived neuropeptide, 39–41
and DA signaling, 225
developmental stages, 5–7
and ECD signaling, 221, 224–225
ecdysone biosynthetic pathway, 39–41
ecdysone synthesis, insects, 12–13
inter-organ signaling coordinates, 49*f*
larval development, 39–41
Manduca sexta and *Anthaerea pernyi*, 222–224
MAPK pathway, 53–54
mechanism, 224–225
and mRNA, 8
MsPTTH gene, 222–224
and PG, 224
PG clock, 13
photoperiodic gating mechanism, 7–8
positive neurons, 12–13
rhythmic release, 12–13
secretion, 23
ultradian rhythmicity, *Drosophila*, 5
PTL. *See* Pancreatic triacylglycerol lipase (PTL)
PTTH. *See* Prothoracicotropic hormones (PTTHs)
Puberty
brain sex differentiation (*see* Brain sex differentiation)
developmental transition
disturbed homeostasis, 300–301
ethnic and geographical factors, 300–301
life-changing phenomenon, 300
physiological mechanisms, 301
postnatal development, 300
metabolic control, 312–317
neurobiology, 301–307
NKB, 307–309
novel mechanisms, 317–320
Pyruvate dehydrogenase kinase isozyme 4 (PDK4), 257

Q

Quercus acutissima, 130, 131*f*

R

Retinoic acid signaling
mouse hypothalamus, 263
retinol deficiency, 263
RXR and RAR, 263
Siberian hamsters, 263
signaling-related genes, 263–264
vitamin A, 263

S

SCN. *See* Suprachiasmatic nucleus (SCN)
Seasonal polyphenisms, diapause
adaptive advantages, 231
Araschnia levana, 228–229, 229*f*
“autumnal/winter” morphs, 227–228

Seasonal polyphenisms, diapause (*Continued*)
B. mori, 229–230
bound morphs, 228–229
bound polyphenisms, 228–229
Brachypterus pupae, 230–231
brain-SOG complexes, 229–230
"diapausing" and "nondiapausing" growth, 228
ECD pulses, 228–229
embryonic diphenism, Bombyx, 231
Macropterus morphs, 230–231
molecular mechanism, 231–233
Orgyia thyellina, 230–231, 230*f*
photoperiodic changes, 227–228
Pieris napi, 228
Shoot maturation, plants
description, 126
heteroblasty, 127–129
miR156 targets, 130–134
postembryonic development, 126
terminology, 126–127
SLC16A1. *See* Solute carrier family 16, member 1 (SLC16A1)
Solute carrier family 16, member 1 (SLC16A1), 258
Suboesophageal ganglion (SOG)
and brain complexes, 229–230
diapause hormone (DH), 217–218
moth nervous system, 217–218
Suprachiasmatic nucleus (SCN)
hibernation timing, 258
hormone melatonin, 253
hypothalamus, 252–253, 258
melatonin, 264
torpor and peaks, 258

T

Target of rapamycin (TOR)
bypasses neuroblast requirements, 57–58
Drosophila (*see Drosophila*)
insulin signaling, 45–46
nutrient-dependent growth (*see* Nutrient signaling)
phenocopies amino acid, 44
Temporal patterning, NBs
embryonic and larval ventral nerve cord, 71–72
epigenetic mechanism, loss, 82–83
feedforward loops
Ap cluster neurons, 84
Cas window, 84, 85*f*
cell fate diversification, 84
NB3-3 lineage, 84–85
integration
temporal and spatial information, 87–88
temporal identity, binary fate choice, 88–90
medulla, optic lobes, 72–73
and neurogenesis
apoptosis, abdominal pNBs, 85–86
mushroom body NBs, 87
Pros-dependent cell-cycle exit, 86–87
olfactory system, 72
restriction, 81
sequence and regulation, TF, 81–82
TFs (*see* Transcription factors (TFs))
transitions
cross-regulations, TFs, 78–80
requirements, cell cycle and asymmetric division, 80
switching factors, 80–81
type II NBs, 73
Terminal growth period (TGP)
nutrient restriction, 41–42
weight and pupariation, 40*f*, 41–42
Tetrodotoxin (TTX), 13, 15, 16–17
TGP. *See* Terminal growth period (TGP)
TH receptors (TRs)
amphibian metamorphosis, 276
anuran metamorphosis, 276
dual function model, 276, 277*f*
frog development (*see* Frog development, TR function)
metamorphic progression (*see* Coactivator recruitment)
metamorphic timing (*see* Corepressor complexes)
natural metamorphosis, 278–279
postembryonic/perinatal period, 276–278
premetamorphic tadpoles, 279
self-renewing system, 276
temporal regulation, 285–291
thyroid gland, 276–278

uterus-enclosed mammalian embryos, 276–278
Thyroid hormone (TH). *See also* TH receptors (TRs)
amphibian metamorphosis (*see* Amphibian metamorphosis)
dependent transformation, 276–278
embryonic plasma, 276–278
premetamorphic tadpoles, 282–283
signaling, 259–260
and TR-RXR, 289
vertebrate development, 276
Thyrotrophin releasing hormone (TRH), 260
Tissues crosstalk, nutrient signaling
gut microbiota, nutrition and developmental progression, 52–53
nutrient signaling and neuroendocrine control
ecdysone synthesis, 53
MAPK pathway, 53
PTTH, 53
nutrient uptake and systemic growth
DILP6 expression, 51–52
expression, *upd2*, 50–51
fat body, 51–52
inter-organ signaling, 48–50, 49*f*
stimulation, systemic growth, 51
TCA cycle, 51–52
TOR cell, 48–50
TOR. *See* Target of rapamycin (TOR)
Transcription factors (TFs)
combinatorial temporal patterning
INPs, 77
type II NB lineages, 77, 78*f*
different sequence, medulla NBs, 76–77, 77*f*
similar/different sequence
Chinmo protein, 75–76
mushroom body, 75–76
pNBs, 76
temporal sequence
birth-order-dependent neuronal identity, 75
cas mutants, 75
embryonic VNC NB7-1, 73–75, 74*f*
NB3-1 lineage, 75
postmitotic progeny, 73–75
Transthyretin precursor (TTR), 251, 260
TRH. *See* Thyrotrophin releasing hormone (TRH)
TRs. *See* TH receptors (TRs)
TTR. *See* Transthyretin precursor (TTR)
TTX. *See* Tetrodotoxin (TTX)

U

Ultraspiracle (USP), 5–7, 23–24
USP. *See* Ultraspiracle (USP)

V

Vegetative phase change, miR156
Acacia colei, 134–135
adult shoot apices, 141
carbohydrates, 142
chlorophyll biosynthetic gene, 142–143
endogenous factors, 141–142
epicormic buds, 140–141
Eucalyptus shoot development, 134
Eucalyptus tenuiramis, 139–140
exogenous sugar, 142
GA-mediated flowering, 141–142
genetic analyses, 144
heteroblasty, 127–129
HEXOKINASE1 (HXK1), 143
juvenile nodes, 140
maize, 141
MIR156A and *MIR156C*, 142–143
miR156 and miR157, 129–130
northern analysis, 134–135
primary shoot apex and lateral buds, 140
and reproductive maturation, 135–139
SBP/SPL genes, 144
SPL gene expression, 141
timing of, 144
tps1 plants, 143
woody species, 131*f*, 134–135
VGF signaling
hibernation, 261
hypophagic behavior, 261
hypothalamic and pituitary anatomy, 254*f*, 260
knockout mice, 260
LTD and LTP, 260–261
synaptic plasticity, 260–261

X

Xenopus laevis
dual function model, 277*f*
gene activation and metamorphosis, 281–282
metamorphosis, 285
N-CoR, 283
premetamorphic tadpoles, 279
pseudo-tetraploid, 280
transgenic expression, 283, 284*f*

Z

Zebra finch *(Taeniopygia guttata)*, 266

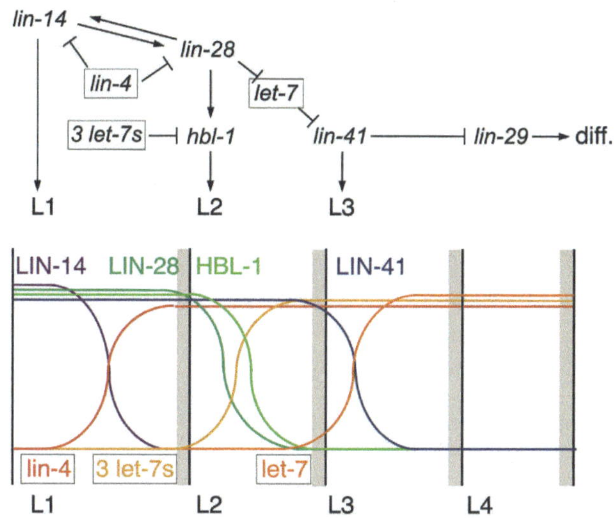

Ann E. Rougvie and Eric G. Moss, Figure 6.2 Genetic and dynamic models for the heterochronic timing mechanism. The top panel shows one model for the established genetic interactions of core players in the timing mechanism indicating approximately when in developmental time they are active. MicroRNA genes are boxed. Stage-specific fates are indicated by larval stage and terminal differentiation by "diff." The bottom panel represents the rising and falling levels of the mature products of key genes. All microRNAs (boxed) begin development in the off state and rise at specific times, and their targets are repressed accordingly, leading to falling protein levels. Not all factors are shown.

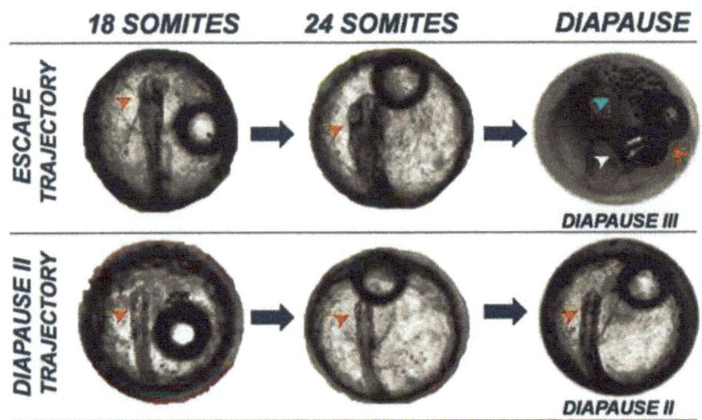

Luca Schiesari and Michael B. O'Connor, Figure 8.6 Developmental growth of the killifish *Austrofundulus limnaeus* in relation to diapause. Model of the two developmental trajectories in embryos of *A. limnaeus*. Diapause is strictly associated with modifications of development. Early embryogenesis normally occurs until 10-somite stage, without any difference between the two trajectories. By the 18-somite stage, nondiapausing embryos (escape trajectory, diapause III) differ morphologically from diapausing ones (diapause II trajectory). At the end of the "escape" trajectory, the fully developed embryos enter obligate diapause III, which occurs directly prior to hatching. Note the morphological differences between the two developmental trajectories at level of trunk and cephalic development (red arrowheads). Blue arrowhead: melanocytes; white arrowhead: eyes. *Modified from Podrabsky et al. (2010) with the permission of The Journal of Experimental Biology.*

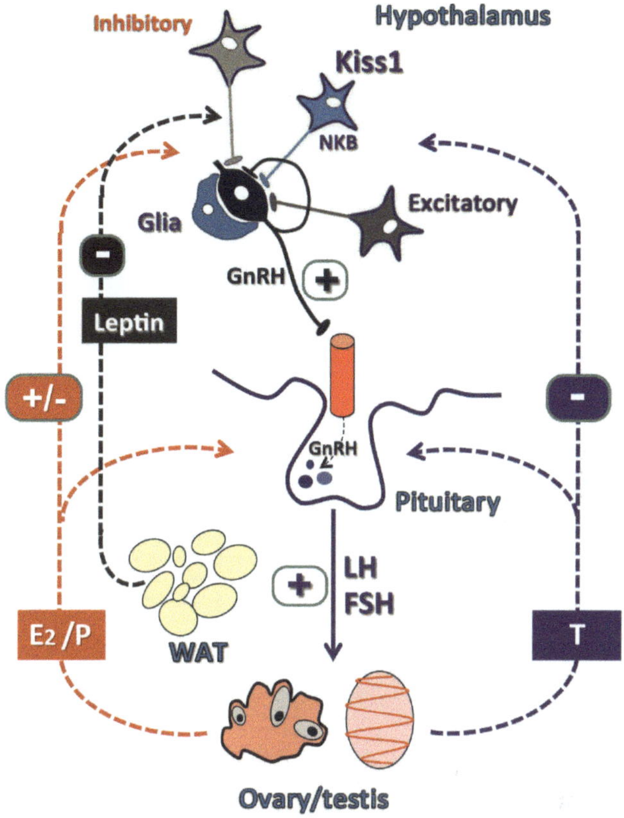

Manuel Tena-Sempere, Figure 11.1 The hypothalamic–pituitary–gonadal (HPG) axis. A scheme is presented of the HPG axis, as master neurohormonal network for the control of reproduction. Hypothalamic GnRH neurons receive transsynaptic inputs (of inhibitory and excitatory nature) and glial-derived signals, which cooperate to modulate the pulsatile release of GnRH. This neuropeptide drives the secretion of gonadotropins, LH, and FSH, from the anterior pituitary, which in turn stimulate the maturation and function of the gonads; note that both the ovary and testis are presented in the figure. These major hormonal elements are connected via feedforward and feedback regulatory loops, in which gonadal steroids play a prominent role: testicular testosterone (T) inhibits GnRH/gonadotropin secretion (negative feedback), while ovarian steroids, mainly estradiol (E2) and progesterone (P), can carry out both negative and positive feedback actions depending on the stage of the ovarian cycle. Other peripheral regulators of the HPG axis mentioned in this review, such as the metabolic hormone, leptin, produced by the white adipose tissue (WAT), are also presented in the scheme. Note that for the sake of simplicity, central inhibitory and excitatory inputs to GnRH neurons are not itemized; only Kiss1 neurons are highlighted. *Adapted from references Pinilla, Aguilar, Dieguez, Millar, and Tena-Sempere (2012) and Roa, Aguilar, Dieguez, Pinilla, and Tena-Sempere (2008), with modifications.*

Kolibri 360 GmbH
Ferdinand-Jühlke-Straße 7
99095 Erfurt, Deutschland
produktsicherheit@kolibri360.de